Nuclear Decommissioning Case Studies

Volume 1

Accidental Impacts on Workers, the Environment and the Public

AF179979

Nuclear Decommissioning Case Studies

Volume 1

Accidental Impacts on Workers, the Environment and the Public

Michele Laraia

ELSEVIER

ACADEMIC PRESS

An imprint of Elsevier

Academic Press is an imprint of Elsevier
125 London Wall, London EC2Y 5AS, United Kingdom
525 B Street, Suite 1650, San Diego, CA 92101, United States
50 Hampshire Street, 5th Floor, Cambridge, MA 02139, United States
The Boulevard, Langford Lane, Kidlington, Oxford OX5 1GB, United Kingdom

Copyright © 2021 Elsevier Inc. All rights reserved.

No part of this publication may be reproduced or transmitted in any form or by any means, electronic or mechanical, including photocopying, recording, or any information storage and retrieval system, without permission in writing from the publisher. Details on how to seek permission, further information about the Publisher's permissions policies and our arrangements with organizations such as the Copyright Clearance Center and the Copyright Licensing Agency, can be found at our website: www.elsevier.com/permissions.

This book and the individual contributions contained in it are protected under copyright by the Publisher (other than as may be noted herein).

Notices
Knowledge and best practice in this field are constantly changing. As new research and experience broaden our understanding, changes in research methods, professional practices, or medical treatment may become necessary.

Practitioners and researchers must always rely on their own experience and knowledge in evaluating and using any information, methods, compounds, or experiments described herein. In using such information or methods they should be mindful of their own safety and the safety of others, including parties for whom they have a professional responsibility.

To the fullest extent of the law, neither the Publisher nor the authors, contributors, or editors, assume any liability for any injury and/or damage to persons or property as a matter of products liability, negligence or otherwise, or from any use or operation of any methods, products, instructions, or ideas contained in the material herein.

Library of Congress Cataloging-in-Publication Data
A catalog record for this book is available from the Library of Congress

British Library Cataloguing-in-Publication Data
A catalogue record for this book is available from the British Library

ISBN: 978-0-12-823700-7

For information on all Academic Press publications visit our website at https://www.elsevier.com/books-and-journals

Publisher: Candice Janco
Acquisitions Editor: Maria Convey
Editorial Project Manager: Alice Grant
Production Project Manager: Debasish Ghosh
Cover Designer: Matthew Limbert

ELSEVIER Book Aid International
Working together to grow libraries in developing countries
www.elsevier.com • www.bookaid.org

Typeset by TNQ Technologies

To my wife Giovanna in gratitude for her constant assistance and encouragement during the drafting of this book

"Collaborating in the very private way of love or the highest kind of friendship..."

Elizabeth Hardwick (1916–2007)

Contents

Foreword

We should all be concerned about the future because we will have to spend the rest of our lives there.

Charles Franklin Kettering (1876–1958)

The long-term considerations in the life cycle of a nuclear facility (the essence of decommissioning) overlap and interact with the very meaning of "sustainability."

According to Brundtland (1987), "sustainable development is development that meets the needs of the present without compromising the ability of future generations to meet their own needs." It contains within it two key concepts:

- "the concept of 'needs', in particular the essential needs of the world's poor, to which overriding priority should be given; and
- the idea of limitations imposed by the state of technology and social organization on the environment's ability to meet present and future needs."

The principle of sustainability is closely related to the polluter pays principle (PPP). One consequence of the PPP is the principle of intergenerational equity (Lindskog et al., 2013).

In other words, sustainability implies environmental protection, economic development, and social progress. A key challenge for implementation of sustainability policies is to integrate these three requirements in a way that does not unduly privilege (or compromise) any of them.

Energy plays a key role among sustainability issues. The generation and use of energy are vital and intrinsic to economic growth and social welfare. These inherent aspects of energy may leave out the environmental impact—and indeed this disregard has been commonplace for centuries. To some extent, all forms of energy generation and use cause environmental impact, often including waste production and resource depletion. Moreover energy forms often involve long-term dimensions (many decades or even much longer periods). But these inevitable consequences cannot be left uncontrolled; on the contrary they require advance planning and mitigation.

A broad assessment of nuclear energy within a sustainability context shows that nuclear energy policies are consistent with the sustainability objective of transferring achievements and beneficial assets to future generations while maintaining environmental protection. The benefits should be ensured for as many years ahead as realistically predictable.

Worldwide, there are thousands of licensed nuclear facilities that will ultimately reach the end of their service life and require decommissioning. They range from large nuclear power reactors (NPP) and fuel processing facilities to small research laboratories, nuclear research establishments, waste storage facilities, and manufacturing plants.

There is a wealth of experience worldwide in the removing from service, dismantling and demolishing of redundant installations. However, those responsible for nuclear and radiological facilities (including, among others, operating organizations, regulators, and waste managers) face the special challenges associated with managing their radioactive inventory over many decades of operation and well beyond it. In recognition of the risks associated with a permanently shut down facility, a safety-driven definition of decommissioning was developed by the IAEA, resulting in the following statement: "Administrative and technical actions taken to allow the removal of some or all of the regulatory controls from a facility" (IAEA, 2014). These actions include complying not only with traditional engineering, financial, and industrial safety objectives but also with legislation/regulations limiting the impact of decommissioning on the health, safety, and radiation protection, and socioeconomic well-being of the workers, public, and environment. It is notable that decommissioning can normally take place from 40 years (immediate dismantling) to 100 years and more (deferred dismantling) after the beginning of nuclear operations: therefore decommissioning-impacting decisions taken at the design and construction stage or during plant operation will show their impacts long after they have been taken and will be felt by an entire new set of stakeholders.

Normal operation of nuclear installations has a low impact on health of the workers and the public, and on the environment. To fulfill sustainable development objectives, nuclear energy must keep high standards of safety throughout all phases of the nuclear fuel cycle, including decommissioning. It is clear that the principle of sustainability of nuclear energy should be best applied to the entire nuclear fuel cycle and to the life cycle of a nuclear installation (in all its phases, from design and construction, operation, to decommissioning and site release): from this viewpoint, decommissioning should not be considered in isolation, as it is the inevitable consequence of having built, operated, and contaminated a nuclear installation. However, it is also true that abnormal (either potential or actual) impacts of decommissioning could in principle affect and alter the broad assessment of nuclear activities, therefore decommissioning events deserve specific consideration *per se*. Many of these aspects belong to environmental sustainability but are not limited to it. For example, decommissioning aspects should be given due consideration during the plant design and operations phases; in the recycle of materials and waste resulting from decommissioning; and in the post-decommissioning reuse/redevelopment of nuclear sites. These aspects are mostly economic and social; therefore they belong to the two pillars of sustainability: economic development and social progress. Specific consideration is given to these two aspects in Adamson and Francis (2009).

Nuclear decommissioning is multidisciplinary, including subjects such as planning (preliminary and detailed), management of radioactive and nonradioactive waste, decontamination and dismantling technologies, radiation protection, industrial safety, record-keeping, security, etc. Therefore to address and confirm the sustainability of nuclear decommissioning different viewpoints are needed—and an entire series of books each covering specific subjects.

All the books of this series on Nuclear Decommissioning make predominant use of case studies (abnormal occurrences, errors, inadequacies during decommissioning). The reader will note that a single case study can be assessed from different viewpoints—and be reported and discussed in different books of this series: for example, one and the same incident can be due to poor planning, result in the generation of abnormal waste, and raise public rage.

According to Press Academia (2018):

- "A case study is a research strategy and an empirical inquiry that investigates a phenomenon within its real-life context.
- Case studies are based on an in-depth investigation of a single individual, group or event to explore the causes of underlying principles.
- A case study is a descriptive and exploratory analysis of a person, group or event.
- A case study research can be single or multiple case studies, includes quantitative evidence, relies on multiple sources of evidence and benefits from the prior development of theoretical propositions.
- Case studies are analysis of persons, groups, events, decisions, periods, policies, institutions or other systems that are studied holistically by one or more methods."

In this series of books events or situations are reviewed in detail within the context of a decommissioning project, cases are analyzed and solutions or interpretations are presented. Generic lessons learned are extracted for the benefits of the international decommissioning community. In this way case studies provide a deeper understanding of a complex subject and help acquire experience about a given situation.

Under the general category of case study there are several subdivisions, each of which can be selected by the investigator depending on the objectives of the research. For the purposes of this series of books Cumulative Case Studies have been selected. "These serve to aggregate information from several sites collected at different times. The idea behind these studies is the collection of past studies will allow for greater generalization without additional cost or time being expended on new, possibly repetitive studies" (Colo, undated).

Through the use of case studies this series of books adopts the inductive approach (as opposed to deductive). "In an inductive approach to research, a researcher begins by collecting data that is relevant to his or her topic of interest. Once a substantial amount of data have been collected, the researcher will then take a breather from data collection, stepping back to get a bird's eye view of her data. At this stage, the researcher looks for patterns in the data, working to develop a theory that could explain those patterns. Thus when researchers take an inductive approach, they start with a set of observations and then they move from those particular experiences to a more general set of propositions about those experiences. In other words, they move from data to theory or from the specific to the general" (https://saylordotorg.github.io/text_principles-of-sociological-inquiry-qualitative-and-quantitative-methods/s05-03-inductive-or-deductive-two-dif.html).

Primarily, the expected readership of these books includes those involved in and responsible for the overall assessment of nuclear applications for their nations (especially, the use of nuclear energy). Decommissioning is an essential, inevitable phase of the life cycle of a nuclear installation and its sustainability should be ensured since the very beginning of a nuclear project. Therefore this series of books is meant to provide informative guidance to the national decision-makers and their technical support organizations in selecting sustainable options.

In addition the books target all decommissioning experts at different technical levels (including nuclear organizations, regulators, waste managers, contractors, university teachers, and stakeholders at large) in providing information about what they can realistically (i.e. based on experience) expect to happen during the decommissioning of nuclear and radiological installations. To this end the books are meant to provide supplementary, up-to-date information and orientation. Once typical events are identified and described, the reader is addressed to specialist guidance in order to systematically check the likelihood of such events and take preventive and mitigating measures, if any, against them.

The decommissioning experts will find in this series of books additional material supposed to build upon their background. Besides, the expert reader will recognize that sustainability—in its three main components or pillars—is the common factor linking all technical, organizational, and human factors highlighted in the series. Actually, many nuclear decommissioning events are representative of events common to other industries as well. Therefore these books could also be of interest to a broader range of professionals, especially safety managers at large. The series can assist in the monitoring, review, and revision of safety work. The descriptive, narrative, occasionally colloquial style adopted (e.g., including literature quotations from the world-famous John Bartlett's "Familiar Quotations") should also make it attractive to the uninitiated.

References to foreword

Adamson, D.W., Francis, J., October 11–15, 2009. The emergence of sustainable practice within decommissioning. In: Proceedings of the 2009 12th International Conference on Environmental Remediation and Radioactive Waste Management, ICEM2009. Liverpool, UK.

Brundtland, G., Chairman, 1987. reportOur Common Future (The Brundtland Report, World Commission on Environment and Development, Oxford University Press, Oxford, United Kingdom. https://sswm.info/sites/default/files/reference_attachments/UN%20WCED%201987%20Brundtland%20Report.pdf.

Colorado State University, n.d. Types of Case Studies. https://writing.colostate.edu/guides/page.cfm?pageid=1290&guideid=60.

International Atomic Energy Agency, 2014. Decommissioning of Facilities. General Safety Requirements, Safety Standards Series No GSR Part 6 IAEA, Vienna.

Lindskog, S., Sjöblom, R., Labor, B., 2013. Sustainability of nuclear energy with regard to decommissioning and waste management. Int. J. Sustain. Dev. & Plan. 8 (No. 2), 246–264. https://www.witpress.com/elibrary/sdp-volumes/8/2/689.

Definition of Case Study, July 9, 2018. https://www.pressacademia.org/definition-of-case-study/#:~:text=A%20case%20study%20is%20a,within%20its%20real%2Dlife%20context.&text=A%20case%20study%20is%20a%20descriptive%20and%20exploratory,a%20person%2C%20group%20or%20event.

Chapter 1

Introduction

Environmental sustainability is the main thread running through this book (the first volume of the series presented in the Foreword). Actual or potential incidents impacting workers, the public, and the environment in the course of decommissioning are highlighted in numbers, severity, and extent as a key indicator of the sustainability of the decommissioning process. However, this book is not a comprehensive catalog of incidents (otherwise its size would grow to unmanageable levels), nor is it aimed at conducting the a priori safety assessment process of expected or potential hazards or imparting guidance on preventive and mitigating measures. Nonetheless, a number of best practices are given in the description of many events reported in this book.

This book is based on facts and return of experience/feedback. It strives to identify a significant number of typical events and come to a judgment on the overall environmental sustainability of nuclear decommissioning. The methodology of applying experience (or lessons learned) to correct inadequacies and/or improve safety and performance is explained in detail in IAEA (2018).

Although information on specific abnormal events in the course of decommissioning projects is available from a number of sources, nowhere are the scattered pieces of this information systematically collected, categorized, and compared on the international scale. It should also be noted that most cases are available only in national or corporate databases and for a restricted readership. The main objective of this book is to fill this gap. As there are literally many thousands of events worth a mention, to cover them all in one database and describe in detail for each of them the context, the circumstances and the causes would be unreasonable: The main body of this book—not including the two Annexes—offers a sample of 170 commented cases (137 fully developed case studies—many of them reporting multiple events—as well as 33 text boxes providing concise descriptions), which is possibly the largest coverage that can be realistically achievable in a book format.

Although guidance on safety assessment methodologies in decommissioning has been promulgated by the IAEA (limited to radiological incidents) and some other organizations, little descriptive material is available internationally on identification, selection, evolution, and root causes of typical incidents. Besides, the implications of decommissioning incidents on the environmental sustainability of nuclear applications (especially, the use of nuclear energy) have never been assessed in depth.

Nuclear Decommissioning Case Studies. https://doi.org/10.1016/B978-0-12-823700-7.00001-1
Copyright © 2021 Elsevier Inc. All rights reserved.

This book provides information and guidance on typical events occurring in the course of decommissioning that actually impacted (or had the potential of impacting) the health and safety of the workers and the public, or the protection of the environment at large. This is an important segment of environmental sustainability. The scope includes both radiological and nonradiological incidents within nuclear decommissioning sites. This book does not address other environmental impacts such as the offsite transport and disposal of decommissioning waste—unless the event is directly related to the decommissioning site. Finally, the book does not deal with events resulting in delays, extra costs, public opposition, and other socioeconomic impacts, although often environmental impacts convey socioeconomic impacts as well—and vice versa. Some reference to socioeconomic impacts is given in Annexes 1 and 2. Further books of the series will address these and other aspects in detail.

The scope of this book is not aimed at decommissioning following severe nuclear accidents (e.g., Chernobyl and Fukushima) although a few examples from those projects have been used in support of statements applicable also to planned, routine decommissioning projects. It is the author's view that severe nuclear accidents could be legitimately used to challenge the environmental sustainability of the whole nuclear fuel cycle, but only to a minor extent the decommissioning sector.

The bulk of this book consists of case studies. Each case study provides information on the following:

- origin, evolution, and conclusion of actual events (or near-misses) directly impacting workers, the public or the environment that have occurred worldwide at different types of nuclear installations during their decommissioning;
- actual or potential impacts from the event;
- analyses and proposed solutions, corrective actions taken in the short and long term; and
- an evaluation of the technical meaning of the event in terms of general applicability (lessons learned).

This detail has never yet been systematically collected and categorized in one publication: In particular, the reader should note that the decommissioning case studies reviewed in this book are internationally based in that they have been drawn from a number of countries including Belgium, Canada, Czech Republic, France, Germany, Italy, Japan, Lithuania, Russian Federation, Slovakia, Spain, the United Kingdom, and the United States.

The main features and contents of this book that are expected to be most valuable to the reader, as listed in a logical sequence, include the following:

- understanding the meaning of environmental sustainability of nuclear decommissioning within the broader scope of sustainable development;

- identifying and understanding the typical incidents/accidents impacting workers, the public, or the environment during decommissioning;
- generically evaluating the overall environmental impact of nuclear and radiological facilities throughout their decommissioning period; and
- confirming the environmental sustainability of nuclear decommissioning as indicated by number and severity of nuclear incidents/accidents.

Reference

International Atomic Energy Agency, 2018. Operating Experience Feedback for Nuclear Installations, Specific Safety Guide No SSG-50. IAEA, Vienna.

Chapter 2

The concept of environmental sustainability as applicable to nuclear decommissioning

I assert that nothing ever comes to pass without a cause

Jonathan Edwards (1703–1758)

Miniver coughed and called it fate, and kept on drinking.

Edwin Arlington Robinson (1869–1935)

It should be recognized that sustainability has been an important element of the nuclear sector for a long time. For example, spent fuel reprocessing was intended—for the countries that used this option—to maintain energy generation with minimal use of fresh uranium. Besides, the safety culture incorporated in the nuclear industry reinforces the sustainability of the workforce, and public acceptance of nuclear power ensures the sustainability of the local communities that draw their economic prosperity from the plant. However, this book considers "sustainability" mainly from the standpoint of the environmental principle of sustainable development, as represented by the occupational and environmental impact of abnormal events.

Antinuclear positions often stress that accidents and radioactive waste downgrade the role of nuclear energy in regard to sustainable development. However, experience shows that properly managed nuclear activities have a small safety risk and modest impacts on the workers, the public, and the environment (in terms of health, environmental emissions, and site contamination).

Abnormal events and errors (which do happen in the best of organizations) should not give rise to impacts to workers, public, or the environment that could compromise sustainability. Such events may happen during decommissioning as well, although normally at a much lower scale than during operations (mostly due to the lack of high temperatures and pressures, which are predominant during nuclear operations). In regard to sustainability, accidents/incidents that could or actually do happen in the course of decommissioning should be

Nuclear Decommissioning Case Studies. https://doi.org/10.1016/B978-0-12-823700-7.00002-3
Copyright © 2021 Elsevier Inc. All rights reserved.

assessed a priori or reviewed after their occurrence as a feedback. In other terms, accidents/incidents occurring during decommissioning are viewed as an indicator of environmental protection, the third pillar of sustainability. This is the very focus of the book.

As prerequisite to considering and reviewing decommissioning accidents/incidents, it should first and foremost be recognized that these events necessarily reflect unsafe practices by a number of parties involved, though occasionally elusive and hard to single out. To resort to "act of God," "bad luck," "force majeure," or "inevitable event" as an easy excuse is unacceptable. Similarly "human error," a term often used—even in this book—as the direct cause of an event, tends to deflect (and hide) responsibilities from a higher managerial level. However, responsibility—a different notion than cause—can be harder to identify, or it can be distributed over such a number of individuals or organizations that the blame can hardly be attributed to anyone. Fig. 2.1 shows an illustrative example to this end: In this book, crew resource management (CRM, identified under inadequate training, lack of communications, disregarded experience, etc.) appears to frequently play an important role alongside personal inadequacies.

The confirmation of accomplished sustainability is not easy. "While there is general agreement on the concept of sustainability, its actual meaning and the principles needed to achieve it in practice are much fuzzier and less well defined. There are many levels at which sustainability principles are currently being set: international organizations, national and local governments, industry

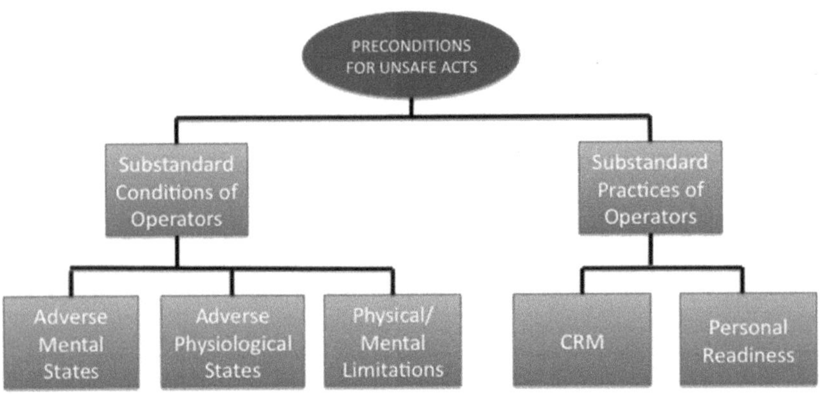

FIGURE 2.1 Preconditions for unsafe acts (SKY BRARY, Wiki Commons). Note: CRM stands for crew resource management (a set of training procedures for use in environments where human error can have grave impacts).

sectors, and individual businesses. The potential for contradictions and inconsistencies is significant, and uncertainty is inevitable given the scale of sustainability. This is particularly relevant for the decommissioning of nuclear facilities where we are dealing with a wide range of issues from the impact of removing jobs from local communities to the transgenerational impacts of managing and storing radioactive waste" (Bonser, 2006).

The following highlights the links between the general concept of sustainability, e.g., as recently summarized in WEF (2014), and the proper approach to decommissioning and site remediation. The reader should especially note the links to the three pillars of sustainability: economic development, social development, and environmental protection. As noted earlier, the focus of this book stays with the latter (often called "environmental sustainability" in this book).

The activities associated with decommissioning a nuclear facility can vary widely. They may include large-scale decontamination works, demolition of massive concrete structures, or enclosing the facility in a safe configuration so as to allow the radioactivity to decay naturally to acceptable levels. On the other end, laboratories in which radionuclides have been used may be fully decommissioned after some modest cleanout activities. In all cases, the decommissioning process addresses the structures, systems, and components of a facility. Additionally, the site (land areas) around a nuclear facility is often contaminated as the result of facility's operation. Therefore, decommissioning of a facility and environmental remediation of its site are typically part of one and the same project: the acronym D&ER is used in this book. Work carried out under decommissioning and remediation programs is accordingly aimed at achieving end states that set the basis for planned or anticipated (future) end uses (i.e., facility and/or site redevelopment). Decommissioning and site remediation programs share resources and several activities. While in the past decommissioning and remediation activities have often been implemented as independent activities, optimization of effort, cost, impacts, and risk reductions can be achieved by carrying out D&ER in an integrated manner (IAEA, 2009).

The concept of life cycle management can be described as the process of managing the entire life cycle of a product from its conception, through design and manufacture to service and disposal. In this book, the "product" is the nuclear fuel life cycle or the operational life cycle of plants and facilities, which have the potential to result in radiological contamination, eventually impacting the decommissioning process (IAEA, 2002).

TABLE 2.1 Applying sustainability principles to decommissioning (elaboration from Bonser, 2006).

Positive aspects	Negative aspects
Generating socioeconomic opportunities (e.g., by developing new productive uses of the site)	Altered socioeconomic conditions (e.g., loss of jobs)
Decontaminating hazardous facilities/ sites	Creating long-term waste management issues
Improving environmental conditions	No guarantee that technology solutions will not create problems (the focus of this book); incidents /accidents during decommissioning test the point
Reducing overall risk	
Developing, using and sharing knowledge (e.g., innovative technologies)	

LCM is a methodology used successfully in various industries to reduce the waste generated from a stream or process to lower costs, optimize production, and increase the value of the business. In addition, LCM can provide an additional benefit for ongoing or planned projects in reducing the extent of end-of-life D&ER. LCM is one typical way of including sustainability in a nuclear project.

While recognizing the link between decommissioning and environmental remediation, this book focuses on the former as a source of relevant events: However, a few events originating from contaminated land around nuclear installations and impacting people or the environment at large have been quoted to confirm the link.

The interactions between sustainability and decommissioning can be represented in Table 2.1. As said before, the focus of this book is the environmental pillar of sustainability, and its representative indicator adopted by this book is the number and severity of incidents during decommissioning insofar as they impact workers, the public, or the environment at large (**boxed entry in** Table 2.1). Fig. 2.2 graphically places the scope of this book within the general context of sustainability and especially environmental sustainability.

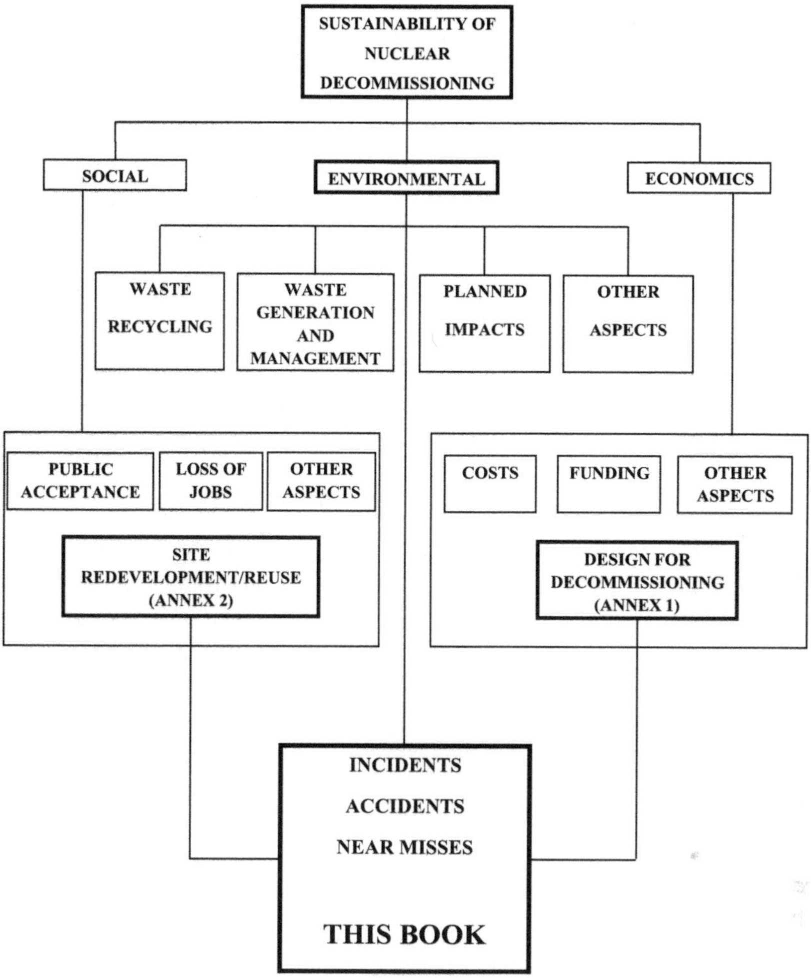

FIGURE 2.2 Relationship of this book to the concept of sustainability.

References

Bonser, D., 2006. Sustainability principles: a practical move towards tomorrow?. In: IBC's 10th Global Conference & Exhibition, Decommissioning of Nuclear Facilities, London 20–22 November 2006.

International Atomic Energy Agency, 2002. Safe and Effective Nuclear Power Plant Life Cycle Management towards Decommissioning, IAEA-TECDOC-1305, IAEA, Vienna.

International Atomic Energy Agency, 2009. Integrated Approach to Planning the Remediation of Sites Undergoing Decommissioning, Nuclear Energy Series NW-T-3.3, IAEA, Vienna.

The World Energy Foundation, August 20, 2014. A Brief History of Sustainability. https://theworldenergyfoundation.org/a-brief-history-of-sustainability/.

Chapter 3

Decommissioning events

Take calculated risks. That is quite different from being rash

George Smith Patton (1885–1945)

There is a pressing need in the industry to identify hazards, manage risks, and avoid or mitigate noxious events. Accidents have multiple causes and involve different levels within an organization, ranging from individual workers to top management. Frequently, accidents are caused by a mixture of intertwined aspects such as insufficient technology, human errors, and organizational failures: Any of these may have more distant causes. Deeply rooted inadequacies, or root causes, can be hard to identify and remove. Root causes in a given organization could be, for example, poor design, missing supervision, and unmanageable procedures. Root causes can be found in traditions, beliefs, perceptions, and priorities of the workforce (safety culture), the environment that surrounds the workers (working environment), and planned operations and specific procedures (safety-related factors). It is essential to identify root causes even before accidents occur, because only hazards and risks that are identified can be tackled (Nevhage and Lindahl, 2008).

A large number of studies have shown that there is a link between the number of near misses and the number of accidental events in a given organization. The relationship is often illustrated as a triangle, where the top section is the number of fatalities and, moving top-down, are the serious accidents, minor incidents and near-misses (the wide base of the triangle) (Fig. 3.1). Identification and assessment of near-misses is important since these events "nurture" grave accidents waiting to happen. Many case studies discussed in this book are near-misses.

"An incident is just the tip of the iceberg, a sign of a much large problem below the surface" Don Brown, Director of Software Products and Services for BasicSafe. In passing the reader should note that incidents and accidents are not qualitatively different and are used interchangeably in this book for "events" or "cases" when their severity is irrelevant. Properly (see Glossary) incidents refer to smaller events than accidents.

Accidents and near-misses can produce severe economic effects including the following (Nevhage and Lindahl, 2008):

- Compensation, care, and rehabilitation of the injured worker (or remediation of the environment)

Nuclear Decommissioning Case Studies. https://doi.org/10.1016/B978-0-12-823700-7.00003-5
Copyright © 2021 Elsevier Inc. All rights reserved.

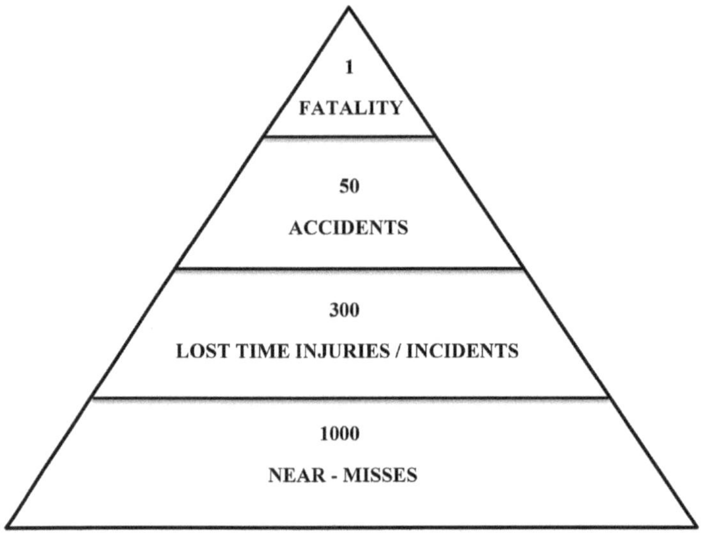

FIGURE 3.1 Triangle of events (aka Heinrich/Byrd's Safety Pyramid). Numbers are arbitrary but highlight a key point: Minor events and near-misses are many, perhaps too many to review individually, and their individual impacts can be negligible or nil, but they can be cues and precursors of less likely, but more serious events waiting to happen (graph and text by M. Laraia).

- Increased production cost (e.g., damaged equipment/supplies, work delays, recruiting new staff, investigation, etc.)
- Higher insurance premiums
- Pressures for additional safety provisions

Although these aspects are occasionally dealt with in this book, the focus is the health impact on people (workers and public) and contamination of the environment.

Certain events have been excluded from the scope of this book. Fuel-related accidents have been excluded in terms of observed causes normally decommissioning activities begin after the fuel has been removed from the plant: Besides, these events are not different from those that may happen during plant's service life and are not specific to decommissioning. Likewise, off-site waste management activities (transport and disposal) are out of scope here because they mostly belong to activities that take place also during regular operations; however, waste management activities (e.g., treatment, packaging, storage) taking place at the plant under decommissioning in preparation to removal off-site are fully part of the scope of this book.

Loss, theft or misuse of radiation sources are also excluded because these very same events can happen any time during a plant's life cycle, including, but not limited to, nuclear decommissioning: However, for sake of completeness, some events of this kind have been mentioned when the sources originated from a nuclear installation being decommissioned.

This book does not try to determine occurrence rates (or percentage ranges thereof) for specified events or categories: A literature overview shows that these numbers are highly variable and site specific, depending on decommissioning activities and phases, organizations' reporting requirements, definitions, and/or priorities.

Reference

Nevhage, B., Lindahl, H., 2008. A Conceptual Model, Methodology, and Tool to Evaluate Safety Performance in an Organization. Master Thesis. University of Lund, Sweden. http://lup.lub.lu.se/luur/download?func=downloadFile&recordOId=1786993&fileOId=1845357.

Chapter 4

Categorization of events

There are more things in heaven and earth, Horatio,
Than are dreamt of in your philosophy.

William Shakespeare (1564–1622), Hamlet.

Decommissioning-related events can be classified in many different ways. The main international classification (Section 4.1) centers on radiation and nuclear safety impacts; other classifications are specific to national regulations and refer to the nature of the events reported regardless of the severity of the impacts. Some classifications have a nation-wide coverage, and others are project-specific. Without being exhaustive, the following sections provide brief descriptions of widely used categories.

4.1 The International Nuclear and Radiological Event Scale

The fundamentals of the International Nuclear and Radiological Event Scale were developed in 1990 by committees of international experts under the aegis of the International Atomic Energy Agency (IAEA) and the Organisation for Economic Co-operation and Development/Nuclear Energy Agency (OECD/NEA). The general objective of these efforts was to establish a reproducible mechanism for communicating the safety impacts of nuclear and radiological events. For over two decades now, the INES has been refined and extended to include any event related to radioactive material and/or radiation sources, including transportation events. The key mechanism for INES to define the nature and seriousness of nuclear and radiation events is a rating scale (from 1 to 7). The upgrading work of INES culminated with the publication of International Atomic Energy Agency, INES, 2013, which is an integrated manual providing detailed guidance for rating any event, as well as examples and remarks. International Atomic Energy Agency, INES, 2013 is structured to assist those in charge of rating the safety impacts of events in terms understandable to the general public. The INES acts as network receiving, processing, storing, and circulating information on events and their INES rating: To this end, the INES Headquarters at the IAEA-Vienna are linked to INES National Officers in several dozen participating Member States.

Nuclear Decommissioning Case Studies. https://doi.org/10.1016/B978-0-12-823700-7.00004-7
Copyright © 2021 Elsevier Inc. All rights reserved.

4.1.1 General description of the International Nuclear and Radiological Event Scale

Events are categorized on the INES scale at seven levels: The scale is roughly logarithmic—that is, the impact of an event is around 10 times higher as one goes up each step (level) of the scale. Levels 4–7 are termed "accidents," and levels 1–3 "incidents." Except when referring to events specifically quoted from INES, this book will tell apart incidents from accidents in a loosely sense; in practice, incidents are defined (see Glossary) as events of smaller significance than accidents. It will be soon shown that events occurring in the course of nuclear decommissioning are only incidents, never accidents. Events without safety significance are categorized as "Below Scale/Level 0" (translated in some Member States as "Not Relevant"). Events that have no safety relevance with respect to radiation or nuclear safety are not classified on the scale; however, many of them are reviewed in this book as indicators of unsafe practices and precursors to more serious events.

It is noteworthy that industrial (nonradiological) events (e.g., falls from heights or exposure to toxic vapors) are not specifically identified in INES or are quoted at level 0 regardless of their seriousness. This book, however, will give considerable space to industrial events occurring in nuclear decommissioning.

INES events are rated in terms of the three following factors:

- People and the environment: This factor evaluates the radiological exposures to people near the event site and the wide, unforeseen dispersion of radioactive substances from a facility.
- Radiological barriers and control: This factor addresses events without any direct impact on people or the environment and only applies inside large installations. It covers unplanned high radiation levels and dispersion of significant amounts of radioactive substances within the installation boundaries.
- Defense-in-depth: This factor also covers events without any direct impact on people or the environment, but for which the range of measures put in place to prevent accidents did not function as intended. As said earlier, this book addresses both actual events impacting humans and the environment, and potential events (e.g., near-misses), which are relevant to the defense-in-depth factor.

The reader should note that a Member State's interest in the event report lies not in its INES rating per se—which may even be unknown at the time of the notification or may be revised later as more detail becomes available—but in the overall safety implications of the event on the country's nuclear safety system.

Comprehensive definitions of the INES levels based on each of these criteria are given in International Atomic Energy Agency, INES, 2013. Simpler definitions can be found in the literature such as follows:

Level 7: Major Accident
Level 6: Serious Accident
Level 5: Accidents with Wider Consequences
Level 4: Accidents with Local Consequences
Level 3: Serious Incident
Level 2: Incident
Level 1: Anomaly

According to the search carried out in this book, decommissioning-related events never exceed level 2. As the IAEA encourages Member States to submit events from level 2 up (in the United States, this is specified in the NRC policy, NRC, 2017), a good deal of decommissioning events goes unreported to INES.

The remaining part of this section describes some arbitrarily chosen events reported according to INES criteria. Several more case studies in this book include the overall INES rating as part of the event descriptions.

4.1.2 Contamination of workers during dismantling of glove boxes (Climatesceptics, 2001; Podlaha, 2020)

4.1.2.1 Decommissioning process; problem encountered: impacts

ÚJV Řež, a. s. (ÚJV) is a leading institution in all areas of nuclear R&D in the Czech Republic and has held a key position in the country's nuclear program since it was formed in 1955 as the Nuclear Research Institute Řež near Prague. ÚJV activities include, among others, nuclear physics, chemistry, and nuclear power. The main issues addressed at ÚJV in the past have included services for NPPs, the fuel cycle, and irradiation services for R&D in the industry and medical sectors. ÚJV also manages most institutional radioactive waste produced in the Czech Republic (around 90%). Around the turn of the millennium, and after several decades of activities in the nuclear field, there were at ÚJV a number of obsolete, excess nuclear facilities to be decommissioned. Many of these facilities were used for radioactive waste (RAW) management, and some of them were or will be reused after decommissioning. The case in question refers to one of ÚJV decommissioning projects.

Glove boxes were extensively used at ÚJV for handling alpha radionuclides (U, Pu, Am, Np) (Fig. 4.1). The boxes became unsuitable for their functions, and it was decided to decommission them. The boxes were heavily contaminated by alpha emitters, and their dismantling represented a serious radiological challenge.

During June–July 2001, several alpha-contaminated dry glove boxes were taken apart. The boxes originated from a Czech installation for production of fire detectors. These boxes were contaminated by ^{241}Am; after dismantling,

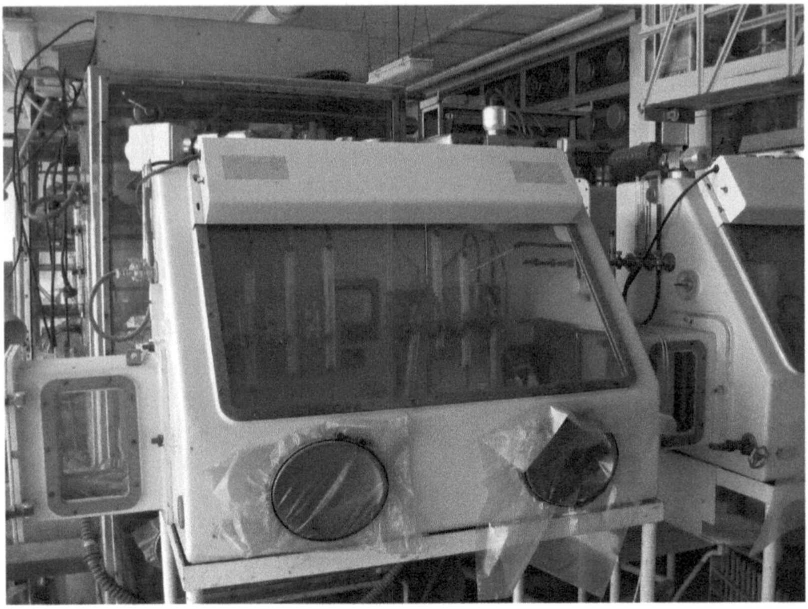

FIGURE 4.1 Glovebox before decommissioning, ÚJV Řež, a. s., Czech R. *(Courtesy of J. Podlaha (2020)).*

the material was placed into containers for storage of radioactive waste. During the dismantling activities, americium particles became airborne and were inhaled by the four dismantling workers. Immediately, after airborne contamination was discovered, the activities were halted. According to early investigations conducted at the Department of Occupational Medicine, Charles University, Prague, and based on total body counting at the State Institute for Radiation Protection, the most exposed individual received a committed effective E (50) dose of about 0.35 Sv, while the committed effective doses of the other three individuals affected were considerably lower. The radioactive releases to the environment were negligible. The event had an INES rating 2. The main features of the events per INES classification include the following:

	No	Yes
Contamination spread within the facility	◉	○
Damage to radiological barriers	◉	○
Degradation of defense-in-depth	◉	○
Overexposure of a worker	○	◉
Person injured physically or casualty	◉	○
Overexposure of a member of the public	◉	○
Release beyond authorized limits	◉	○
A continuing problem	○	◉
Was a press release issued	◉	○

4.1.2.2 Analysis, solutions

Based on learning from previous experience, the decision was taken that the glove boxes would not be further dismantled, but disposed of as one piece in nonstandard, purposely built units after approval by the organization responsible for operation of the radioactive waste disposal facility. The boxes were disconnected from their supports and sealed. Before that, the internal space of the boxes was filled with other radioactive waste to optimize the volume of the boxes as waste containers. The boxes were then placed into the disposal units, grouted in concrete, and shipped for disposal (Fig. 4.2).

4.1.2.3 Lessons learned

By its very nature, dismantling consists of the gradual removal of barriers that during operations confined radioactive products and prevented their release to

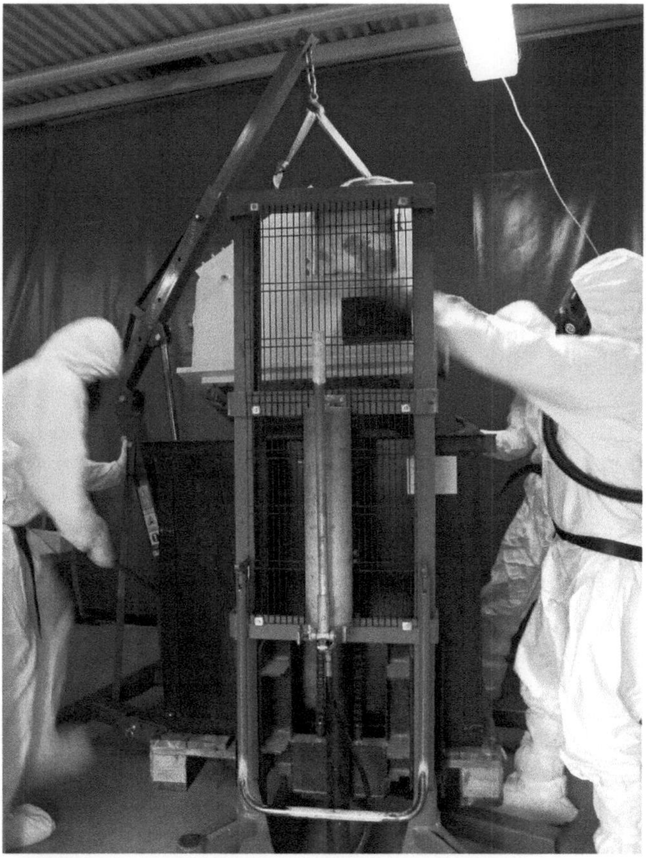

FIGURE 4.2 Insertion of a glove box into nonstandard disposal unit, ÚJV Řež, a. s., Czech Republic. *(Courtesy of J. Podlaha (2020)).*

the closer or more distant environment. Therefore, the opening and removal of barriers conveys an intrinsic potential for inadvertent, unexpected release of radioactive products and occupational contamination. In this case, the defense-in-depth worked, and remaining barriers prevented contamination from spreading beyond the immediate surroundings of the affected equipment. Considerations for any removal plan should include, but not be limited to, one or more of the following:

(1) characterization of the mobility of radioactive species (e.g., loose vs. fixed contamination, presence of liquids or dust);
(2) ease of extraction of radioactive items/products without creating airborne contamination;
(3) in situ conditioning or sealing of items/products before extraction;
(4) doing removal from a distance (e.g., remotely operating equipment);
(5) workers wearing proper PPE; and/or
(6) intact removal with no dismantling (the solution finally adopted by ÚJV).

More incidents involving release and inhalation of ^{241}Am can be found in Sections 6.5, 6.9, etc.

4.1.3 Flooding event at the EUREX facility (Climatesceptics, 2000)

EUREX was a pilot fuel reprocessing plant, which was shut down in 1984. During a flood, on October 16, 2000, water invaded the ventilation system and cable room some 3 m below grade, causing the system to be inoperable for a few days. The water in the room became slightly contaminated by the activity embedded in the ventilation filters. The room was cleared by removing the water to the ponds. The contamination levels were much lower than the discharge limits. The ventilation system was returned to service after some days.

The event was initially rated at level 0 and then upgraded to level 1 in consideration of the fact that a similar event had already occurred at EUREX in 1994 (with minor impacts) and at another installation on the same site (with similar consequences as those described here).

The main features of the event include the following:

	No	Yes
Contamination spread within the facility	◉	○
Damage to radiological barriers	◉	○
Degradation of defense-in-depth	○	◉
Overexposure of a worker	◉	○
Person injured physically or casualty	◉	○
Overexposure of a member of the public	◉	○
Release beyond authorized limits	◉	○
A continuing problem	◉	○
Was a press release issued	◉	○

4.1.4 Plutonium uptake at a reprocessing facility (International Atomic Energy Agency, INES, 2013)

Four workers went into a controlled radiation zone to carry out work on a ventilation component. The work consisted of the removal of a baffle box in a building that formerly housed a plutonium processing plant. The plant had been idle since 1957 in preparation for decommissioning.

The workers were wearing PPE and monitoring equipment. Cutting of the baffle box went on for 1 h 40 min when dust was seen falling from the box. When the workers halted work and evacuated the area, personal contamination monitors revealed clothing contamination on all of them. Work restrictions were promptly placed on involved personnel; a dose assessment was initiated through bioassay methods. Exposures were initially estimated to be lower than 11 mSv effective doses. More accurate estimates gave 24—55 mSv effective doses for the workers affected. The annual limit in force at the time was 50 mSv.

Rating criteria: numbers refer to relevant sections of (IAEA, 2013a)	Explanation
2.2. Activity released	Not applicable. No release to the environment.
2.3. Doses to individuals	One worker received a dose greater than the annual limit. The number of individuals receiving such a dose was less than 10, so the rating is not increased due to the number of individuals affected. Rating level 2.
3.2. Radiological barriers and controls at facilities	The contamination occurred during the decommissioning of a specific item in an area, which had been prepared for the potential contamination (i.e., an area "expected by design"). The criteria are therefore not applicable.
Rating for actual consequences	Level 2.

4.1.5 Fire in decontamination equipment at former nuclear manufacturing plant of Bosco Marengo, Italy (Climate sceptics, 2012)

On December 22, 2011 SOGIN, the operating organization responsible for the decommissioning of Fabbricazioni Nucleari (Nuclear Manufacturing) plant at Bosco Marengo, Alessandria, Italy, notified the Italian Regulatory Body (ISPRA) of an abnormal event that had occurred at a shot blasting decontamination machine while it was in operation inside the plant. This machine is used to decontaminate for free release very low-contaminated materials, arising from decommissioning activities.

A statement given by SOGIN reported that during the machine operation, a bearing of the fans of the final filtration system had overheated. This caused a fire in the filtration section with smoke being released locally. Immediately after the event, emergency procedures were activated, and the fire team stepped in. The fire was soon extinguished. The Alessandria fire brigade confirmed that the fire was put out and the area was safe again. The emergency case was declared closed. Radiological measurements done by the Health Physics team ruled out any contamination of offsite areas, or contamination of personnel and equipment in the controlled area. The measurements confirmed that the levels of surface and air contamination in the controlled area were the same as those measured before the event; similarly, the contamination levels of the chimney filters were unchanged.

From the beginning of the event, regional authorities monitored the environmental contamination in cooperation with ISPRA. Environmental data show that the concentrations of α and β activity in airborne particles inside and outside the plant are comparable with background values. The event was rated at INES level 1.

Rating criteria	No	Yes
Contamination spread within the facility	◉	○
Damage to radiological barriers	◉	○
Degradation of defense-in-depth	○	◉
Overexposure of a worker	◉	○
Person injured physically or casualty	◉	○
Overexposure of a member of the public	◉	○
Release beyond authorized limits	◉	○
A continuing problem	◉	○

Many events in this book have been assessed and rated according to INES criteria, including those reported in Sections 4.1, 4.2, 5.36, 5.41, 5.46, 6.8, 6.9, 6.17, 6.18, 6.23, 6.26, 6.28–6.31, 7.13, 7.20, 7.21, 7.23, 7.24, 7.29, 7.35, 7.36, 7.38, 7.39, and 7.46.

4.2 UK Office for Nuclear Regulation

Depending on the objective the UK Office for Nuclear Regulation (ONR) gives to its reports, various classifications have been adopted for nuclear and radiological events. A very detailed classification is given in ONR (undated). This reference puts ONR's categorization in the light of INES. A decommissioning-oriented elaboration of ONR categories is given in Table 4.1: Categories unrelated to decommissioning have been highlighted, and explanations are given for the lack of coverage in this book.

TABLE 4.1 ONR categorization of decommissioning events.

Category	Event description	Relevance to decommissioning and coverage by this book
1	Administration matter	Relevant
2	Automatic or manual reactor trip following plant anomaly	Not relevant (operating state)
3	Conventional safety event	Relevant
4	Environment event regulated by the Environment Agency or the Scottish Environment Protection Agency	Not relevant (UK specific)
5	External event	Relevant
6	Flask or package movement or transport event	Relevant (covered only from the viewpoint of decommissioning organization and receiving party)
7	Unplanned leak or spillage of active or potentially active process liquor or material	Relevant
8	Unplanned leak or spillage of contaminated or potentially contaminated coolant	Relevant
9	Lifting event with nuclear safety implications	Relevant
10	Media interest only	Not relevant (not specific to safety and health impacts)
11	Nonadherence to a safety-related procedure	Relevant
12	Noncompliance or inadequacy associated with arrangements made in the interests of safety	Relevant
13	Noncompliance with conditions and limits necessary in the interest of safety	Relevant
14	Radiation or contamination event affecting personnel	Relevant
15	Safety-related plant defect or degradation	Relevant
16	Safety-related plant requirement incorrectly implemented	Relevant
17	Smouldering/smoking material or fire	Relevant

A simplified categorization is offered by ONR (2016). Regulatory functions in response to categorized events include the following:

Nuclear and radiological safety
Fire safety
Nuclear security
Transport
Conventional health and safety

ONR (2016) covers the period 2001–15 and includes 3866 events at both operating and shutdown facilities in the United Kingdom. Therefore, it can be considered a representative picture of the nuclear situation in a developed country with many old facilities. However, during this time, ONR's (or its predecessors) responsibilities have been broadened to encompass fire safety at UK nuclear sites, nuclear security (2007), and transport of certain materials (2011). This restricts the applicability of numerical comparisons of reporting rates from year to year. Security events are not mentioned in ONR (2016). Conventional safety events are listed separately.

Of the 3866 events, 3141 were rated on the INES scale as being of no nuclear safety relevance (INES level 0 or not rated), and 716 were rated at INES level 1, which is the lowest level of INES relevance. There were eight events rated at INES level 2 (incident) and one event—at the operating THORP plant—rated at INES level 3 (serious incident). No events were given a higher INES rating during the reporting period. It appears that no incidents above level 2 ever occurred at a decommissioning plant or can be related to conditions typical of decommissioning plants (although this is somehow a subjective judgment). This ample database confirms the general statement—supported also by other sources—that no decommissioning-related events have yet been given an INES rating higher than 2. After ONR (2016) was published, two level 2 events occurred at Sellafield, one in 2017 (Section 6.26) and the other in 2019 (Section 7.13).

A cursory analysis of the data indicates a significant rising trend in the number of events of "no nuclear safety relevance" over the years. This trend should be regarded as positive because near-misses and other minor events contain significant lessons and improvement opportunities. Besides, a higher frequency of minor events, while the number of more significant safety events does not increase, appears to be consistent with a more conscious, proactive, and transparent safety culture embraced by the operating organizations.

None of the INES level 2 cases had any adverse impact on the public or the environment.

Another observation is that the categories associated with non–nuclear-specific events (conventional health and safety, fire events, events originating outside of the nuclear site, and transport events) constitute a major share (some 24%) of the total events reported over the period 2001–15.

4.3 German regulations

AtSMV (2010) provides official guidance on reporting criteria for nuclear facilities in decommissioning. The events are grouped under the following categories:

1. Radiology and radiation protection
- **1.1** Planned, operational discharges of liquids, gases, or airborne particles on designated ways
- **1.2** Unplanned, accidental releases into the environment, e.g., caused by component failure or human error
- **1.3** Contamination
- **1.4** Spreading of radioactive substances outside of designated areas

2. Plant technology and operation
- **2.1** Malfunctions, damage, or failures in safety-relevant facilities
- **2.2** Damage or leakage to pipes or containers
- **2.3** Falling of loads; events during handling, storage, or transport of radioactive substances
- **2.4** Other events

3. External and internal events
- **3.1** External events
- **3.2** Internal events

Each category includes several reporting criteria. GRS (2013) offers a statistical analysis of 237 events (NPPs and research reactors) submitted from 1995 to early 2013 in compliance with 14 reporting criteria: Of these, five reporting criteria include 80% of all events, namely:

2.1.1 Malfunction in safety-relevant systems

2.1.2 Failure, damage, or finding, indicating a systematic error in safety-relevant systems

2.1.3 Failure/damage to fire protection equipment

2.2.1 Leakage

2.4.1 Security specification change required.

Only 9 events out of 237 are classified "E" (urgent notification by telephone and writing within 24 h after the event has come to the attention of the operating organization). All other events are not subject to the same stringent requirement.

As decommissioning progresses, the (few) events reported show an increase of causes belonging to category 2.1.3 and a decrease of causes under category 2.2.1.

4.4 US Department of Energy

A comprehensive classification of decommissioning incidents is given in DOE (1998).

This publication lists 658 occurrences for the period June 1990 to August 1997. Each case is identified through brief occurrence description, work type (e.g., characterization, decontamination, demolition, etc.), hazard (radiological, chemical, physical exposure/trauma, biological), environmental, safety, and health (ES&H) consequences (see the following), significance ranking (in classes, see the following), date of occurrence, and identification code and number.

Subcategories for ES&H consequences include the following:

- Radiological hazards
  - Radiological exposure
  - Internal inhalation/injection
  - Clothing/skin
  - Environmental contamination
- Chemical hazards
  - Chemical exposure
  - Internal inhalation/injection
  - Clothing/skin
  - Environmental contamination
- Physical trauma hazards
  - Impact/trauma
  - Puncture/cut
  - Slip/trip/fall
- Physical exposure hazards
  - Burn
  - Fire/explosion
  - Electric shock
  - Asphyxiation

The significance ranking is structured in four classes as follows:

Class 1: Occurrences resulting in an injury or exposure requiring hospital treatment or confirmed, severe environmental effect; also occurrences with potential to cause these effects with all safety barriers down, except, for example, that workers were not nearby to be injured or exposed, or escaped in time, or the climate conditions were favorable: Very significant

Class 2: Occurrences characterized by minor injury (e.g., first aid/safety and health personnel in attendance) or exposure, minor environmental damage, or temporarily uncontained hazard excursions into the environment; also near-misses where one additional safety barrier remained to prevent consequences to those in class 1: Relatively significant

Class 3: Potential precursors to occurrences in class 1 or 2: Minor significant.
Class 4: Minor but potentially frequent occurrences, such as leaks, spills, releases and excursions frequently with no ES&H consequences or impact: Insignificant

As detailed in DOE (2017), there are 10 reporting criteria groups as follows:

Group 1—Operational Emergencies
Group 2—Personnel Safety and Health
Group 3—Nuclear Safety Basis
Group 4—Facility Status
Group 5—Environmental
Group 6—Contamination/Radiation Control
Group 7—Nuclear Explosive Safety Group
Group 8—Packaging and Transportation
Group 9—Noncompliance Notifications
Group 10—Management Concerns and Issues

Each reporting criterion is identified by Group, Subgroup (if applicable), Sequence number, and Report Level (RL). RLs indicate the impact associated with a given occurrence in terms of health, safety, and security to workers, the public, the environment, and the operational mission. The three RLs are high, low, and informational. The detailed list and description of occurrences given under (DOE, 2020) follow the classification given in DOE (2017).

4.5 US Nuclear Regulatory Commission

The US Nuclear Regulatory Commission (NRC, 2002) includes a comprehensive list of decommissioning incidents/accidents based on licensing documents submitted by the 20 reactors reviewed (many events are partly overlapping or are plant-specific descriptions). With the exclusion of fuel-related and offsite transportation events (out of scope in this book), the NRC (2002) list has been elaborated by the author and is given in Table 4.2. Please note that the events listed in the following are not actual occurrences, but assumptions (plant-specific or established by the safety assessor).

In some cases, the accident could easily be placed under more than one heading. For example, one licensee reviewed an explosion and/or fire in the ion exchange resins. This accident could have been placed under "Explosions," "Fires," or "Spent Resin Accidents." Another example would be the "oxyacetylene explosion and release of HEPA filter contents," which was considered by several licensees. This accident could have been listed under either "Explosions" or "Loss of HEPA filters."

TABLE 4.2 Accidents involving radioactive materials (non-fuel-related), elaboration from (NRC, 2002 App I).

Decontamination-related accidents
Spray release during in situ decontamination of systems Gross leak or accident during in situ decontamination (spray and liquid) Decontamination of liquid spill Accidental spraying of contamination with high-pressure jets
Radioactive material (non-fuel) handling accidents
Waste container drop and rupture (containing activated concrete rubble) Dropping of filters or packages of particulate material Dropping of contaminated components Packaging events Materials-handling event Steam generator drop inside/outside containment Dropping the reactor pressure vessel
Dismantling-related accidents
Contamination release during accidental cutting of contaminated piping Loss of engineering controls during dismantling of reactor cavity Contamination release during dismantling of main coolant system loop Dismantling of reactor coolant system and safety injection piping without or with loss of local engineering controls Absence of blasting mat during explosive removal of activated concrete
Loss of HEPA filters
Rupture of contamination-control envelope; release of contamination on HEPA filter Loss of integrity of portable filtered ventilation enclosure Temporary loss of local airborne contamination control during blasting Temporary loss of local airborne contamination control during scarfing of contaminated concrete surfaces with jackhammer Loss of contamination-control envelope during oxyacetylene cutting of reactor vessel
Radioactive gas waste system leaks
Radioactive liquid waste releases
Liquid waste storage vessel failure Liquid radioactive waste release to lake through earthquake-induced cracks in building Rupture of spent fuel pool, contents released to bay Liquid waste discharge pumped to river without sampling Leaks and failures in radioactive liquid waste system Condensate storage tank contents pumped into ground during in-service leak test
Containment breach (open penetration to containment)
Containment vessel breach, subsequent loss of contents to air/water Open penetration—unfiltered pathway from containment

TABLE 4.2 Accidents involving radioactive materials (non-fuel-related), elaboration from (NRC, 2002 App I).—cont'd

Spent resin accidents
Spent resin handling accident (exothermic reaction during dewatering) Dropped resin vessel during removal from containment building Release of resins from makeup and purification demineralizer Storage of spent resins Explosion and/or fire in ion exchange resins
Vacuum filter bag
Vacuum filter bag rupture during decontamination of spent fuel pool floor Vacuum filter bag rupture during cleaning of the reactor building floor Vacuum canister failure
Loss of electric power
Loss of offsite power affecting HEPA filters, etc.
Loss of compressed air
Fire
Fire inside reactor vessel Fire inside stairwell Fire in reactor building or fuel handling building Fire in storage facilities Fire in intermodal container of waste Fire in combustible waste piled in yard Fire in low-level radioactive waste storage building Combustible waste fire in 200-L drum container Contaminated clothing or combustible waste fire Contaminated sweeping compound fire (sawdust with oil and other additives, used to enhance collection of loose surface contaminants) Fire initiating residual sodium release
Explosion
Explosion of liquid propane gas leaked from front-end loader in containment Oxyacetylene explosion in the containment building while cutting reactor coolant system piping and release of HEPA filter contents within portable enclosure Explosion of oxyacetylene during segmenting of reactor vessel Explosion inside warehouse Explosion of large fuel-oil storage tanks Detonation of unused explosives in reactor cavity Sodium interaction with water caused by water inflow through a crack in a tank Onsite transportation accident
Aircraft crashes
Water level alterations
Site flooding
Flood, lake level oscillations, and tsunamis
Minimum water level, lake surges or waves

Continued

TABLE 4.2 Accidents involving radioactive materials (non-fuel-related), elaboration from (NRC, 2002 App I).—cont'd

Winds
Wind and tornado missiles
Other accidents initiated in external events
Earthquakes Volcanoes Lightning Forest or brush fire Freezing temperatures, loss of plant heating Physical security, intruder event
Others
Hazardous non-radiological chemical events Chemical combustion (from sodium-water interaction) and dispersal

4.6 European Commission

Give me a fruitful error any time. Full of seeds, bursting with its own corrections.
Vilfredo Pareto (1848–1923)

In an unofficial EC report (EC, 2014), NPP decommissioning events have been reviewed and classified in the following groups:

- Leakage of radioactive liquids
- Breach of technical specifications
- Contamination and radiation protection issues
- Inadvertent activation of equipment
- Control of nuclear and radiological substances
- Others

The reader will note that this classification is somehow different from all others given in this book. Some additional explanation on this classification is given in the following to provide informative detail on a number of events described in this book.

Leakage of radioactive liquids: Leakages are typically caused by design inadequacies, human errors, and malfunction of instrumentation. Near-term actions consist of stopping the leakage, cleaning and decontaminating of the affected areas, and fixing of the leaking equipment.

Breach of technical specifications: This category includes also operations beyond the design basis and designated operating ranges. These events are primarily originated by human errors, due to inattention to or misunderstanding of the procedures (however, not always are the procedures wholly adequate or easy to follow). Secondly, many such events are due to the specific configuration of the installation being decommissioned. In these cases, the violations are due to technical specifications not being updated consistently with plant modifications resulting from decommissioning (for instance, the technical specifications could request tests and maintenance on a component that is no longer used).

A good deal of these events is violation related to the handling of heavy loads, improper lifting and securing of the loads, prejob tests, etc. The recurrence of these events at several nuclear facilities is due to the fact that heavy loads are handled more often and on different, unfamiliar paths during decommissioning than during former operations.

Corrective actions in this regard include enhanced training; drafting, review, or updating of procedures and safety assessments; and organizational improvements.

Contamination and radiation protection issues: This group typically includes improper handling of contaminated materials and the unexpected discovery of radioactive substances outside of the controlled area. The main causes are inadequate planning, poor encapsulation of contaminated equipment, errors in dose estimates, and neglect of the radiation protection program (for cases of activity discovered outside the controlled area). Corrective actions include the amending of wrong estimates, improving procedures and training, and upgrading the radiation protection program. Any active substances identified beyond the controlled area are removed, and the affected area is barricaded and then cleaned.

Inadvertent activation of systems or equipment: These events include the inadvertent activation of plant systems, or, vice versa, the failure to duly activate plant systems. Direct causes are component faults or false signals, caused in turn by material deterioration, operational errors, or electric disruptions. Corrections include the following sequential actions: check why the activation occurred, replace the equipment, and fix the system. Improvement of the procedures and reinforced training are often implemented in consequence of these events.

In some cases, it turned out that the faulty system was unnecessary for plant decommissioning, which highlighted that insufficient attention was given to the removal of some equipment or systems that had become redundant in the operation-to-decommissioning transition. The corrective actions in this case include the deactivation and removal of the affected equipment or systems and the abrogation of all related surveillance and maintenance requirements.

In other cases, still-necessary systems are inadvertently deactivated typically due to of human errors. The corrective actions include the reinstating of the wrongly deactivated systems and improvements in planning, training, and supervision of nonroutine activities. These events normally do not present risks, nor noxious impacts, but indicate weaknesses in the planning, management, and monitoring of the project, which can harbinger more troublesome events.

Control of nuclear or radioactive substances: Missing of nuclear or radioactive substances defines another group of decommissioning events. Such events are typical during surveys, when written records are cross-checked against the physical identification of substances, or when the decommissioning operations are meant to manage such substances. In some cases, the direct cause is operational (e.g., equipment that becomes contaminated and is not controlled); in other cases, the cause is historical, and old nuclear or radiological substances are involved (legacy material). The control requirements for these substances have evolved over the years and have become much stricter. Besides, the record-keeping media for nuclear or radiological substances have become much more efficient (e.g., from paperwork to electronic data storage). Corrective actions commonly include a search for the missing substances, the assessment of the impacts of a potentially inadvertent shipment of these substances, and the upgrade of control and surveillance procedures (in both the licensee and regulators' realms).

Others: Several other events cannot be categorized as described earlier. Although these events are quite dissimilar (design flaws, operational errors, loss of offsite power, fires, lack of coordination of different teams, etc.), a common element is the need to adjust the organizational structure and practices to the decommissioning objectives.

Corrective actions include the interruption of the works until the situation is stabilized, the upgrading of training and coordination, and enhanced supervision of the works by the decommissioning organization.

4.7 Others

Needless to say, the simplest classification would be based on facility type, e.g., nuclear power plants; research and test reactors; nuclear fuel cycle facilities (enrichment, conversion, fabrication, reprocessing, etc.); and small medical, research, and industrial installations. This book will showcase studies from all of these facility types.

Other classifications are given in papers and reports. For example, Iguchi and Kato (2010) offers a classification based on experience of incidents (called "troubles") and specific examples at decommissioning nuclear power plants, as follows:

Fire
Drop
Leakage
Exposure
Failure
Human error

Security problem
Natural phenomenon
Chemical hazard

The ASME (2004) offers a list of possible accidents (and associated causes) as follows:

Falls from heights
Being hit by a vehicle
Being hit by a falling object
Tripping
Collapse of structures and machinery
Contact with moving objects or machinery.
Manual handling
Contact with electricity
Ingestion of hazardous substance
Failure of pressurized systems
Working in confined spaces.
Exposure to excessive noise
Electrocution
Shorting and flash burns
Lack of oxygen

Lists of decommissioning events are also given in plant-specific studies and reflect plant-specific assumptions. For example, the decommissioning of the Barsebäck NPP in Sweden is subject to the prescriptions of the Article 37 EURATOM (potential impacts on neighboring countries). In SSM (2019a), the following D&D external events have been assessed in regard to potential discharges of activity to the environment (and local environmental conditions affecting their impacts):

- Heavy rain, snow, or winds
- High and low sea level
- High sea temperature (water supply)
- Pollution and other transients in the water supply
- Lightning
- Earthquake
- Airplane crash

Highly unlikely D&D events are not assessed in this report unless they are supported by evidence (at least one case study).

The following D&D internal events have been assessed in regard to potential discharges of activity to the environment:

- Internal fire
- Internal flooding
- Lifting accidents (load drop)

- Faults in the power supply
- Faults in the ventilation system
- Faults regarding instrumentation and control
- Events that can cause missiles
- Unplanned activity emission to water

To identify the most relevant events for D&D, the aforementioned events have been evaluated further on the basis that the two Barsebäck reactors no longer stored any spent fuel or reactor internals located in the reactor pressure vessels and that there were no leading mechanisms such as high pressure or high temperatures.

In SSM (2019a), it is stated that, except highly unlikely events, internal fire can cause higher releases of radioactivity to the environment than all the external events. The limiting event is therefore a fire in the area containing the most combustible radioactive material.

In view of its future dismantling, the small reactor at Ågesta, Sweden, has been subjected to a similar study for the purposes of Article 37 EURATOM (SSM, 2019b). At Ågesta, the nuclear fuel has been long removed, and the plant has been kept in a safe enclosure mode for many decades. For Ågesta, the study states that, for example, a fire may cause a release of activity to the environment: However, the radioactivity is fixed in the plant structures and components (no combustible material remains), and even if a large fire occurred, the activity would not be easily released. Events potentially resulting in a radioactive discharge to the environment include the following:

- Fire
- Internal events (e.g., internal flooding)
- External events (e.g., external flooding, extreme weather or earthquake)
- Human errors (e.g., illicit removal of radioactive waste or supposedly valuable component)

Fort St Vrain (FSV), CO, US, was a 330 MWe high-temperature gas-cooled reactor. A study on postulated decommissioning accidents at FSV is given in Iacovino et al. (1992). According to this study, the risk of decommissioning accidents is considerably less than during operation, mostly due to fuel being removed prior to decommissioning. Six accidents were assessed in detail:

- Dropping of contaminated concrete debris
- Heavy load drop
- Fire
- Loss of prestressed concrete reactor vessel (PCRV) shielding water: The dismantling strategy employed at FSV consisted of filling the PCRV with water for shielding while internal components were taken away
- Loss of power
- Tornado

The fire event gave the largest dose of the six accidents, although well below all applicable legal or regulatory limits.

Founded on a database of 653 occupational accidents in demolition work documented in 1984–2012 in the US Occupational Safety and Health Administration (OSHA), a statistical study (ERTAS and Erdogan, 2017) highlighted the most frequent hazards as presented in the following (elaboration from ERTAS and Erdogan, 2017, Table 1). For the purposes of this book, it should be noted that industrial demolition events represent only a part (albeit significant) of the full range of nuclear decommissioning events.

1. Asbestos exposure
2. Overturn/collapse of buildings or structures
3. Electric shock
4. Slipping/tripping/falling/stumbling
5. Cuts/scratches/jamming/hitting/puncturing/manual operation
6. Machine accidents
7. Hit by falling objects/flying debris
8. Traffic accidents
9. Fire
10. Falling from heights
11. Control of chemicals hazardous to health
12. Dust
13. Noise
14. Personal protective equipment

4.8 International Atomic Energy Agency

The IAEA has not promulgated an official list of incidents/accidents occurring during decommissioning: However, a generic list of safety-related decommissioning hazards is given in Appendix VIII of (IAEA, 2013b) (Table 4.3).

In this table, there is no direct connection or "one-to-one" relation between a hazard and an accident/incident as described in this book and in other classifications.

If there are no events reported of a given kind, it does not mean there are no hazards potentially causing these events. The lack of reports only means that the responsible organizations controlled the hazards well in their embryonic phase and prevented them from occurring. However, the organization responsible for the decommissioning safety assessment should analyze/assess all potential hazards regardless of their actual occurring.

Decommissioning safety assessment deals first with normal operations (even if everything goes as planned, people will be getting some exposures, will have to work at heights, will be exposed to heat, etc.). In addition, the safety assessment deals with unplanned incidents and accidents that may happen during the work and proposes ways to prevent them and to mitigate

TABLE 4.3 Generic checklist for hazard identification (IAEA, 2013b).

Radiological hazards
Direct radiation sources Improper removal of shielding Radioactive material, including form (solid, liquid, gaseous) Criticality Contaminated liquid, material Other radioactive sources (smoke detectors, lightning rods)
Fire/explosion hazards
Oxygen Sodium Explosive substances Flammable gases (e.g. oxyacetylene, propane gas), liquids, dust Combustible/inflammable materials Compressed gases Hydrogen generation Overheating or fire, caused by, for example, portable heaters, overload of electrical circuits, application of cutting techniques
Electrical hazards
High voltages Power overload and shortcuts, power failures Inadequately disconnected circuits/prevention against inadvertent connection
Non-ionizing radiation hazards
Non-ionizing radiation sources, including lasers Electromagnetic radiation (e.g. microwaves) High intensity magnetic fields
Chemical/toxic hazards
Chemo-toxic material Spills Chemicals (aggressive chemicals) Accidental mixing/combination of chemicals (e.g. in sewage systems, decontamination work) Asbestos and other hazardous materials, such as lead or beryllium Pesticide use Biohazards
Physical hazards
Kinetic energy Potential energy (springs, Wigner energy in graphite) Degraded or degrading structures, systems and components Steam Temperature extremes (high temperatures, hot surfaces, cryogenics) High pressure (pressurized systems, compressed air)

TABLE 4.3 Generic checklist for hazard identification (IAEA, 2013b).—cont'd

Working environment hazards
Working at heights (e.g. ladders, scaffolding, man baskets)
Excavations, formation of underground cavities (subsidence) from rain, waste degradation, etc.
Vehicle traffic
Heavy lifts, material handling, heavy equipment, manual lifting, overhead hazards, falling objects, cranes
Inadequate illumination
Inadequate ventilation
Noise (high noise areas and tools)
Dust
Pinch points, sharp objects
Confined space
Dangerous equipment, e.g. power tools, compressed gas cylinders, welding and cutting, water jet cutting/decontamination, abrasive decontamination techniques, grinding, sawing
Remote work area
Obstruction of passageways or exits
Human/organizational hazards
Human error
Safety culture aspects
Assigning inadequate training for work steps
Assigning inadequate protective measures for work steps
External hazards/initiating events
Ambient temperature extremes
Aeroplane crash
Storm and adverse weather conditions
Earthquakes
Flooding
External explosions and fires
Other hazards
Degraded/corroded barriers, ageing of materials
Unknown or unmarked materials
Spills

their consequences (failure of equipment, loss of power, loss of ventilation, loss of confinement, fire, mechanical injury).

Safety assessment is prepared in advance, on the basis of the decommissioning plan, characterization results, and some other inputs (facility construction data, meteorological data, etc.). It looks at what has been planned and also tries to anticipate potential unplanned situations and to assess the safety of decommissioning activities on the basis of detailed analysis of planned activities and identified scenarios.

The list of hazards provided in IAEA (2013b) is supposed to be a checklist for those developing a safety assessment to go through it and ask such questions as follows: Is a given hazard relevant for my decommissioning project? Does this hazard appear during activities that go as planned (no surprises)? What can go wrong? So the list is a tool for hazard identification.

Then all the identified hazards are screened out to decide which ones need special attention and more detailed analysis, and which ones are acceptable as they are or are well controlled by existing safety measures. The last phase of the safety assessment includes detailed analyses using formulas, models, and computer codes, to calculate the consequences and to define in detail what safety measures are needed if any (time limitation, shielding, distance, respiratory protection, ventilation, ad hoc tools or procedures, remote operations, special covers, nonflammable materials, etc.).

A number of decommissioning incidents are reported by the joint IAEA-NEA Incident Reporting Systems for Nuclear Installations (IAEA, undated). The IAEA manages three distinct incident reporting systems (IRS) covering, respectively, nuclear power plants, research reactors, and nuclear fuel cycle facilities. Events refer to all phases of a nuclear installation's lifetime (decommissioning has only a fraction of cases).

Participating Member States submit event report to the IRS. Reports illustrate the event, analyze its causes, and indicate corrective actions and lessons learned. The national organizations involved include regulators, utilities, operators, technical support, vendors, research centers, and universities. Each participating Member State appoints authorized users responsible for submitting event reports and for circulating information from the IRS. The IAEA organizes regular meetings of authorized users. It should be noted that access to IRS is restricted to registered parties.

The NPP-IRS was established in 1980 and had 33 member countries in 2018. The Research Reactors—Incident Reporting System for Research Reactors (IRSRR) was established in 1997 and had 59 participants in 2018. The Fuel Cycle Facilities—Fuel Incident Notification and Analysis System (FINAS) was established in 1992 and had 31 participating countries in 2018.

The Report on Safety Lessons Learned from Nuclear Power Plant Operating Experiences Worldwide (2015–17) was recently published (NEA, 2020). The report groups events into three categories: human performance, equipment issues, and management and oversight.

4.9 Categorization employed by this book

The use of traveling is to regulate imagination by reality, and instead of thinking how things may be, to see them as they are Samuel Johnson (1709–84)

There are two basic categories of safety indicators:

1. Leading indicators, which are related to preventive actions (before the fact)
2. Lagging indicators, which are related to occurrences and impacts (after the fact)

Most organizations measure safety performance by lagging indicators such as number of reportable events or lost time injuries. Such statistical indicators may highlight the type, the occurrence rate, and the severity of the event, but they tell little about the causes. To prevent an accident, it is vital to identify its cause. Lagging indicators tend also to have a significant uncertainty, due to a large number of variables affecting the findings such as the willingness to report and how to report an event. Another disadvantage of lagging indicators is that they are reactive measurements. That is, an accident or incident must happen before the responsible organization can identify and implement corrective actions. Instead, leading indicators are proactive measurements associated with event prevention. Examples of leading indicators could be training courses and walk-downs. With the use of leading indicators, the organization can improve safety performance and promote organizational changes. Inadequacies in leading indicators can be associated with latent conditions (e.g., poor design, insufficient monitoring of processes, and unworkable procedures), precursor to coming events.

Being an overview of occurrences, this book does not aim at directly measuring safety performance or suggesting solutions to prevent incidents/accidents. However, it makes use of event descriptions (potentially resulting in lagging indicators), event analyses, corrective actions, and lessons learned (potentially resulting in the identification of causes and the development of leading indicators) (Nevhage and Lindahl, 2008).

IAEA (2013b) classification of hazards has been used as the basis of this book—albeit from a different viewpoint. The approach taken by this book assumes the main rationale for excluding or including hazards in a checklist should be their credibility (i.e., having occurred). Credibility can only be ascertained by facts. Otherwise, a list of hazards may tend to be just an exercise of imagination.

Included among cases (incidents/accidents) reported by this book are also potential cases and near-misses, precursor to events that were averted only by lucky circumstances. In the overview given in this book, these events are quoted as well, along with actual impacting incidents/accidents. Potential cases and near-misses are frequently reported in the decommissioning literature. Additionally, some events that occurred (or were close to occurring) in nonnuclear industries are quoted as by their own nature they could readily reappear in nuclear facilities.

Table 4.4 builds upon Table 4.3 (IAEA, 2013b)—with some deletions (when no applicable events were identified in the technical literature), additions, merging, and reformulations; some events that were not quoted in IAEA

TABLE 4.4 Generic Checklist for Hazard Identification (this book).

Fire/explosion hazards
Sodium (5.21, 5.44, 7.20) Combustible/inflammable materials, explosive chemical solutions, thermal reactions (5.29, 5.33, 5.45, Box 5.45, 5.47, 6.18, 6.19, 6.31, 7.42, 7.45) Hydrogen generation (5.21, 5.44) Overheating or fire, caused by heaters, cutting techniques, improper operations, malfunctions, etc. (4.1.5, Box 5.46, 6.18, 6.19, 6.31, 6.35, 7.20, 7.34, Box 7.35, 7.40)
Electrical hazards
High voltages (5.11, 5.25, 5.50) Power failures (5.39, 7.38, 7.40, 7.42, 7.43) Inadequately disconnected circuits, cables inadvertently cut or removed, circuits left energized (5.4, 5.18, 5.19, 5.25, 5.39, 5.48, 5.50, Box 7.21, 7.30, 7.34)
Chemical/toxic hazards
Chemotoxic material (5.20, 5.22, 5.27, 5.31, Box 5.31, 7.3) Spills of chemical solutions, chemical reactions (5.31, 5.44, Box 5.44, 5.47, 6.19, 7.37, 7.46) Hazardous materials, such as asbestos (5.23), lead, arsenic, copper (Box 5.22), mercury (7.28, Box 7.28), beryllium (5.38, Box 5.38, 6.19), PCBs (7.12) Biohazards (5.2, 5.14, 7.28, 7.35) Smoke (5.45, Box 5.46, 6.31, 7.20, 7.40)
Physical hazards
Degraded or degrading structures, systems and components, degraded/corroded barriers (5.13, 5.15, 5.17, 5.34, 6.20, 7.5, 7.19, 7.27, 7.28, 7.29, 7.31, Box 7.34, 7.38, 7.42) Structural collapse, demolition hazards (5.17, Box 5.23, 6.27, 7.19, 7.28) Workplace temperatures (high/low temperatures, hot surfaces, thermal cycles, cryogenics) (5.17, 6.33, Box 6.33, 7.22, 7.24, 7.29) High pressure (pressurized systems, compressed air, overpressure) (5.3, 5.21, 5.26, Box 5.31, 7.24, 7.34, 7.46)
External hazards
Storms, high winds, freezing, other adverse weather conditions (5.15, 5.17, 5.34, 7.22, 7.27, 7.29, 7.31, 7.47) Flooding (4.1.3, 7.2, 7.29, 7.30, 7.42) Lightning, earthquakes (7.43, Box 7.43, 7.47) External explosions and wildfires (7.43, 7.47) Gas/oil pipelines and tanks (7.45)
Working environment hazards
Working at heights and falls (e.g., ladders, scaffolding, elevated structures) (5.1, 5.5, 5.7, 5.8, Box 5.8, 5.17, 5.49) Slips, trips, falls, hazardous contacts (where the height of the fall is not the major factor) (5.30, Box 5.32, 5.34, 5.39) Excavations, formation of underground cavities (subsidence), waste degradation, etc.

TABLE 4.4 Generic Checklist for Hazard Identification (this book).—cont'd

(7.5, 7.11, 7.15, 7.18, 7.19, 7.44) Vehicle incidents, traffic control (5.2, 5.6, 5.24, 5.25, 5.32, Box 5.32, 5.39, 5.46, 7.43, 7.44, 7.47) Manual lifting, overexertion (5.37) Falling objects, heavy loads, cranes (5.12, 5.13, Box 5.13, 5.15, 5.16, Box 5.16, 5.36) Inadequate lighting (5.27, 5.49, 6.28) Inadequate ventilation (5.14, 5.27. 5.38, Box 6.5, 6.9, 6.18, Box 6.24, 6.30, 6.34, 7.29, 7.42) Noise (high noise areas and tools) (5.28), vibration (Box 5.28, 7.38) Dust (Box 5.22, 5.35) Pinch points, sharp objects (6.24, 6.26, 6.28) Confined/congested spaces, obstruction of passageways or exits, accessibility issues (5.2, 5.11, 5.34, 5.38, 5.41, 5.50, 6.28, 6.29, 6.33, 7.26, Box 7.28, 7.29, 7.40) Proximity to sensitive areas/components (5.16, 5.36, 5.39, 5.44, 5.45, 6.1, 6.13, 6.27, 6.32, 7.9, 7.22, 7.27, 7.28) Potentially dangerous or improperly used techniques, e.g., power tools (5.28), welding and cutting (5.28, 5.29, Box 5.46, 7.43), cutting torch (5.2, 5.18, 5.29, Box 5.22, 6.16, 6.18, 6.35) water jet cutting/decontamination (6.14), abrasive techniques, grinding, hot fragments, sparks (5.28, 5.33, 5.45, Box 5.46, 6.18, 6.23, 6.31, 7.43), sawing (5.9, 5.28, 5.43), explosives (Box 5.23), fixatives (6.30, 6.31, 6.32, 6.33, 6.34, 7.1, 7.7) Unknown or unmarked materials/components/contamination found (5.20, 6.4, 6.7, 7.29, Box 7.29) Loss of containment (5.44, 6.18, 6.34, 7.1, 7.2, 7.24, 7.27, 7.39, Box 7.42, 7.46)
Radiological hazards
External irradiation (6.25, 7.26) External contamination (Box 6.11, 6.15, 6.23, 6.24, Box 6.25, 6.29, 6.32, 6.33, 6.34) Internal contamination (4.1.2, 4.1.4, 6.1, 6.5, 6.9, 6.10, 6.11, 6.24, 6.26, 6.28, 6.29, 6.30, 7.1) Leakages, spills, spread of radioactive materials (solid, liquid, gaseous) (6.1, 6.2, Box 6.2, 6.3, 6.19, 7.3, 7.6, 7.15, 7.16, 7.17, 7.21, 7.22, 7.24, 7.26, 7.27, 7.32, 7.34, 7.37, 7.38, 7.39, 7.42) Unplanned releases of radioactive materials to the environment (6.32, 7.1, 7.5, 7.10, 7.13, 7.14, 7.20, 7.23, 7.25, 7.29, 7.33, 7.34, 7.35, 7.36, 7.41) Criticality risk and analysis (6.8, Box 6.8a, Box 6.8b, 6.20, Box 6.20, 6.21, 6.22) Contaminated land/water (6.7, 6.15, 6.32, 7.1, 7.4, 7.5, 7.8, 7.9, 7.11, 7.15, 7.16, 7.17, 7.18, 7.23, 7.25, 7.26, 7.28, 7.31, 7.32, 7.39) Nonionizing radiation sources, including lasers (5.40)
Human/organizational causes of reported events
Design deficiencies (5.17, 5.22, 5.27, 5.36, 6.30, 7.2, 7.13, 7.21, 7.35, 7.46, 7.47, Annex A-1 *passim*) Human error, violation of procedures (5.1, 5.2, 5.3, 5.6, 5.9, 5.17, 5.24, Box 5.32, 5.33, 5.39, 5.46, 6.2, 6.8, 6.16, 6.17, 6.26, 6.28, 6.30, 6.31, 7.1, 7.10, 7.21, 7.22, 7.24, Box 7.24, 7.37) Complacency, low priority given to decommissioning (5.3, 5.32, 5.40, 6.2, 6.26, 7.11, 7.21, 7.22) Lack of qualifications (5.6, 5.39, 6.16, 7.6)

Continued

TABLE 4.4 Generic Checklist for Hazard Identification (this book).—cont'd

Safety culture aspects (5.30, Box 5.30, 6.1, 6.8, 6.19, 7.22)
Inadequate records or as-built drawings (5.4, 5.16, 5.19, 6.3, 6.7, 6.25, 7.3, 7.6, 7.7, 7.9, 7.17, 7.46)
Inadequate planning, missing/unclear/conflicting procedures, lack of monitoring/supervision/surveying (5.7, 5.8, 5.17, 5.21, 5.25, 5.27, 5.29, Box 5.32, 5.39, 5.41, 5.42, 5.43, 5.45, 5.49, 6.1, 6.5, 6.6, 6.8, 6.12, Box 6.12, 6.13, 6.17, 6.19, 6.21, 6.24, 6.29, 6.30, 6.31, 6.32, 6.34, 7.1, 7.4, 7.6, 7.7, 7.10, 7.14, 7.17, 7.19, 7.23, 7.30, 7.35, 7.41, 7.44, 7.46, 7.47)
Inadequate/faulty instrumentation and other components (Box 5.10, 5.19, 6.22, 6.32, 7.3, 7.4, 7.6, 7.21, Box 7.22, 7.24, 7.38, 7.40, 7.41)
Insufficient maintenance (5.24, 5.34, 5.40, 7.5, 7.8, 7.21, 7.30, 7.32, 7.34, Box 7.34, 7.47)
Assigning inadequate training for work steps (5.1, 5.2, 5.8, 5.9, 5.12, 5.27, 5.33, 5.39, 5.40, 5.41, 5.42, 6.19, 6.35)
Assigning inadequate protective measures, e.g., PPE or wrong plant configurations (5.8, Box 5.8, 5.9, 5.10, 5.12, 5.15, 5.27, 5.29, 5.33, 5.40, 5.49, 6.1, 6.5, 6.9, 6.10, 6.11, 6.12, 6.28, 6.29, 6.30, 6.33, 6.34, 7.40)
Poor health of worker (5.32)
Insufficient communications (5.17, 5.39, 5.40, 5.41, 5.42, 5.48, 6.12, 6.13, 6.25, 6.35, 7.1, 7.10, 7.22, 7.32, 7.36, 7.47)
Time pressure (Box 5.32, 5.40, 5.45, 5.49, 6.16, 7.1, 7.14, 7.44)
Missing incorporation of lessons learned (5.6, 5.21, 5.41, 5.43, 6.19, 6.21, 6.35, 7.47)
Poor medical assistance, inadequate emergency preparedness/response (5.6, 5.27, 5.43, Box 6.1, 6.5, 6.26, 6.31, 6.35, 7.47)

Numbers refer to the book sections; Text Box numbers are given separately.

(2013b) come up in Table 4.4. To be comprehensive, certain categories of events are extracted from other categorizations and are also inserted in Table 4.4.

The table refers to the section(s) of this book describing applicable case studies. The reader will note that some case studies apply to more than one hazard as also indicated in the same table.

Categorizing accidents is not a simple, direct process. Frequently, an originating event can result in more than one type of accident. For example, the loss of electric power could cause a crane failure, resulting in a heavy load being dropped on a highly contaminated surface and spread of contamination. The same loss of electric power could affect the ventilation system and cause the loss of HEPA filtration and consequent release of airborne contamination. Alternatively, a single accident could be caused by different types of originating events. For example, the loss of spent fuel pool water could be caused by the loss of offsite power, a break in a pipe (e.g., from shearing the wrong pipe or shearing with wrong tools), or an external event (such as pipe damage from freezing temperatures or rupture of the pool structure due to an earthquake). Therefore, each category of events indicated in Table 4.4 can be referred to by one or more sections of this book.

The reader seeking to identify events of a similar nature or a similar cause should consult Table 4.4, where he will be addressed to relevant case studies. A point of caution: the number of events cited in each category has no reference to the frequency of actual events; however, the reader will note that most events are linked with administrative, procedural, or human errors. It is a rare event that happens regardless of identifiable responsibilities due to human action (or omission).

A number of external site-specific events during decommissioning for which no concrete cases were identified in the literature include earthquakes, volcanic activity, aeroplane crash, sabotage, and war. Likewise, certain internal events such as electromagnetic radiation (e.g., microwaves) or high-intensity magnetic fields apparently have caused no impacts of any interest. Brief mention has been made to some of these events in the text. However, as stated in Section 4.8, those responsible for the decommissioning safety assessment may wish to consider on a site-specific basis some of these unlikely events as potential hazards impacting the safety of workers and the public, and the protection of the environment.

4.9.1 The presentation of events in this book

Draw from others the lessons that may profit yourself; Terence (190–159 B.C.) The Self-Tormentor

Following introductory chapters and event categorizations orienting the readers, the bulk of this book is structured in case studies.

Case studies identify a research method to gain a better understanding of a subject (in this series of books, abnormal events in nuclear decommissioning). Case studies are intended to reveal the weaknesses of plans, technical and administrative processes, and human behaviors. Case studies are stories, ranging from anecdotal evidence to scientific investigation. In this book, they tell how a misguided approach or strategy caused (or went close to causing) abnormal impacts to health and safety of humans, or to the protection of the environment: in other words, how a misguided approach or strategy failed to achieve the project objectives. However, the outcome of the set of case studies presented proved that the environmental sustainability of nuclear decommissioning was not compromised by actual impacts on humans or the environment caused by decommissioning activities.

In this book, case studies are presented in accordance with a specified template and sequence:

- an outline of the decommissioning process in question and problems/difficulties/impacts experienced by either workers or the public, or affecting the environment;
- an analysis of the event and solutions considered or enforced to move the project ahead more safely; and
- lessons learned of general applicability.

The situations described are typical incidents/accidents/near-misses occurring during the execution of nuclear decommissioning projects. The information is not a full catalog of events: Those described in this book are a selection mostly based on completeness and identification of important issues. Although the events covered are not intended to be exhaustive, readers are encouraged to evaluate the applicability of these events to their own projects. In this way, case studies allow to evaluate methods and techniques based on others' experience—a sort of calibration.

Within a case study, lessons learned draw conclusions based on the past experience and provide for better behavior. The lesson includes recommendations, dos and don'ts. Organizations must learn; in many organizations, especially in the high-tech industries, and where safety is a major concern, acquisition and processing of lessons learned is a routine, managed process. The reader will note in many case studies cited in this book that inattention to lessons learned from previous events is generally viewed as a serious omission and blame on the organization(s) responsible for the decommissioning project. The various phases of a lesson learning process are described in Rowe and Sikes (2006). "A man learns more quickly and remembers more gladly what he scorns than what he approves and esteems" (Horace, Epistles).

This book groups all case studies under three major headings: nonradiological impacts (Chapter 5), radiological impacts on workers (Chapter 6); and radiological impacts on the public or the environment (Chapter 7). A cursory look to Table 4.4 will be enough to show that a single category of events—or even one and the same event—can concern the workers, the public, or the environment (or all three together) and can cause both radiological and nonradiological impacts simultaneously.

Finally, it should be mentioned that conditions leading to accidents and innovative solutions are an active research field. A broad review of safety-driven innovations is given in Jilcha and Kitaw (2017). As one specific example among hundreds, deterioration or collapse of structural concrete is mentioned in several case studies (e.g., Sections 5.17, 6.15, 7.28), and a related research project is indicated in Fig. 4.3.

Only a few studies have evaluated the safety issues involved in decommissioning nuclear installations, especially from a structural and industrial safety standpoint; the focus is generally on the radiological risks regardless of the fact that the most serious decommissioning risks are nonradiological. A research work illustrated in Byeol et al. (2020) highlights the inadequacy of

FIGURE 4.3 ORNL and the University of Tennessee work on ultrasonic linear arrays to detect degradation in nuclear concrete (Oak Ridge National Laboratory, Wiki Commons).

current deterministic risk assessment techniques, which fail to take account of the uncertainty and complexity of occupational hazards. Byeol et al. (2020) offers a fuzzy logic–based safety assessment method that analyzes and compares alternatives by means of step-by-step risk quantification and multidimensional approach. This approach allows to assess nuclear decommissioning risks that cannot be quantitatively assessed owing to a lack of data. Kwan Seong Jeong (2016) describes a virtual reality application to reduce the probability, and mitigate the impacts, of incidents/accidents during the decommissioning of nuclear facilities.

References

American Society of Mechanical Engineers (ASME), 2004. The Decommissioning Handbook. Two Park Avenue, New York, NY 10016-5990, US.

Federal Ministry of Justice and Consumer Protection Ordinance on the Nuclear Safety Officer and the Reporting of Incidents and Other Events (Nuclear Safety Officer and Reporting Ordinance - AtSMV), 2010. Appendix 4 Reporting criteria for reportable events in decommissioning facilities according to § 7 paragraph 3 of the Atomic Energy Act (Reference: BGBl. I 2010, 773 - 776) (German original: Bundesministerium der Justiz und für Verbraucherschutz, Verordnung über den kerntechnischen Sicherheitsbeauftragten und über die Meldung von Störfällen und sonstigen Ereignissen (Atomrechtliche Sicherheitsbeauftragten- und Meldeverordnung - AtSMV); Anlage 4 Meldekriterien für meldepflichtige Ereignisse in Anlagen in Stilllegung nach § 7 Absatz 3 des Atomgesetzes (Fundstelle: BGBl. I 2010, 773–776). https://www.gesetze-im-internet.de/atsmv/BJNR017660992.html.

Byeol, K., Joo Sung, L., Yong Han, A., 2020. Fuzzy based risk assessment for decommissioning concrete bioshield structures in nuclear power plants: structural risks and worker safety. Appl. Sci. 10, 2614. https://www.mdpi.com/2076-3417/10/7/2614.

Climate Sceptics, 2000. Flooding Event at the EUREX Facility, November 22. http://www.climatesceptics.org/event/690.

Climate Sceptics, July 16, 2001. Contamination of Workers during Dismantling of Alpha Dry Glove Boxes. http://www.climatesceptics.org/event/711.

Climate Sceptics, January 1, 2012. Start of a Fire in Decontamination Machine of Former Nuclear Manufacturing Plant of Bosco Marengo (AL) under Decommissioning. http://www.climatesceptics.org/event/958.

European Commission, Joint Research Centre, 2014. European Clearinghouse: Nuclear Power Plants Decommissioning Related Events. Report EUR 26559. https://publications.jrc.ec.europa.eu/repository/bitstream/JRC88897/summary_report_decommissioning_agm_mmr_online_version.pdf.

Ertaş, H., Erdoğan, A.S., 2017. An analysis of occupational accidents in demolition work. Civil Eng. Archit. 5 (2), 37–51. http://www.hrpub.org/download/20170330/CEA1-14808185.pdf.

GRS, Eich, P., Kaulard, J., Strub, E., 2013. Analyse von meldepflichtigen Ereignissen in der Stilllegung (in German), Gesellschaft für Anlagen- und Reaktorsicherheit, Schwertnergasse 1, 50667 Köln (Analysis of Reportable Events during Decommissioning, Association for Plant and Reactor Safety, Schwertnergasse 1, 50667 Cologne).

Iacovino, J., Augustine, T., Johns, J., March 1–5, 1992. Development of Accident Scenarios for the Decommissioning of the Fort St. Vrain Nuclear Generating Station. Waste Management Symposia, Tucson AZ, US.

International Atomic Energy Agency, 2013. Safety Assessment for Decommissioning, Safety Reports Series No. 77. IAEA, Vienna.

International Atomic Energy Agency, Incident Reporting Systems for Nuclear Installations https://www.iaea.org/resources/databases/irsni.

Iguchi, Y., Kato, M., October 2010. Risk-informed approach for the regulation of decommissioning of nuclear facilities. J. Eng. Gas Turbines Power 132.

International Atomic Energy Agency, INES, 2013. The International Nuclear and Radiological Event Scale- User's Manual, 2008 Edition. IAEA, Vienna.

Jilcha, K., Kitaw, D., 2017. Industrial occupational safety and health innovation for sustainable development. Eng. Sci. Technol. 20, 372–380. https://reader.elsevier.com/reader/sd/pii/S2215098616308497?token=CDB9992673BB3F69311A50CE24210F33E63DB04F5FB01FD5703B4422AB238BFEFCF9779222E6B64E49D88F55CB83362E.

Kwan Seong Jeong, et al., 2016. A training system based on virtual environments to prevent incidents and reduce accidents during decommissioning of nuclear facilities. Int. J. Saf. Secur. Eng. 6 (2), 330–340. https://www.witpress.com/elibrary/sse-volumes/6/2/1176.

OECD Nuclear Energy Agency, 2020. Nuclear Power Plant Operating Experiences from the IAEA/NEA Incident Reporting System 2015-2017. NEA, Paris. http://www.oecd-nea.org/nsd/pubs/2020/7482-npp-operating-experience.pdf.

Nevhage, B., Lindahl, H., 2008. A Conceptual Model, Methodology and Tool to Evaluate Safety Performance in an Organization. http://lup.lub.lu.se/student-papers/record/1786993.

Office for Nuclear Regulation (ONR), The INES System, Nuclear Directorate (ND) Event Records and Descriptions of Events Notified to ND. http://www.onr.org.uk/ines/ines-context-and-event-explanation.pdf.

Office for Nuclear Regulation (ONR), February 2016. Events Reported to the Nuclear Safety Regulator in the Period of 1 April 2001 to 31 March 2015. http://www.onr.org.uk/documents/2016/events-reported.pdf.

Podlaha, J, 2020. Personal communication from Josef Podlaha, ÚJV Řež, Czech Republic, to M. Laraia, 18 May 2020.

Rowe, S.F., Sikes, S., October 2006. Lessons Learned: Taking it to the Next Level, PMI Global Congress. Seattle, WA, US, pp. 21–24. https://www.pmi.org/learning/library/lessons-learned-next-level-communicating-7991.

Swedish Radiation Safety Authority (SSM in Swedish), April 07, 2019. General Data in Accordance with the Requirements in Article 37 of the Euratom Treaty, Decommissioning of the Barsebäck Nuclear Power Plant in Sweden, 2019. https://www.stralsakerhetsmyndigheten.se/contentassets/b99546b3a17a4b5593688b3ab1abaca8/2019-07-general-data-in-accordance-with-the-requirements-in-article-37-of-the-euratom-treaty.pdf.

Swedish Radiation Safety Authority (SSM in Swedish), May 09, 2019. General Data in Accordance with the Requirements in Article 37 of the Euratom Treaty, Decommissioning of the Nuclear Reactor Ågesta in Sweden 2019. https://www.stralsakerhetsmyndigheten.se/contentassets/c6cedb11efdd4f5f87b944ae2fa0dbcf/201909-general-data-in-accordance-with-theequirements-in-article-37-of-the-euratom-treaty.pdf.

US Department of Energy, August 1998. Statistical Evaluation of Department of Energy D&D Occurrences, DOE/EH-0578. US DOE, Germantown, MD, US.

US Department of Energy, January 17, 2017. Occurrence Reporting and Processing of Operations Information, DOE Order 232.2A. https://www.directives.doe.gov/directives-documents/200-series/0232.2-BOrder-A/@@images/file.

US Department of Energy, April 30, 2020. Occurrence Reporting and Processing System (ORPS) updated. https://data.doe.gov/MS/asp/Main.aspx.

US Nuclear Regulatory Commission, November 2002. Generic Environmental Impact Statement on Decommissioning of Nuclear Facilities, NUREG-0586, Supplement 1 Volume 1. App. I. https://www.nrc.gov/docs/ML0234/ML023470323.pdf.

US Nuclear Regulatory Commission, August 28, 2017. International Nuclear and Radiological Event Scale (About Emergency Preparedness) last update. https://www.nrc.gov/about-nrc/emerg-preparedness/about-emerg-preparedness/emerg-classification/event-scale.html.

Chapter 5

Nonradiological events

5.1 Worker fell 95 m to his death inside nuclear chimney (McWhirter, 2017)

5.1.1 Decommissioning process; problem encountered; impacts

In 1957, a severe fire occurred in Pile 1, one of two plutonium production reactors at Windscale (now in the Sellafield site); after the accident, both reactors were permanently shut down. Active decommissioning began in the 1990s (Fig. 5.1).

As Pile 1 chimney was heavily contaminated, it was decided to develop the dismantling process at Pile 2, where the contamination levels were acceptably lower. The chimney was 95 m tall, with an inner diameter of 10 m.

The case study described in the following is a tragic, but all too common, example of nonradiological accidents during nuclear decommissioning (or any construction/demolition project for that matter).

A worker was carrying out decommissioning work on a chimney at Sellafield nuclear plant in Cumbria, when he plunged inside the chimney on January 9, 2003.

A working platform was designed and positioned inside the chimney, and an acceptable work procedure was prepared. Contractor workers were required to operate from this platform, and as an added precaution, they were equipped with a lanyard (a fall arresting harness), which was attached to the edge of the platform. However, a worker deviated from the approved procedure when he left the safety of the platform. This deviation was not checked or noticed by anyone. Fig. 5.2 shows an industrial worker equipped with safety harness and fall arrestor.

The worker was standing on the working platform when a steel girder fell on to a ledge that was adjacent to the platform. Finding it difficult to retrieve the girder, he stepped off the platform and moved on to the ledge. As the worker was moving the girder, it tipped upward and fell through the gap between the ledge and the platform. A metal bracket cut the worker's harness. The weight of the girder took him over the edge, and the worker, no longer clutched by the harness, fell to the bottom of the chimney and suffered fatal injuries.

Nuclear Decommissioning Case Studies. https://doi.org/10.1016/B978-0-12-823700-7.00005-9
Copyright © 2021 Elsevier Inc. All rights reserved.

FIGURE 5.1 Historic view of the Windscale Piles, ca.1956. *Credit to DOE, Wiki Commons.*

FIGURE 5.2 Worker with safety harness and fall arrestor. *Photo by CEphoto, Uwe Aranas, Wiki Commons.*

Due to high contamination inside the chimney, the body of the worker had to be decontaminated before it could be evacuated by medical staff and health physics in personal protective equipment (PPE).

The investigations and the judicial process took nearly 6 years after the tragedy to complete. This period of time caused a similar delay to the decommissioning project and extra decommissioning costs. The damage to the credibility of the operating organization, the contractor (and perhaps of the whole nuclear sector), is likely to have been serious.

Sellafield Ltd. (then operating as British Nuclear Fuels Ltd.), the operating organization, pleaded guilty and was fined £150,000 ($185,000) and ordered to pay costs of £50,500 ($62,000).

PC Richardson & CO Ltd., the contractor tasked to carry out the chimney work at the site, pleaded guilty for failing to ensure safety of an employee. They were fined £100,000 ($123,000) and had also to pay costs of £25,000 ($30,000).

5.1.2 Analysis; solutions

The direct cause of this event was the deviation from the approved, safe work procedure: However, it was established by the regulators (UK's Health and Safety Executive) that if the work had been adequately monitored, the responsible organization would have realized that unsafe working methods were being adopted, and would have conducted a new risk assessment.

This tragic case highlights that when nonroutine operations (typical of nuclear decommissioning) are being carried out, assumptions about the level of understanding of the procedure on the part of the workers cannot be taken for granted, and additional monitoring is essential.

At the inquest, it was witnessed that workers would often abandon the platform to go out on to the ledge to retrieve any beams that had ended up there. Although safe procedures were in place, they were not always adhered to in practice.

The key point is that, regardless how detailed and rigorous procedures and risk assessments are, if those who carry the work are not well trained, their understanding of the procedure is uncertain, and their compliance with it is not monitored in situ, accidents are likely to occur. Safety management requires feedback, in the form of monitoring compliance with safe working practices.

If the approved procedure is found to be unsafe, uncomfortable, or hardly applicable, the workers (and their supervisors) should stop the work at once, report the problem, and demand that the drafters of the procedure revise it.

The use of mockups or 3-D models may often help.

5.1.3 Lessons learned

The following lessons learned have been extracted from this accident evaluation report.

1. A formal process should be prepared to ensure that both the operating organization and the contractor have sufficient expertise to safely execute the decommissioning work.
2. To implement this, decommissioning-related qualifications and skills should be clearly defined, enabling competent people to be involved in all phases of a decommissioning safety case.
3. The safety case should include dismantling plans and detailed techniques and procedures.
4. Hazards and risks should be identified and related responsibilities assigned to those who are best able to bear them.
5. Pretender health and safety plans should take due care of identifying situations and factors that could impact safety and requirements for enhanced management, supervision, and other worksite controls.
6. Procedures and project management plans must avoid misinterpretations and deviations.
7. Training documents of contracted workers must be routinely checked against stipulated training requirements before the start of work.
8. Operating organizations should use formal standards for performing work and challenging deviations communicated to contracted workers.
9. Prejob briefings should be carried out engaging contracted workers to reinforce the key requirements and highlight any expected problems.
10. Postjob debriefings should be arranged to identify emerging problems and promote improvements to work procedures.
11. Any changes to work procedures should be assessed to ensure the following:
    - consistency with existing approved documents including the safety case;
    - control of radiological and nonradiological risk; and
    - consistency with contractor's pre-ascertained competence and expertise.
12. Local surveys must be performed with a focus on working arrangements in use to identify the appearance of inadequate behaviors and procedural deviations.
13. Local audits should be performed to ensure that the contract is consistent with stipulated procedures.

Another case study also related to Windscale stack decommissioning is given in Section 6.50.

5.2 Working in confined spaces

5.2.1 Problem encountered 1; impacts: forklifts in confined areas need spotters, Nevada National Security Site (away from the principal area of activity), United States (DOE, 1999a)

A skilled mechanical technician, who was fully trained and qualified, was operating a forklift in support of a test. The task required him to operate the forklift near chemical containers. During the task, the technician was seen leaning out of his forklift. The technician was asked to cease working on the task pending an investigation.

The operator was restricted from operating forklift until repeating a Forklift Training course, including both the academic and the practical evaluation. The project manager met with facility employees to discuss the need to use a spotter near chemicals, when transporting chemicals, in congested areas, during critical lifts, and at all other times that they believe using a spotter will result in a safer operation. See Fig A-1.4 as an example of congested environment.

5.2.2 Analysis; solutions

The size of the fork truck precluded the technician clearly viewing the gas bottles he was attempting to move. The technician was leaning out of the fork truck in an effort to see where to place the prongs. There was the potential for losing control of the fork truck in an area that contained hazardous chemicals. The operator was trained to engage in the use of a spotter as required by company procedures and federal safety guidelines. The operator did not use a spotter for the part of the task that required operation near the chemical containers.

5.2.3 Lessons learned

This near-miss occurred in a congested environment, which is typical of decommissioning activities.

Skilled workers may become complacent as they repetitively operate familiar equipment within the same environment and therefore do recognize a situation that is potentially hazardous.

5.2.4 Problem encountered 2; impacts: personnel considerations when working in confined spaces, mound facility, Miamisburg, Ohio, United States (International Atomic Energy Agency, 2005)

The on-site workforce at the Mound facility stack decommissioning project was reluctant to working at the base of the stack to remove contaminated debris. Working within the confined space and less controlled environment of the stack was a new challenge to the workers. Many of them used sick and vacation leaves to dodge the work, and the 1-day task took 2 weeks.

5.2.5 Analysis; solutions

Supervisory personnel successfully managed to assemble a crew to complete the task, using focused communications and escorting the workers into the stack.

5.2.6 Lesson learned

It should not be presumed that decommissioning workers will readily adapt to new or changing working environments.

Examples of confined spaces that may be present in demolition sites include narrow passageways, silos, tanks, sewers, chambers and ducting, unventilated or poorly ventilated rooms, and access shafts. Pits, trenches, and other excavation works may also be viewed as confined spaces especially when access or egress is difficult.

The hazards associated with confined spaces include, but are not limited to:

- Toxic atmospheres causing such acute effects as impairment of judgment, unconsciousness, and death. Simple asphyxiants (i.e., gases that are so concentrated that they displace oxygen in the air) include argon, nitrogen, carbon monoxide, etc.
- Lack of oxygen. Low oxygen levels (below 19.5%) can cause rapid breathing, accelerated heart rate, clumsiness, emotional upset, and fatigue. As less oxygen becomes available, nausea and vomiting, collapse, convulsions, coma, and death can ensue.
- Flammable or explosive mixtures.
- Flowing liquids or free-flowing solids (typically in excavations).
- Lack of natural light.
- Dusts in high concentrations.
- Excessive heat.
- By-products of the task being undertaken, for example, cutting torch fumes.
- Visibility (e.g., smoke particles).
- Biological hazards—viruses, bacteria from fecal matter and sludge, or molds.
- Symptoms of claustrophobia, e.g., fear of losing control, fear of fainting, trembling.

Working materials and support systems also should be taken into account, including

- fire or spark generation,
- waste removal,
- ventilation, and
- space needed for tool storage and handling.

In general, there should be a regular monitoring of the working environment to ensure a safe, viable workplace. This monitoring should take place before decommissioning work commences, and during work, to ensure the workplace changes (indeed the gist of decommissioning) do not make conditions unsafe. The working environment should facilitate safe operations, taking into consideration physical, physiological, and mental skills and limitations of workers, e.g, claustrophobia or awkward movements.

5.3 Contractor employee injury, Savannah River Site, South Carolina, United States (DOE, 2011a)

5.3.1 Decommissioning process; problem encountered; impacts

The R-Reactor (Building 105-R) is the largest of the five SRS reactors. It was also the first to start operation in 1953 with the mission of producing weapons material (e.g., tritium and ^{239}Pu). A few years later, it was evaluated that the production resources of the reactor were no longer needed: The reactor was permanently shut down in 1964 and remained in an inactive state for many years since. Option and end-state analyses were completed for Building 105-R in 2006. The structural analyses estimated that the sturdy, reinforced-concrete structures would remain unharmed for 500 years provided (1) the overhead cranes for the shield doors were taken away, (2) the stacks were demolished, and (3) concrete roofs were installed over specified areas.

Deactivation began in August 2008 and was practically ended by September 2010. The decommissioning scope included, among others, removal of above grade structures, removal of the stack, grouting of the reactor vessel and other below grade structures, and sealing of the entombed structure. The in situ decommissioning project (following the stack demolition referred to in this case study) was completed in 2011 (Serrato, 2013).

On October 12, 2010, a subcontractor employee was injured when the concrete hose that he was holding whipped out of his hand hitting him in the neck/head area. The employee and his crew were working on the 105-R building roof. They were involved in work activities associated with the preparations for the demolition of the building stack (Fig. 5.3).

The concrete pumping operation had been suspended due to a blockage at the pumping truck. Clearing of the blockage at the concrete pumping truck resulted in an air pocket forming in the line. When the air pocket reached the end of the hose, it caused the line to whip out of the hose man's hand resulting in the injury.

FIGURE 5.3 R Reactor as built (left), dismantling stack (right). *Credit to DOE, Wiki Commons.*

The injured employee was taken to a local hospital via an ambulance where he was examined. The employee was provided pain medication and returned to work later that same shift with no restrictions.

Description of direct cause: Based upon discussions with the subcontractor management, the hose man failed to comply with the worker protection/task-specific plan requirements as well as his training and experience when he handled the concrete hose before the air had been purged from it.

5.3.2 Lessons learned

According to the subcontractor management analysis, the root cause was attributed to complacency.

Personnel should not allow previous successful operations with a particular method or tool to decrease their awareness or focus on the task at hand.

They should not allow themselves to become so comfortable with the task that the real hazards associated with it end up being perceived to be diminished or unlikely.

5.4 Inadequate as-built drawings lead to contact with underground electrical line, Hanford Site, Washington, United States (DOE, 2006a)

5.4.1 Decommissioning process; problem encountered; impacts

On June 20, 2006, a Washington Closure Hanford (WCH) Field Remediation (FR) subcontractor was performing road maintenance on a gravel haul road with a motor grader. This was a work activity performed routinely since the beginning of backfill operations. While grading the road, the motor grader caught the edge of a splice box/hand-hole enclosure buried under the haul road.

The tug on the enclosure caused the live line that ran from the enclosure to a junction box to disconnect from the junction box conduit. The act of pulling the cable out of the junction box placed tension on the fuse connection, pulling and breaking the lower fuse block from the main panel, and dislodging a fuse in the panel. During the event, the wires had melted the insulation and arced to the metal conduit.

The enclosure was not identified on any drawing, nor was it clearly marked in the field per National Electric Code. There was another enclosure nearby that had been located during a field walk-down and was marked with a construction candle.

No one was injured and there was no damage to the grader.

5.4.2 Analysis; solutions

The electrical system involved in this event was installed several years earlier by another remedial action subcontractor. As-built information was provided and filed upon completion of the job. As-built data were retrieved from the files and incorporated on drawings provided to the WCH subcontractor. The original as-built information was not sufficient, however, to accurately reflect the features installed. WCH and their subcontractor relied upon the data and did not verify and mark the physical location of the underground utilities: This additional step could have prevented this incident.

5.4.3 Lessons learned

As a point of caution, utility information on drawings should be a priori considered inaccurate or missing important detail, and appropriate field verification should be conducted in planning for activities that would potentially interfere with utilities.

Electrical hazards are of concern during decommissioning. During the operations-to-shutdown phase, licensees often rewire the site to eliminate redundant electrical circuits or repower certain operations from outside the buildings. During safe enclosure periods, monitoring equipment may have to be installed, and some systems will have to be depowered. All these activities, and others (maneuvering trucks and cranes near power lines (Section 5.25), inadvertent digging near buried cables (Box 7.42), etc.), pose electrical hazards to workers.

If underground cables are damaged, workers can be injured or even killed by electrocution, electrical arcs (and resulting explosion), and flames, which can cause serious burns to hands, face, and body. Such events occur when a cable is

- cut through by a sharp object or a tool or
- crushed by a heavy truck or excavator.

Underground electrical cables can be particularly hazardous because they often look like pipes, and it is impossible to visually tell if they are live.

Excavation work should be properly managed to control risks, including the following:

- Accurate planning
- Using cable maps
- Cable locating devices
- Safe excavating practices

Several relevant case studies are given in https://www.hse.gov.uk/electricity/excavations/index.htm.

5.5 Operational safety incident at BR-3 reactor, SCK-CEN center, Mol, Belgium (International Atomic Energy Agency, 2006a)

5.5.1 Decommissioning process; problem encountered; impacts

The BR-3 reactor was the first PWR (pressurized water reactor) in Western Europe to become critical (1962); it was shutdown in 1987. Within the umbrella of the European Commission's 5-year programs for research and technological development for the decommissioning of nuclear facilities, BR-3 was selected, together with three other European facilities, as a pilot project for the testing/validation of decommissioning strategies and methods for nuclear installations. These projects aimed to develop the scientific bases and technical know-how for planning and execution of the decommissioning of commercial, large-scale installations. According to current plans, BR-3 will be completely decommissioned by the end of 2023.

During BR-3 decommissioning, an employee at work in a radiologically controlled area was worried about the potential contamination of his ladder. This ladder was made of aluminum to facilitate decontamination, but the worker was worried that the rubber pieces at the ends of the ladder would get contaminated. The worker then secured those parts against contamination by wrapping them with plastic bags. When the ladder was used in the controlled area, it slipped. The worker fell and broke a leg, which kept him away from work for 3 months.

5.5.2 Lessons learned

In the optimization of decommissioning work, attention should be paid to both radiological and nonradiological risks. The interrelations between these two aspects can be significant and may demand an expansion of the ALARA concept to a broader scope than radiation protection.

5.6 Subcontractor fatality at the Pond B Dam Upgrade Project, Savannah River Site (SRS), South Carolina, United States (DOE, 2004a)

In the discovery of secret things and in the investigation of hidden causes, stronger reasons are obtained from sure experiments and demonstrated arguments than from probable conjectures and the opinions of philosophical speculators of the common sort.

William Gilbert (1540–1603).

5.6.1 Decommissioning process; problem encountered; impacts

The SRS is located halfway between Augusta, Georgia, and Aiken, South Carolina. Except for the installations operated for the National Nuclear Security Administration, many of the original SRS production installations are being decommissioned, and large waste management operations associated with the decommissioning program are conducted under the jurisdiction of the US Department of Energy's (DOE) Office of Environmental Management.

The site is managed by a team led by Westinghouse Savannah River Company, LLC (WSRC), which is responsible for SRS nuclear facility activities, environment, safety, and health (ES&H), quality assurance, and all administrative functions and the Savannah River National Laboratory. Among major subcontractors to WSRC one should mention in this context Bechtel Savannah River, Incorporated (BSRI), which is in charge of environmental restoration, project management, and engineering and construction work.

The dam referred to in this case study is managed under the Federal Energy Regulatory Commission (FERC) guidelines. WSRC and FERC inspections confirmed that the prejob conditions of the Pond B Dam did not fulfill FERC guidance and were in need of repair. In December 2003, BSRI granted a subcontract to Grade South, Inc., to make the needed upgrades to the SRS's Pond B Dam to bring the dam into compliance with FERC guidelines. For part of the work, Grade South had to use an excavator rented from Metropolitan Tractor. The deceased driver was an employee of Guthrie Grading and Hauling, which had been hired to return the excavator to Metropolitan Tractor. The driver's accident occurred at the Pond B Dam toward the end of the upgrade project.

On July 26, 2004, a truck driver was crushed by an excavator at the Pond B Dam while he was loading the machine onto a low-bed trailer for return to the rental company. The accident occurred when a driver from Grade South, Inc. and a driver from Guthrie Grading and Hauling tried to rearrange an outrigger beam on the low-bed trailer. The beam had been displaced when the excavator was being positioned on the trailer. The Grade South driver employed the excavator's arm to raise one of the excavator tracks off the trailer so that the other driver could rearrange the beam under the track. The excavator dropped while the driver was rearranging the beam, and the driver was crushed (Fig. 5.4).

5.6.2 Analysis; solutions

Two days after the accident, the Acting Assistant Secretary for Environment, Safety and Health appointed a Type A Accident Investigation Board to analyze causal factors, identify root causes, and define Judgments of Need (JON) to

FIGURE 5.4 John Deere excavator similar to the one involved in the SRS accident. *SRS*, Savannah River Site. *Credit to Torben Hansen, Wiki Commons.*

prevent future accidents of this type. The Board completed their investigations on September 1, 2004. Results from these investigations allowed to perform a causal factors assessment as specified by the DOE Workbook, Conducting Accident Investigations. The investigation process dictated by DOE rules is meticulous and referred to in (DOE, 2004a) and other references describing DOE Type A and B investigations: However, the investigation process itself is not described in this book, only its conclusions. In general, a Type A investigation is performed for the more serious accidents and is appointed and managed by DOE's Office of the Assistant Secretary for Environment, Safety, and Health; a Type B investigation is appointed and managed at the site level. However, the components of the investigation process and the report format are the same. Categorization and investigation criteria are detailed in DOE Order 225.1A, 1997 (https://www.emcbc.doe.gov/Content/Office/doe_o_225.1a.pdf).

According to DOE terminology, causal factors are events or circumstances that produced or contributed to the event and consist of root, observed, and contributing causes. The root cause of the accident is the basic cause that, if fixed, would prevent the accident from happening again. The observed cause is the immediate event or circumstances producing the event. Contributing causes are events or circumstances that, together with the other causes, increase the probability of the event but, singly considered, did not produce the event.

The Board stated the following root causes of the event:

- The DOE, WSRC, and BSRI ignored the program inadequacies in the delivery and enforcement of safety requirements for subcontracted work at the Pond B Dam Project.
- The subcontractors' unsystematic approach failed to ensure that safety requirements were incorporated into work control specifications; did not enforce compliance with such requirements; nor did they define their safety objectives prior to starting work.
- The WSRC did not incorporate the causal factors stemming from previous occurrences into a corrective process.

The Board further concluded that the observed causes of the accident included the following:

- The deceased driver was working under or close to the excavator.
- The Grade South driver was not qualified to act on the excavator.
- Grade South, Inc. failed to exercise control over their staff.
- Stop-work actions were not taken.

The Board also described contributing causes: Among them, one should especially quote deficiencies in medical response. First, the emergency medical response was hindered by the lack of a detailed description of the incident due to insufficient communications. Second, the emergency responders did not rigorously follow the SRS Medical Protocols for trauma transport. The use of a helicopter could have lowered transport time to the Medical College of Georgia by at least 30 min.

In summary, the Board concluded that this accident was avoidable. Based on their assessments, the Board defined the proper JON in responding to this event. JONs are managerial tools and safety controls intended to prevent the likelihood of a recurrence or mitigate its impacts. They are also meant to assist managers in establishing corrections. All DOE Accident Investigations reports described in this book include JONs.

5.6.3 Lessons learned

Loading and unloading equipment (a common activity at construction and demolition/decommissioning sites) can produce a serious hazard should the equipment fall off of the trailer. In addition to excavators, equipment posing such a hazard includes, among others, risers, loading machines, and compactors. Typically, the weight of an excavator can be as high as 20 t.

In the United States, Title 10 Code of Federal Regulations Part 851 (Appendix A/Construction Safety) imposes a hazard assessment prior to start of work, which is meant to highlight predictable hazards and protective arrangements to avoid or reduce such hazards (DOE, 2019).

The Board revealed weak points in the SRS' enforcement of overarching safety management policies into work practices as implemented by subcontractors.

Although internal and external supervision and a range of operational events had systematically shown safety-related difficulties in similar cases, a lack of diligent causal analysis did not enable the SRS operating organizations to extract lessons learned and recurrent weaknesses and to launch a program of serious corrections. The reader may note that in 2018 at SRS, there was a similar accident: A contractor was loading a compactor onto a trailer, and the compactor fell from the trailer ramp. The contractor suffered slight injuries.

5.7 Fall injury accident, Savannah River Site, South Carolina, United States (DOE, 2011b)

Happy the man who could search out the causes of things

Virgil (70-19 B.C.), Eclogues

5.7.1 Decommissioning process; problem encountered; impacts

The K-Area Complex (KAC) is centrally situated within the SRS. The KAC main operation building is Building 105K, formerly housing the K Reactor, one of the numerous reactors built and operated in the United States in the 1950s to produce nuclear material for military purposes. In 1996, the DOE mandated the site contractor to deactivate the K Reactor plant to a permanent shutdown condition. The KAC was mandated to initiate plutonium storage in 1997. At the time of the accident described in the following, the Purification Area Vault (PAV, a name still associated with former operations) was being modified to accommodate the storage of excess plutonium.

On July 1, 2011, a worker fell from portable scaffolding onto a concrete floor while he was doing plant modifications in the PAV. The worker was urgently attended to by coworkers nearby, and emergency action was requested and obtained via phone and radio within very few minutes. He was transported via SRS helicopter to the MCG Health Medical Centre, Augusta, Georgia, and hospitalized with head trauma and broken ribs. As this accident met DOE's Accident Investigation Criteria (i.e., hospitalization of the worker for more than 5 days, beginning within 7 days of the accident), the DOE's Office of Environmental Management nominated an Accident Investigation Board, which completed the investigation by the end of July.

Based upon the outcomes of the investigation, the Board stated that this accident was avoidable. The Board identified the root cause of this accident as the inadequacy of SRS's operating organization, which failed to identify and rectify hazardous work practices in use before or during the work that led to the accident. In addition, the Board singled out 21 contributing causes.

5.7.2 Analysis; solutions

When the accident occurred, the worker was carrying out the dismantling and removal of gypsum board and metal studs from the PAV. Following approval of the work package and the first prejob briefing, work began on June 24 and went on through July 1, the day of the accident. The work carried out on July 1 was considered "backshift" as it was performed outside the regular weekly timetable. The work required general hand and power tools and was conducted at heights using portable scaffolding ("towers"). PPE included helmets, heavy work shoes, gloves, safety eyewear, and more protective aids as needed.

Normally, gypsum board was broken up by using pry bars to detach it from the metal studs and using saws to cut smaller pieces. Then the debris was removed using pallets and pallet trucks to a large disposal container outside the building. Before moving the scaffolds to new work locations, the floor was freed of debris and other obstacles.

On the day of the accident, a worker moved a tower to a new location and climbed the scaffold. A little later, he was seen on the scaffold in the middle of the work platform. At one point, a coworker heard sort of a "chain noise" and saw the other worker falling and landing on the floor. First aid was immediately provided by the coworkers, and the injured worker was air-transported to the hospital. His conditions were serious, and he could not be interviewed by the Board.

At the conclusion of their investigations, the Board stated the following:

- Ambiguities in the requirements, inconsistencies, and lack of full safety professional involvement in planning produced a work order (WO) that did not precisely determine all the needed controls to safely carry out the work.
- The work authorization process did not verify that facility conditions continued to support the ongoing work.
- The required hazard controls did not fully adhere to the work activity, putting the workers at higher risk.
- On the day of the accident, the work team adopted impromptu methods for gypsum board removal and handling that were inconsistent with the stipulated WO hazard controls; no consideration was given to reanalyzing the hazards.
- The configuration, surveys, and operation of scaffolding did not comply with US Occupational Safety and Health Administration (OSHA) requirements.
- Six of eight individuals at work during the accident were not trained on the specific type of scaffolding used.
- At SRS, adequate supervision is conducted of higher risk operations. Unfortunately, the supervision of lower-risk activities is not devoted the same focus. Programmatic support staff should enhance in-field supervision. The communications between the in-the-field people and the support staff (the technical subject specialists) should allow better share emerging difficulties and incorporate supervision resources.

5.7.3 Lessons learned

- Many enterprises rely on contractors to carry out work. There should be a system to ensure that contractors observe the same or consistent requirements, policies, and procedures as the enterprise that awarded the contract. Contractors should receive training and direction under the same conditions as the employees of the operating organization.
- Responsibility of activities remains with the operating organization/licensee and cannot be delegated to a contractor. Accurate supervision of the former on the latter is essential at all times.
- Full consistency between work objectives, overarching plans and detailed procedures should be ensured.
- Procedures should be comprehensive and unambiguous. Controls should be clearly specified. Freedom of interpretation should be restricted or subject to ad hoc approval. Personnel training should be as realistic as possible.
- Communications between all involved parties should be open, timely, and comprehensive.

See a similar case in Section 5.17.

5.8 Fall fatality, Hanford 200 East Area, Washington, United States (DOE, 2004b)

No snowflake in an avalanche ever feels responsible.

Stanislaw Jerzy Lee (1909–1966).

5.8.1 Decommissioning process; problem encountered; impacts

The Hanford Site encloses some 1500/km along the Columbia River in Washington State, United States. The site was formerly a plutonium production complex with nine reactors and related plutonium processing plants and is currently the home of the world's broadest decommissioning and environmental remediation program. Fluor Hanford, Inc. (FHI) manages Hanford's Asset Disposition Program for the USDOE Richland Operations Office (DOE-RL) under the direction of the DOE Office of Environmental Management.

On July 15, 2004, an employee of All Mobile Transporting & Repairs (AMTR) was found at the foot of a ladder with a serious head injury. Although there were no eyewitnesses to the apparent fall, it was apparent that the employee was working on the ladder about 1.5–2 m high taking away metal screws from the aluminum trim of an office trailer in preparation for its transport off-site. The employee was in critical conditions: He underwent surgery, reportedly collapsed twice, and passed away.

The Accident Investigation Board reviewed the mandate, roles, and responsibilities of the DOE and its contractors, and their management schemes as possible contributors to the accident. DOE-RL and FHI stated they had only minimal involvement and responsibilities—except permitting safe access—for work carried out at Hanford by nongovernment contractors. The work originating this incident was being carried out following DOE-RL transfer of ownership of a redundant office trailer to the Community Reuse Organization (CRO). In turn, the CRO sold the office to a commercial company before taking the office. The commercial company then contractually agreed with AMTR to remove the office from DOE property at Hanford and transfer it to a new location under CRO's supervision. Although the appearance of AMTR on DOE property may be viewed only for non-DOE commercial purposes, in fact DOE benefited from the removal of redundant assets that it would have otherwise had to effect contractually.

The terms of agreement between DOE Headquarters and DOE-RL about ownership transfer were considered only for guidance and were not inserted into the FHI contract or DOE-RL contractual procedures. If these provisions had been enforced, AMTR would probably not have been at Hanford on the day of the accident, as the office trailer would have been removed to a CRO-supervised location by Hanford staff and the work carried out by AMTR on the day would not have been needed.

Integrated safety management (ISM) was not applied to this work because the work was not conducted according to a subcontract under the Project Hanford Management Contract (PHMC). Therefore, the key aspects of safety management incorporating environment, safety, and health into work planning, hazard assessment, controls, and work details were not applied to this work in regard to its risk and complexity. If CRO and/or AMTR work had been subject to the provisions of ISM, this accident may have been avoidable.

Under the provisions of the PHMC Integrated Environment, Safety and Health Management System, DOE-RL expects that all types of work will be conducted safely on the Hanford Site. To fulfill this objective, FHI had formal and informal procedures in use to safeguard visiting organizations by such processes as site orientation and hazard assessment. These procedures were not rigorously enforced for the planning and implementation of the work carried out by AMTR.

DOE-RL and FHI did not produce and enforce formal procedures specifying responsibilities in controlling the CRO asset disposition schemes. This use of such formal procedures would have offered opportunities to avoid the accident.

Following the accident, DOE-RL and its direct contractors took actions to establish and implement controls to render a similar accident unlikely. Letters ordered the contractors to urgently review all work underway that involved industrial hazards to confirm timely and effective transmission of requirements, proper supervision of subcontracted work, and subcontractor safety performance assessment.

The Accident Investigation Board also concluded through interviews, medical records, and specialist consultations that the AMTR employee had medical conditions that may have been conducive to his fall. Moreover, the high temperatures on the day of the accident and the worker's problematic heat adaptation may have worsened his medical conditions. AMTR did not have, nor was it mandated to have, an occupational medical checkup program. The worker did not reveal some of his medical problems to AMTR, possibly fearing that he would not appear qualified for the job.

The Board also observed inadequate work control practices in AMTR's operations. Ladder issues were noticed concerning training, load carrying, maintaining three point contact while going up or down (possible OSHA noncompliances), and ladder labeling/binding (noncompliances with Washington Industrial Safety and Health Act, WISHA). Ladder use could have been improved by using a step ladder or a utility belt. At AMTR, heat stress was considered informally and went unregulated. Selection of PPE was not commensurate with work operations. Prejob safety briefings were not held. All controls should have been formalized through a formal job-specific hazard assessment and instructions, but AMTR did not have any such mechanisms: Basically, they counted only on the experience of their workers. Any or all of these issues could have contributed to the accident.

5.8.2 Lessons learned

This accident clearly delineates a whole range of issues that may occur through a chain of responsibilities being passed on from a licensee to contractors and from contractors to subcontractors.

For example, it is essential that the basic components of ISM be transmitted and enforced through subsequent players including work plans, hazard and risk analysis, controls, and work implementation.

An ISM is a strategy based on the overall safety policy. The ISM provides a structured approach to managing plans and provisions intended to ensure a good safety performance. The ISM should define an ambition target that the operating organization deems appropriate for its business. The minimum ambition target is to follow regulations: Moving from there to more ambitious targets should be instilled in the strategy (the "constant improvement" concept). To be a good driver, it is not enough to follow the traffic rules.

It is also essential that mandates, roles, and responsibilities be clearly allocated for each step of work down to contractors and subcontractors. In this regard, the scope of activities should be specified for each responsible party leaving only a minimum of personal interpretation.

A critical aspect is related to ownership transfer of site areas and other assets in the course of decommissioning: This process implies a transfer of accountability, time and space boundaries, and old and new obligations/constraints, which should be clearly detailed in the legal act of transfer.

BOX 5.1 Occupational fatality; Fukushima Daiichi Nuclear Power Plant (NPP), Japan

In March 2011, waves from a tsunami flooded the Fukushima Daiichi NPP, impaired the plant's cooling systems, and caused the meltdown of three reactors. The catastrophic event resulted in the massive contamination of the plant itself, the plant site, and the surroundings, forcing the evacuation of some 80,000 residents.

In early 2015, some 7000 workers were at work to stabilize the plant and contain large amounts of contaminated water.

A middle-aged worker died after he fell into the 10-m-high water tank, which was empty at the time of the fall. He had been surveying the tank with two co-workers at the top.

"He was wearing a harness, but the hook was found tucked inside the harness. This means the harness was not being used. We are investigating whether safety measures were appropriately observed," a spokesman from Tokyo Electric Power Co. (the licensee) stated (BBC, 2015).

Unfortunately, contractual ambiguities in this type of transactions are all too common. Another fall fatality was reported during decommissioning work at Fukushima Daiichi Nuclear Power Plant (NPP), Japan (see Box 5.1).

5.9 Serious finger injury during machine shop work, Argonne National Laboratory, Illinois, United States (DOE, 2016a)

Training is everything. The peach was once a bitter almond: cauliflower is nothing but cabbage with a college education.

Mark Twain (1835–1910).

5.9.1 Decommissioning process; problem encountered; impacts

While the event described in the following is not reportedly attributed to a specific decommissioning project, it can be representative of a wide range of events that are likely to happen in the course of nuclear decommissioning. Cutting of metal components is a regular decommissioning operation.

This case study refers to the USDOE investigation into inadequacies in University of Chicago (U of C) Argonne, LLC's enforcement of machine guarding requirements at ANL and especially into the August 13, 2015, event when a machinist had a finger nearly amputated while he was using a band saw (Fig. 5.5).

When a worker switched on the band saw, and before the cutting operation began, the worker's left index finger went in contact with the moving,

FIGURE 5.5 Metal cutting band saw. *Photo by M. Laraia, 2010.*

unguarded blade, resulting in a near amputation of the finger: The injury caused a bone fracture and digit nerve damage demanding surgical repair. It was just a matter of luck that the injury had not suffered worse consequences.

The DOE scrutinized the injury event, and extensive machine guarding inadequacies raised serious safety issues. In particular, the incident disclosed inadequacies in (1) management accountability, (2) hazard identification and evaluation, (3) hazard preventive and mitigation measures, and (4) training and communications.

5.9.2 Analysis; solutions

The DOE's investigation disclosed several noncompliances by U of C Argonne with occupational safety and health requirements. Several significant noncompliances are extracted from the investigation report (DOE, 2016) and are summarized in the following.

(1) Management accountability

Notwithstanding the requirements in place, U of C Argonne did not fully enforce procedures regarding major risks related to machining activities and did not verify that the work was being carried out safely.

a. The in-shop machining of common metals and plastics using milling machinery, shapers, compactors, shears, saws, etc. was classified as low-risk activities by the responsible ANL Division Management. In other words, this work could be performed relying only on the skills of the worker. Legally speaking, the consequences of incidents in machining operations are not necessarily minor, including severe injuries such as amputated fingers, lacerations, head traumas, etc.

b. The responsible ANL Division management allowed the use of a horizontal band without appropriate guards on both sides of the point of attack.

c. The responsible management permitted the in-shop machining of common metals and plastics using milling machinery, shapers, compactors, shears, saws, etc. by workers who did not have the formal, approved qualifications for those activities.

(2) Hazard identification and evaluation

Notwithstanding the requirements in place, U of C Argonne did not verify that machine guarding inspections were capable of identifying inadequacies or removing inadequately guarded machines from use.

(a) The Central Shops Safety Team runs monthly safety inspections of the machine shops. During the semester preceding the near-amputation incident, these surveys noted only one machine guarding inadequacy. That inadequacy event was not recorded in the ANL database. No machine guarding inadequacies had been noted through inspections in the building where the near-amputation event occurred, including even the inspection carried out on the very day of the event. However, the review U of C Argonne conducted in the Central Shops shortly after the event spotted 30 machines with machine guarding deficiencies, a significant share of all machines surveyed.

(b) U of C Argonne's work planning was unable to recognize the hazards inherent to the operation of machining tools and keep them under control through guarding and safe work procedures. Some 180 of the 800 pieces of equipment that were in use at the time of the incident did not have adequate guarding, shielding, and/or fastening.

(c) The ANL Nuclear Engineering Division promulgated procedures for a range of machining operations but did not specify the hazards of moving machine components and related machine guarding requirements.

(3) Hazard preventive and mitigation measures

Notwithstanding the requirements in places, U of C Argonne did not implement an effective program to reduce machining hazards in a timely manner or protect workers from dangerous impacts resulting from poorly guarded machines

Many machine guarding issues were noted during a 2003 inspection by OSHA and again during a 2007 assessment by an external consultant hired by U of C Argonne. Following each inspection, U of C Argonne considered most of the noted issues; however, reviews were not capable of avoiding recurrence; in fact, machine guarding issues came again to light and mostly were not solved. Thus, the postevent review identified a lot of remaining machine guarding inadequacies.

(4) Training and communications

Notwithstanding the requirements in places, U of C Argonne did not verify that personnel were sufficiently trained to understand the hazards and controls regarding the safe operation of machining tools or communicate hearing conservation issues to the supervisors responsible for the safety of the workers.

(a) Workers authorized to operate machining equipment were mandated to follow a dedicated training course, which delivered general information on guards but did not fully cover OSHA machine guarding requirements, ANSI standards, or guarding requirements specific to individual types of tools or equipment. Therefore, many machines remained in use without appropriate guarding, indicating that training had been ineffective.

(b) Industrial hygiene survey reports, which are meant to highlight noise exposure findings, comparisons to requirements, and directions for supervisors, included misleading or confusing details.

5.9.3 Lessons learned

Inadequate appreciation of guarding requirements due to insufficient review of all operating conditions of a machine and tolerance of unguarded machine hazards wrongly founded on perceived skills and experience; time pressure cumbersome; and insufficient guards are all factors producing a complacency toward full guarding and causing severe incidents. Complete guarding is vital to protect workers against absent-mindedness (DOE, 2015).

5.10 Respirator reduction aids ALARA programs at the DOE and utility nuclear facilities, US sites (Pentek, 2014)

5.10.1 Decommissioning process; problem encountered; impacts

The DOE's SRS is a 842 km^2 industrial site located at Aiken, South Carolina, which produced nuclear materials for military purposes. Over 1000 areas will be decommissioned at SRS over the 30 years following the cessation of production. SRS has implemented a Respirator Usage Reduction (RUR) program and has sought out more effective engineering controls to reduce reliance on respirators as well as reduce exposures. Among other strategies, a contractor has been selected that provides decontamination services while minimizing the generation of airborne contamination and so effectively reducing the need for respiratory protection.

The contractor's dustless decontamination system eliminates the health concerns of respirator overusage and allows the decontamination work proceed in longer shifts. Therefore, teams employing the contractor's equipment typically work without the burden of respirators or the need for expensive containment structures. The success of RUR program relies heavily on the engineering controls used to keep down exposures.

Holtec International, a nuclear manufacturer and services firm, will decommission the three reactor units at the Indian Point NPP in as few as 12 years. Current owner (Entergy) hopes to finalize the transfer of the NPP operating license to Holtec by May 31, 2021, so the company can begin decommissioning as soon as the 1080-MWe Unit 3 ceases operations on April 30, 2021. The same-output Unit 2 was shut down on April 30, 2020, and Unit 1, 615-MWe, was shut down in 1974. The accelerated decommissioning plan proposed by Holtec will cut 60 years off the schedule initially set by Entergy.

Indian Point NPP presents another application of dustless decontamination techniques. They tasked the contractor's decontamination team to remove paint and concrete substrate in 16 areas totaling about 1600 m^2 for release of these areas from controlled status.

The majority of the flooring was flat and unobstructed ideal for the contractor's MOOSE (Fig. 5.6), a remotely operated, floor scabbling robot with onboard VAC−PAC high-efficiency particulate air (HEPA) vacuum that collects and packages waste in a single-step dustless procedure. The MOOSE is capable of high production rates and complies with ALARA program goals by allowing the operator to remotely guide the scabbling robot from a vantage point in a low radiation area. In parallel, the decontamination team attended to harder-to-access areas with handheld Corner-Cutter needle guns (Fig. 5.7) and SQUIRREL-III floor scabblers. Once the utility's health physics concluded

FIGURE 5.6 Application of Pentek's MOOSE to an intractable asphalt floor sealant. *Credit to DOE, Wiki Commons.*

FIGURE 5.7 Hand scabbling using a Pentek corner-cutter. *Credit to DOE, Wiki Commons.*

that respirators were no longer needed due to the strict emissions controls built into the contractor's system, the use of burdensome, full-face gear was soon discontinued.

The reader should note that vacuum systems are not immune to incidents. Box 5.2 describes an incident associated to the failure of HEPA vacuum system.

5.10.2 Lessons learned

Concerns about excessive respirator usage refer to the risk of lung damage. Studies indicate that standard air purifying respirators increase breathing resistance from 20-mm to 50-mm water pressure. Dead air space is also increased, forcing a worker to breathe deeper to obtain oxygen, which contributes to dizziness, accelerated fatigue, and heat stress. Recognizing the health and safety problems associated with respiratory protection, the Institute

BOX 5.2 Failure of filter on high-efficiency particulate air vacuum system (DOE-RL, 2011)

Problem encountered; impacts

During floor decontamination, a high-efficiency particulate air (HEPA) filter system used to vacuum debris and dust failed. This work was being carried out in a containment tent with a portable radioactive air emissions unit that discharged the air to the environment through HEPA filtration. Based on survey and air monitoring results, the vacuum failure caused no environmental release.

The decontamination tool was a scabbler associated with a HEPA vacuum. The debris and dust from the scabbler operation was pulled through a knockout pot that collected the large particles and then into a HEPA vacuum which discharged into the Operations Room of the tent. After the fifth pass of the scabbler, a mist spread in the Operations Room. Work stopped, and the area was secured. Radiological surveys on exiting workers showed no contamination. Radiological surveys carried out on the outside of the containment tent indicated no loss of containment. Likewise, the air samples taken at the discharge point indicated no loss of containment. Therefore, the containment system as a whole operated as designed to control contamination releases.

Analysis; solutions

The event cascade can be described as follows. The HEPA vacuum probably failed because enough fine concrete dust passed through the 1-μ cloth bag filter to clog and break the HEPA filter. The paper collection bag may have been breached by sharp debris or overfilling, allowing unfiltered aerosols to get to the 1-μ cloth filter. The cloth bag filter may have breached, letting particulates build up on HEPA filter. HEPA filter in turn may have clogged due to buildup of particulates. Suction exceeded the HEPA filter differential pressure limit, and the filter got broken. Note: The HEPA vacuum has no visual filter stage capacity indicators.

Lessons learned

Filter failure may lead to the spread of contamination.

HEPA filter change-out rate should be based on operating conditions, i.e., the amount of particulates expected for collection in one filter.

Have a contingency plan for filter failures, e.g., a containment tent and emergency response.

of Nuclear Power Operations (INPO) published their Guidelines for Radiological Protection at Nuclear Power Stations stating: "Because respirators can increase external exposure by affecting worker efficiency, their use should be minimized. Too often, respirators are used when engineering controls could be used to eliminate or reduce the source of airborne radioactivity" (INPO, 2005).

5.11 Arc flash accident at Los Alamos National Laboratories Technical Area 53, New Mexico, United States (Roberson, 2015)

5.11.1 Decommissioning process; problem encountered; impact

Strictly speaking this event does not refer to a specific D&D project, but it is believed to be representative of a range of events that could occur during decommissioning, especially for laboratories and other small nuclear facilities.

On May 3, 2015, at Los Alamos National Laboratories (LANL), during preventive maintenance on an electrical substation, a worker (W1) entered a cubicle on the energized portion of the switchgear to clean it with a commercial spray cleaner. The observed cause of the accident was the cleaning fluid sprayed into the air gap between an energized switchgear bus and the grounded cubicle wall. The aerosolized fluid created a path to ground, resulting in an arc flash and blast. The force of the arc flash/blast ejected W1 from the cubicle. In addition to suffering significant burns, the worker lacerated his head when he fell backward and hit nearby equipment. In total, this electrical accident affected nine employees: Seven were soon released, two of them required hospitalization, but only W1 was in critical conditions. The root cause of the accident was the inadequate implementation of control.

It was also disclosed that the spray cleaner was to be used only on non-energized surfaces because it has no established insulating properties to avoid the conducting of electrical current.

5.11.2 Analysis; solutions

Two specific root causes were identified by the Review Team and can be merged into the single root cause of control implementation: the failure to implement zero-voltage checks and the lack of establishing physical barriers.

Basic training and process requirements for electrical work require "zero-voltage" checks on equipment before beginning manual work. The hazard analysis process for this specific work did not consider changes in the work environment during the work period: Unfortunately, in this case, changes that occurred from beginning to end of the work left a mix of similar equipment partially energized. As the hazard analysis of the work plan was not revisited, no new controls were considered to differentiate between the energized and deenergized equipment. Prudent work controls would install sturdy barriers to mitigate errors due to crew entering energized cubicles. A physical barrier preventing W1 from entering the energized cubicle would also have avoided this accident.

The Review Team identified the five actions shown in the following that clearly are broader in scope than this electric incident and definitively involve decommissioning-related aspects:

- Management needs to reinforce expectations regarding work scope as well as task-level work and hazard analysis.

- Operating organization needs to establish consistent and stringent implementation of safety requirements especially when work involves mixed teams (in the LANL case staff and contractors).
- Operating organization needs to implement human error prevention tools in work planning and hazard analysis.
- Management needs to reinforce and clarify expectations and implementation of zero-voltage verification requirements in the course of all electrical work.
- Operating organization needs to improve its capability to take action on, and verify implementation of, corrective actions from previous reviews and incidents.

5.11.3 Lessons learned

Review of the management processes relevant to this work showed adequate procedures were in place to conduct electrical maintenance work. However, this and other recent events indicate that these procedures are often applied to a small extent, and sometimes not used at all.

Without fixing the recurrent inadequacies in implementing these procedures, it is likely that incidents of this type will reoccur. To prevent this, it is vital that top management and all managerial levels cooperate to achieve the rigor needed for hazardous work. In the case of the LANL accident, either a zero-voltage check or a strong barrier to restrict entry would have avoided serious consequences; the former is mandated by LANL processes, and the latter should be a routine practice.

5.12 Grinder fell from scaffolding, Idaho Cleanup Project, Idaho, United States (DOE, 2011c)

5.12.1 Decommissioning process; problem encountered; impacts

The DOE's Idaho Cleanup Project (ICP) involves the safe environmental remediation of the Idaho Site, including dispositioning TRU waste, spent fuel storage and HLW processing. Fluor Idaho began site cleanup work in June 2016, with project completion scheduled for 2021. In April 2011 a 4.5 -kg grinder, placed on a softening mat erroneously positioned over a 10−15 cm gap between the scaffolding border and the working surface, fell and almost hit a worker some 7 m below. The cord of the grinder hit the worker's shoulder. The worker was wearing a helmet and protective eyewear. After the event, the worker was escorted to the medical department, after which he could safely return to unrestricted work.

It could not be ascertained who placed the mat over the gap or why the grinder was there. The scaffolding was built in accordance with specific requirements; however, softening mats were stacked to a height higher than the top of the scaffold side barrier, which did not meet the requirements.

5.12.2 Analysis; solutions

A. Training on the task was not included in the worker training instructions.

Specific requirements for the use of scaffolding include the following (among many others): (1) the user is instructed to keep garbage and redundant material off the scaffolding to leave the widest room available to walking or working; (2) the piling of brickwork on the scaffold surface is limited to a specified height; and (3) the scaffolding should be free of rubble and garbage.

Other regulations detail methods to shield workers from falling objects during scaffolding work (e.g., side barriers, screens, guardrails, catch nets or platforms, or awning). Of particular relevance to this case study are (1) the use of panels or screens protruding from the side barrier or the deck to the top of the guardrail where tools or miscellaneous items can be safely stacked to height higher than the top of the side barrier and (2) the use of an awning, net, or platform strong enough to resist the impact of falling objects.

During a course on Environmental, Safety, and Health (ESH), the workers had been introduced to requirements for scaffolding (e.g., prework inspection; do not use scaffolding if floorboards are slippery; do not load the scaffolding with too great a burden), but no advice had been provided about protection from falling objects.

B. Progress/status of work insufficiently monitored

The work being carried out from scaffolding in this project was enormous. The professionals in charge of the supervision of this work area did not notice the piling and position of materials on the scaffolding deck as a hazard for which no protection had been considered.

5.12.3 Lessons learned

Although scaffolding may be constructed in accordance with requirements it is not necessarily risk free. Minor gaps present in scaffolding may let tools and other items fall through and endanger the workers below. Moreover, timely and effective housekeeping practices should prevent the accumulation of unused materials or waste. The grinder referred to in this case study was not in use when the incident occurred and should have been removed rather than sit on the scaffolding.

Falls from heights, e.g., from ladders, scaffolding, and partly constructed or demolished structures are among the commonest causes of fatal injury or permanent disability at construction or decommissioning sites. A fall protection plan should be enforced, which includes one or more of the following training and use strategies (IFC, 2007):

- Fall prevention devices, such as rails or other barriers able to support a weight of 100 kg, when working at heights equal or greater than 2 m or at any height if the hazard includes falling into machinery in motion, into liquids, into noxious substances, or through a surface opening.

- Fall arrestors, e.g., whole body harnesses and energy-absorbing lanyards able to support 2.5 t. Rescue procedures for workers whose fall has been successfully arrested. The connection point of the fall arrestor to the scaffolding should also support 2.5 t.
- Installation of control zones, monitoring, and warning systems in the vicinity of fall hazard zones. Securing, marking, and labeling covers for openings in floors (Sections 5.16 and 5.49), roofs, or walking surfaces (Section 5.17).

5.13 Facility degradation introduces new hazards demanding advance review, Hanford Site, Washington, United States (IAEA, 2018)

5.13.1 Decommissioning process; problem encountered; impacts

On September 1, 2006 during decommissioning activities within an abandoned tank pit at the Plutonium Finishing Plant (PFP), a 1.0 cm × 1.2 m threaded rod broke loose from the ceiling after being inadvertently struck by a ladder moving within the pit. This made the stud fall to the bottom of the tank pit (around 5 m), hitting the bottom some 1.5 m behind a worker.

Activities in progress within the tank pit had taken away most of piping and components from the grating level. Several protruding bases of the piping (others were previously sliced off for collision protection), at a level higher than a worker's heads, were left standing on the ceiling.

5.13.2 Analysis; solutions

Preliminary characterization and hazard review of the tank pit highlighted a possible corrosion of its metal structures. This corrosion was assessed by mechanical engineers who tried to better determine the risk the corrosion posed to the workers. However, to provide sufficient protection to the workers, the analysis focused on structures that the workers would surely contact (ladders, grating, surfaces, etc.). During that review, the corrosion to the piping supports was not deemed significant. Therefore, the work plan addressed only the removal of the collision hazards; the integrity of the remaining structures was not evaluated as workers were not supposed to get in contact with those structures.

The threaded rod within the tank pit was heavily corroded from its long operational history. When bumped by the ladder, it could not withstand the side shock and broke loose. Although this corrosion was known, the work plan did not see this condition as an occupational hazard.

5.13.3 Lessons learned

New hazards are constantly introduced in old plants due to the growing deterioration of materials. Even when the work plan has been reviewed for

BOX 5.3 Fall of contaminated pipe, Bugey-1 Nuclear Power Plant, France; December 15, 2014 (IRSN, 2015)

At Bugey-1 reactor (currently being decommissioned), a vertical pipe section (weight about 3t), whose interior was slightly contaminated, fell vertically along its handling axis from a height of 20 m and penetrated two floors during a lowering maneuver prior to being segmented. The pipe section had been grabbed in a handling collar consisting of two steel half-shells bolted together, holding the pipe by its upper end and from which it was hanging. This fall luckily did not cause any injuries, although several workers were manning the area.

The contact between the handling collar and the pipe was lost while the pipe was being handled, resulting in its slide and fall.

hazards specific to the planned operations, other hazards may remain hidden. When conducting job hazard analyses for work inside old, abandoned areas (especially those having overhead structures), the work plan must take into account the potential deterioration and collapse of these structures. Any structure threatening to fall always poses occupational risks. When planning work, ensure that no potential hazards are disregarded. Similar events are given in Box 5.3 and in Sec. 5.16.

5.14 Biooccupational hazards during facility deactivation, Oak Ridge, Tennessee, United States (IAEA, 2018)

5.14.1 Decommissioning process; problem encountered; impacts

The Department of Energy Oak Ridge Operations (DOE-ORO) initiated an investigation based on preliminary notice from the K-25 decommissioning contractor (Bechtel Jacobs Company) that there had been a chain of respiratory illnesses that could be associated with the workers' occupations. The K-25 Building has been unused since the 1960s. Maintenance had lessened over time. Ventilation systems were idle, and metal sheets had been welded over vault openings on grounds of security. The roof needed repair, and water leaking into the vaults was a recurrent issue. The moisture buildup coupled with the lack of ventilation created the ideal environment for the proliferation of mold, fungi, and bacteria. Most vaults were radiologically controlled areas. Being aware of the conditions inside the K-25 Building, management required the Deactivation Plan to include data supporting strategies aimed to mitigate conditions in the building. Previous versions of the Deactivation Plan had not considered occupational illnesses such as those related to exposures to airborne mold, fungi, and bacteria.

In response to workers' concerns, Industrial Hygiene conducted a review of conditions in the impacted vaults and work activities in K-25. Evaluations also included sampling for bioaerosols such as airborne fungi and bacteria. Airborne samples were taken both from within the affected vaults and outside K-25 to establish a comparison. The sampling findings indicated that measurable bioaerosol levels were present in the building.

5.14.2 Analysis; solutions

Recommendations were issued to supervisors to improve working conditions, e.g., by increasing ventilation in biocontaminated areas, reducing air moisture, and evaluating chemical treatments to remove the mold. Disposable respirators were made available, provided that wearers had to be medically checked and found fit for the use of respirators. Sampling data and reviews by external industrial hygienists confirmed that high fungal and bacterial levels were present.

5.14.3 Lessons learned

In addition to general guidelines and safety and management requirements, a facility Deactivation Plan should address specific hazards associated with that facility. Key elements of the plan are the nonoperational functions of the facility, such as waste storage, and the configuration the facility's structures, systems and components (SSC) should have when being transferred to safe enclosure. All teams expected to conduct activities in the deactivated facility should be consulted and their views incorporated into the final Deactivation Plan. The Plan should specify how to ensure adequate protection to those going in and out of the facility and identify what building systems must be maintained and to what extent. The plan should also address biocontamination (to bioaerosols, bird droppings, rodents, etc.) and other unique hazards associated with the facility. The Plan needs to be regularly reviewed to incorporate lessons learned, the adoption of corrective actions, and follow-up evaluations.

Regarding biological hazards, ordinary sanitation practices can eliminate most disease bearers. Having potable water, marking nonpotable water, and making cleansing areas available are the most important components of a sanitation plan.

Because many nuclear facilities under decommissioning are old, there is a higher potential for workers to be exposed to molds and other bioorganisms that proliferate in and on buildings. Inhalation of molds and fungi can cause minor to significant lung problems. Skin contact can cause eruption or irritation. A full survey of the plant should be carried out prior to execution of decommissioning operations, and accurate cleaning and PPE should be used upon discovery of bioorganisms.

5.15 Falling objects, East Tennessee Technology Park, Tennessee, United States (IAEA, 2018)

5.15.1 Decommissioning process; problem encountered; impacts

A team was performing sprinkler tests inside the K-309-1 and K-309-3 units of the K-25 Building. Two Fire Specialists at work on the operations floor had to open an inspector test valve, while another Fire Specialist took readings at the sprinkler riser on the cell floor below. The movement of the Fire Specialists on the operating floor resulted in a few small pieces of concrete floor dropping near the Fire Specialist taking readings on the floor below (a near miss).

5.15.2 Analysis; solutions

The direct cause of this event was the deterioration of the operations floor panels. This deterioration resulted from water leaking through the deteriorating roof of the main K-25 building and ending on the operations floor. Since the operations floor is open to ample weather variations, the panels are exposed to recurrent freeze and thaw cycles. These conditions, and the fact that the roof of the K-25 building has far exceeded its service life and maintenance activities to prolong its service life have not materialized, have worsened deterioration of the floor panels. Deterioration will not stop unless the roof is fixed or a new roof is installed.

5.15.3 Lessons learned

Demolition activities may lead to the fall of materials or tools, as well as projection of solid pieces from abrasive or cutting operations, which can result in head, eye, and leg injury to workers, surveyors, inspectors, etc. Preventive and mitigation techniques include the following:

- Using an access-restricted waste chute or discharge areas, and/or a designated path for safe transfer of waste from upper to lower elevations
- Carrying out segmenting, cutting, grinding, etc. with adequate barriers around the machinery and firmly securing tools
- Keeping traffic routes clear of loose scrap and debris to avoid the bumping of heavy equipment/machinery over these materials
- Using interim fall protection measures in scaffolds and on borders of elevated platforms, such as hand rails and vertical barriers attached to work surface to avoid uncontrolled displacement of materials
- Evacuating work areas during blasting operations, or using blast mats or other containment barriers to minimize the risks from projection of missiles if work is carried out in the vicinity of individuals or sensitive structures

- Establishing exclusion zones and hard-hat areas, clearly signaled with barriers
- Using covered gangways
- Using high-reach machines
- Shielding machine cabs to protect the drivers
- Training and supervising site workers
- Wearing industrial PPE, such as safety goggles, face shields, helmets, and safety boots (Fig. 5.8)

Site managers should remain at all times aware of the structural status of site installations, especially those that are in a safe enclosure or on standby. Workers at work in these installations should report at once any unusual or evolving conditions. Administrative and technical controls should strive to maintain workers' safety and offset structural deficiencies. Adequate funding is needed to ensure a safe status of the structure and protect men at work inside the installation. To prevent uncontrolled collapse, periodical structural survey should consider

- the age of the structure,
- its former and current use,

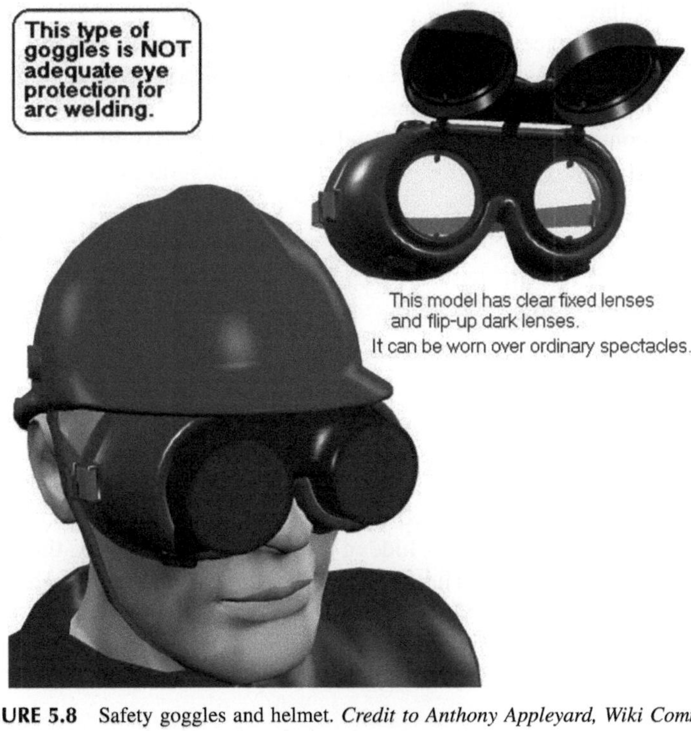

FIGURE 5.8 Safety goggles and helmet. *Credit to Anthony Appleyard, Wiki Commons.*

- the construction material, features, and geometries,
- contiguous structures,
- the weight of removed components, waste, or machinery standing on floors above ground level.

5.16 Pipe falls through ceiling hole and results in near hit, Nevada Site, United States (IAEA, 2018)

5.16.1 Decommissioning process; problem encountered; impacts

Before the event in question, a maintenance team was removing components and materials from the second floor of Building A-1, including wall panels, in a beryllium remediation program. On the day of the incident, the crew removed the panels between Rooms 4113 and 4120, exposing a formerly hidden, unsupported vent pipe, which was not included in the removal program.

Later on the same day, the pipe fell from the second to the first floor, Room 4525. The pipe halted about 1 meter above the first floor and less than 50 cm short of hitting a worker seated at his work station. The pipe had passed through a clearance hole between the second and the first floor. The pipe stopped when a section of the pipe with a 45 degrees curve bumped and set on the border of the clearance hole.

The worker at once informed the removal team of the incident, and the team barred the area as well as all ceiling/floor penetrations in the area. They also checked that all remaining penetrations were identifiable, and there were no other second-to-first floor gaps and utility dead ends in the area. The removal team crew guarded the scene so an investigation could begin right away.

5.16.2 Analysis; solutions

The modifications to this old facility (construction year 1975) including the vent pipe rendered permanently nonoperational were not recorded on any of the drawings available at the time of the remediation program. After the vent pipe event, interviews revealed that in 1984–1985 a sink and a section of its vent pipe, between the first and second floors, was discontinued from service. The second-floor section of the pipe was not removed but left in place. The remaining section of pipe extended down from the Building A-1 roof, through the inside of an inner wall (so hidden from sight), and partly through the second floor clearance hole. The first-floor workers at the time of the incident had been at work in room 4525 (where the pipe fell) for many years; they had noticed the hole in the ceiling but had never seen anything inside the hole and, therefore, did not feel endangered by the hole.

Past incidents and other safety concerns identified during the remediation program led to the introduction of regular (two or three times a day) job safety briefings. If a new hazard was found in a given area, the team would stop working there until the issue was solved. Moreover, two employees (one from Safety and the other from Industrial Hygiene) were seconded to the remediation program to assist with the identification and mitigation of the building hazards, e.g., electrical issues. However, when the workers removed the wall panels from the second floor, they assumed that the pipe was fully operational, supported, and not "abandoned in place" since other walls containing hidden pipes had been readily removed with no problem.

Although the general aim of the remediation program was occupational safety, it did not adequately ensure the safety of occupants on the first floor of the building in this specific case. On one hand, an inspection of the first floor was carried out to ensure asbestos exposure, and other work controls were enforced, but on the other hand, it was expected that because of its cast concrete build, the demolition on the second floor would in no way affect the first-floor workers.

5.16.3 Lessons learned

Working in old buildings may require a specific focus on hazards not readily identifiable in modern buildings. Construction codes and standards have much improved over the past 30+ years. Maintenance and demolition activities in older buildings require prejob reviews and surveys that are not normally considered by work planners and implementers. A maintenance team was at work in such an old building, when a near miss occurred. The root cause analysis concluded that all realistic safety measures had been taken. However, an undocumented 35-year-old building modification regarding a vent pipe that ran from the building roof through the second floor to a first-floor drain resulted in a near miss when the vent pipe fell through a second- to first-floor penetration into a manned area. A broader hazard analysis including lessons learned and hazards already identified in the remediation program will be used for future work involving older buildings.

Two fatalities due to falling objects are described in Box 5.4.

5.17 Employee fall and injury at the K-25 Building, East Tennessee Technology Park, Oak Ridge, Tennessee, United States (DOE, 2006b)

Like dead, remembered footsteps on old floors

Edwin Arlington Robinson (1869–1935)

BOX 5.4 Two fatalities during the dismantling of German Nuclear Power Plants (NPPs)

On June 6, 2019, a fatal accident occurred in the Isar Unit 1 NPP during dismantling work on a large container. A contractor was caught by a falling part and fatally injured. The criminal police took over the investigation. According to the licensee, the accident did not result in the release of radioactive material (KRIPO, 2019).

On December 21, 2016, there was a fatal accident during dismantling activities at Mülheim Kärlich NPP.

For reasons that had not yet been clarified at the time of reporting, a 60-year-old worker died in the reactor dome of the plant. According to witnesses, the man was hit by a falling component. He was injured so badly that he died at the scene of the accident (RPR1, 2016).

5.17.1 Decommissioning process; problem encountered; impacts

Uranium enrichment at the K-25 Site Gaseous Diffusion Plant (now called East Tennessee Technology Park [ETTP]) started operation in 1945 and continued until 1964, after which modified operations went on until 1977. After production ended, the building remained closed until total demolition was completed in 2017.

Since the late 1990s, the roof had been left to deteriorate unrepaired, and many fragments of gypsum and debris fell onto the operations floor below (see Section 5.15). The concrete panels of the operations floor had been exposed to rain and freezing and thawing cycles. Weather exposure corroded the steel reinforcement in the panels and caused the concrete to break apart. Since 1995, inspections of the bottom of these panels have revealed the deterioration of floor panels, and the faulty panels had been painted on the upper surface with specified markings. In March 2005, the wheels of a cart "sank" through the operations floor, and site management declared the operations floor unsafe for personnel entry.

The Radiation/Criticality Accident Alarm System power supply panels and fire protection equipment were installed on the operations floor and necessitated surveillance and maintenance (S&M). Because the structural resistance of the operations floor was uncertain and management had declared the floor unsafe, a network of steel grating walkways had been installed by August 2005. Later on, the network had several branches added to allow workers to reach the equipment without wearing fall protection. The installation of lateral grating in building unit 309-1 was part of the job when the incident occurred.

The fall incident occurred on January 3, 2006, when a worker fell almost 10 m to the cell floor below after a deteriorated floor at K-25, Oak Ridge National Laboratory (ORNL) collapsed. Even though he suffered serious fractures to his right leg and right arm, he was very lucky to have survived this fall. This was an accident that could easily have turned into a fatality.

5.17.2 Analysis; solutions

The investigation disclosed inadequacies in work planning and control, engineering design, job, and safety supervision. Multiple barriers failed, which caused a combination of conditions, deviations, and events that resulted in this accident. For instance, there were violations of the Work Instructions, such as the worker carrying a piece of grating exceeding the weight limit of 25 kg, but the supervisor did not impose this requirement. Safety supervision missed to see the crews installing the grating did not use fall protection devices. The design of the lateral branches of the walkway grating did not comply with the OSHA regulations for walking/working surfaces. Communications among the main players responsible for the planning and execution of the Work Package relevant to the accident were ambiguous, misinterpreted, unofficial, and unrecorded. The work instructions and the activity hazard analysis (AHA) contained contradictory positions.

Specifically, the work instructions contained the following warning:

"Personnel shall not stand on operating floor panels without fall protection. To avoid the use of fall protection personnel may stand on the previously installed decking and slide the one being installed in place."

Instead, the AHA states the following:

"All Operations Floor panels have suspected integrity problems. 100% fall protection required when walking on Operations Floor until plate/grating established."

This confusion led to informal work changes whereby workers would forego the use of fall protection during grating installation. The formal change control process was not adhered to, and the safety organization was not informed. The work instructions did not include criteria for telling apart established safe walkways from walkways being installed but not yet established, inspected, or certified safe.

During its barrier analysis, the investigation noted that a validation checklist was made in response to the BJC Operational Readiness Review (ORR) for the K-25/K-27 High-Risk Equipment and Other Process Gas Equipment Removal, which was carried out in August–September 2005. This checklist was not applied to the operations floor grating installation Work Package, which had been initially issued in April 2005, well before the ORR. Had this checklist been applied to review the grating installation work package, the conflict between the work instructions and the AHA would probably have been observed and clarified. This inattention shows that the grating installation job enjoyed a low management priority and limited safety supervision. In fact, the main focus was high-risk equipment removal and nuclear safety.

5.17.3 Lessons learned

The accident also prompted a major change in the demolition strategy at K-25, which was reformulated to make it safer for workers. Instead of moving on with the same strategy that had been used earlier to remove equipment from three old uranium-enrichment buildings, Bechtel Jacobs Co. (BJC) devised an innovative approach to limit worker access to at-risk areas and leave most of the process equipment there—but foaming it or sealing it, etc., to make sure that it was safe when the building came down.

Asked if his company would have changed strategies if there had not been a near-fatal accident, the BJC President said: "I don't know. There was enough deterioration in the building that I think we would have ultimately concluded that we had to do something different, whether we had the accident or not. But certainly the accident made us stop doing anything until we really reassessed the integrity of the structure and the like" (Knoxblogs, 2011).

5.18 Electric shock during underwater plasma arc use, EBWR, Argonne National Laboratory, Illinois, United States (Fellhauer, 1998)

5.18.1 Decommissioning process; problem encountered; impacts

During EBWR decommissioning, plasma arc underwater segmenting of the reactor core was carried out by a crew of three workers: one executed the cutting, a second held the piece to be cut with clutches mounted on an extension device, and the third watched the proceedings and monitored the plasma arc torch. During the cutting, the third worker noticed the second worker convulsing and at once deenergized the torch and went to help. An emergency call was made. The worker was not injured and did not necessitated medical treatment.

5.18.2 Analysis; solutions

The clutches and extension device were metallic, and the worker was wearing protective clothing wet with sweat.

Moreover, he was in contact with a metal guardrail during the cutting. It is assumed that when the piece being cut was removed from the reactor core assembly, electrical ground moved to the segmented piece and the electrical current passed from the extension device through the worker's body to the guardrail.

5.18.3 Lessons learned

1. Tools used in plasma arc underwater cutting should be constructed of nonconductive material.

2. Metal guardrails and the work area floor should be coated with a rubber layer to impede inadvertent grounding.
3. The attendance of an observer in charge of cutting power immediately can avoid a potentially fatal accident. In this case, the observer's fast reaction may have saved the life of his mate.

5.19 Unexpected cutting of energized wires, JANUS reactor, Argonne National Laboratory, Illinois, United States (Fellhauer, 1998)

5.19.1 Decommissioning process; problem encountered; impacts

The JANUS reactor was operated as a neutron source to investigate the biological and genetic impacts of acute and chronic exposures to neutrons. The reactor was in operation during 1963–1992 at a power up to 200 kW. The JANUS decommissioning began in October 1995 and was basically completed in September 1997. An independent verification survey was performed in November 1997. The facility was released for unrestricted use with the exceptions of two areas, which remained under restricted access.

During the removal of the JANUS control panels, two live wires were cut. The issue was identified later when it was found that the stairway exit lights were off. Work was stopped, and an inquiry was launched.

5.19.2 Analysis; solutions

All known circuits had been placed in "lock out tag out" (LOTO) mode and confirmed as deactivated by reading the voltage at all terminals of the control panels. Workers cut the wires at the panel inlet point instead of disconnecting the wires, pulling them off the panel and then cutting them. One of the workers, believing to take an extra safety measure, was rechecking each wire after he had cut it to confirm it was deactivated. This caused two live wires being cut and topped with wire nuts. The wires had been positioned in the early 1990s during an emergency light upgrade. The circuits had been mistakenly routed via the panels instead of in their own duct. Because the wires did not terminate inside the panel, they were not identified as active in preparation of the work.

5.19.3 Lessons learned

1. Use inductive electrical detection instruments to improve voltage measurements during LOTO verification. This type of instruments would have spotted active circuits.

2. Workers should be repeatedly alerted to stop work and notify any unforeseen condition. The worker cutting the wires told his supervisor that he had discovered the first live wire. After some minutes, he told his supervisor he had found another. The supervisor was not watching the operation and assumed the worker was doing voltage checks, not wire removals. The work should have been halted as soon as the worker discovered the first live wire.
3. Never assume the old facility drawings to be 100% accurate.

5.20 Unexpected chemical hazards, East Tennessee Technology Park (ETTP), Oak Ridge, Tennessee, United States (DOE, 2001)

5.20.1 Decommissioning process; problem encountered; impacts

Building K-33 was a gaseous diffusion uranium enrichment facility at the ETTP: Built in 1954, it was in operation until 1985. The multistory building covered over 130,000 m^2 (Fig. 5.9). Within the frame of a reindustrialization strategy established in 1997, a number of decontamination and dismantling activities were carried out around the turn of the millennium. The case study given in the following refers to this initial D&D phase.

The partly decontaminated plant still had radiological and chemical contamination, hazardous materials, asbestos, and polychlorinated biphenyls (PCB). It was totally dismantled by July 2012.

On December 19, 2000, at the Building K-33, prework sampling of four large tanks in the Holding Drum Room disclosed lethal concentrations of chlorine trifluoride ($ClFl_3$) gas. The tanks were assumed to be empty;

FIGURE 5.9 The interiors of K-33. *Credit to DOE.*

therefore, hazards and related accidents posed by the residual gas were not considered in the work authorization basis. The contractor notified an Unreviewed Safety Question Determination on January 25, 2001, and filed an Unusual Occurrence Report on February 5, 2001. The sampling of the tanks avoided the risk of an uncontrolled gas release of the gas and any resulting harm to workers tasked with the dismantling of the tanks and their piping.

Project workers were checking the contents of the four tanks in preparation for decommissioning. The workers opened the interlinking piping between the four tanks to equate gas concentrations and took air samples. The sampling indicated gas concentrations of the order of 1000–4000 ppm. The Immediately Dangerous to Life and Health limit for ClF_3 is only 20 ppm. In this case, workers were not harmed.

Chlorine trifluoride was used in the uranium enrichment process. This gas and other hazards in Building K-33 were assumed to have been removed during earlier postshutdown phases. Moreover, earlier samplings of other ClF_3 containers in Building K-33 had showed no evidence of the gas. However, based on previous discovery of other unexpected hazards during ETTP decommissioning activities (e.g., lubricants), the contractor routinely employs confirmatory sampling. In this case, the sampling might have saved lives.

Following discovery, the contractor at once isolated valves and vents, locked the Holding Drum Room, and installed danger signs.

A similar incident occurred at ETTP on December 13, 2000. In this case, fluorine leaked from the ETTP K-1302 Fluorine Storage Facility. This facility was in a deactivated state, and its fluorine tanks and piping were assumed to have been emptied and purged of the hazardous substances. However, remaining fluorine corroded and perforated a weld joint, causing gas discharges. Two workers were hospitalized with nausea, and others suffered minor impacts.

5.20.2 Lessons learned

The unexpected discovery of ClF_3 in Building K-33 confirms the need and benefits of preventive confirmatory sampling before dismantling tanks and other components. More generally, predecommissioning assessments and plans should not take historical characterization for granted. See a less significant case in Box 5.5.

BOX 5.5 Some personal experiences—hidden contamination

"During an apparently simple job of clearing boxes from a store, one box was opened and was found to contain contaminated tooling (unwrapped). Investigation and remedial actions took 2 weeks to complete. In the same store unmarked bottles were found on a shelf that contained mercury. This again added to delays until such time the all-clear was given that there were no remaining mercury issues (Kerr, 2008)."

5.21 Sodium reaction causes rupture of piping and release of asbestos, Idaho National Lab, Idaho, United States (DOE, 2020b)

5.21.1 Decommissioning process; problem encountered; impacts

This accident occurred in the Sodium Boiler Building (Fig. 5.10), which was next to the shutdown (now entombed) sodium-cooled experimental reactor (EBR-II).

While decommissioning workers were processing passivated sodium, a pressure excursion occurred in the piping and forced a dead-leg, 30-cm pipe section running outside the building to get broken and detach from the rest of the pipe. This treatment solution spread from the piping and displaced pipe insulation to the area between buildings. Workers evacuated the building upon hearing loud noises and alerted the fire unit. Asbestos had also been released in amounts that required reporting. One worker was exposed to a haze of treatment solution; doctors found no evidence of injury and soon released the worker.

The most likely cause of the incident was a sodium−water reaction, worsened by the differential pressure on a large volume of water in the 30-cm header and the complex geometry of the system. The major hydrogen production from the sodium−water reaction caused a detached mass of water to gain high speed and impacted and ruptured the dead leg outside the building;

FIGURE 5.10 Idaho National Laboratory, piping in the east boiler basement of the sodium processing building was color-coded for easy identification. Orange indicates sodium, and green identifies cooling water. *Credit to DOE, Wiki Commons.*

when the pipe section separated from the system, it dispersed to the environment a large amount of treatment solution, water, bicarbonate, and hydrogen.

5.21.2 Analysis; solutions

The hazard analysis did not realize that water could take unexpected directions or paths or that a pipe section was outside a controlled area. It also did not realize that a major quantity of water remaining in the system could react with the sodium and create a significant hazard. The Engineering Design File (EDF) established process controls to take account of the unknown quantities and location of sodium by dosing limited quantities of water to react with sodium. However, the EDF failed to assess the unexpected movement of the dose water to unmonitored sections of the piping, allowing level increases and reactions resulting in large movements of water. Moreover, disregarding the volumes occupied by the sodium bicarbonate, even with a proper dosing, could have caused a full line to react with sodium and subsequent overpressure.

The response to management self-assessment (MSA2) recommendations was insufficient: The work order did not include the MSA2 recommendation to avoid undetected nonpassivated sodium from being wholly dipped in liquid dosing solution.

The procedures were extremely hard to follow and occasionally conflicted with the EDF requirements. The procedures allowed unnecessary use of "Engineering direction" and did not include all lessons learned from the MSA2. These inadequacies resulted in hard-to-manage situations for the workers. The written instructions were obscure, inconclusive, or ambiguous. Different procedures about the same work had different requirements. Contradictory or incongruous requirements were set in different steps of a given procedure. Requirements were set in different units (i.e., different numbers).

5.21.3 Lessons learned

The root cause analysis (barrier analysis) illustrates a lost opportunity to avoid the incident: an MSA2 observation about unexpected movement of water into spaces with unreacted sodium. An observation does not require corrective action, and the response of the decommissioning organization to this observation was to lower the rate for dosing treatment solution. The a posteriori assessment showed that controlling dosing did not prevent the resulting water hammer. The decommissioning department, independent reviewers of the EDFs, and third-party assessors failed to see the impacts and likelihood of a water hammer. If more attention had been paid on the MSA2 observation, the work planners may have realized the water hammer risks and establish preventive controls.

This incident stresses the need of carrying out a complete hazard assessment addressing the widest range of undesirable system interconnections. Lessons learned that can be drawn from this incident and generalized include the following:

- Procedures must be written clearly and must not include ambiguous or contradictory statements. Regarding this event, the procedures were very hard to follow and occasionally contradicted requirements given elsewhere. The procedures allowed vague use of "Engineering direction" and missed to include lessons learned from an MSA2. These inadequacies induced the system workers to make mistakes.
- It is vital that work orders incorporate recommendations and lessons learned from specific assessments or other events that are relevant to the work in question. Regarding this event, the work order did not include the MSA2 recommendations that specified actions preventing undetected nonpassivated sodium from being totally dipped in liquid dosing solution.
- The reader should note that this case study refers to an uncontrolled sodium reaction similar to the one that occurred at Rapsodie (Section 5.44). Never underestimate the difficulties and the unpredictability of handling sodium, and have contingency plans at hand.

5.22 Occupational impacts of toxic vapors released by underground waste storage tanks, Hanford Site, Washington, United States (WRPS, 2017)

In the multitude of counsellors there is safety.

Bible, Proverbs 24, 6

5.22.1 Decommissioning process; problem encountered; impacts

During the past 10 years or so of the past century, the main mission at the DOE Hanford site shifted from producing plutonium for national defense purposes to decommissioning, waste management, and environmental remediation, including treatment and disposal of more than 200,000 m^3 of high-level waste (HLW). This waste is stored in 177 aging underground storage tanks (USTs) on-site. Of these tanks, 149 are single-shell tanks (SSTs) built during 1944–1964, and 28 are double-shell tanks (DSTs) built during 1968–1986. All the SSTs long exceeded their design lives, and 67 of them are known or suspected to have leaked 4000 m^3 of HLW into the ground. None of the DST is known to have leaked, but a number of them will reach the end of their design lives before the waste can be retrieved and processed. The DOE is legally obliged to remediate the waste tanks. To manage the maintenance and cleanup of the tank waste, the Tank Waste Remediation System (TWRS) scheme was defined in 1991. The current program for remediating the Hanford tank waste consists of waste retrieval, pretreatment, immobilization, and disposal. The HLW is planned to be immobilized in a borosilicate glass matrix; the resulting glass canisters will then be disposed in a deep repository offsite. The low-activity waste resulting from the pretreatment also is planned to be immobilized in a borosilicate glass matrix and disposed of at a Hanford disposal facility (NRC, 2001).

Hanford's huge USTs range in volume capacity from 200 m^3 to some 4000 m^3. For their construction, large sunken landforms were cleared; the tanks were installed in arrays named "tank farms." The tanks were then enveloped with soil, leaving the upper parts of the tanks a few meters below ground to shield the radiation.

Pipes, named risers, provide access to the tanks. None of these risers were designed to facilitate the emptying of the tanks.

Tank waste retrieval is extremely difficult. Because of the limited access and the highly radioactive and toxic environment retrieval, equipment must be used inside the tank with the assistance of video monitors and remotely controlled. Radiation and chemicals inside the tanks damage electronics and other components. Workers must be very careful to avoid exposure to chemical and radioactive waste. Furthermore, a number of pipes and rods congest the tanks and present barriers to handling equipment.

The waste retrieval program includes transfer of waste from SST to DST through hose-in-hose lines. The removed waste then goes through an evaporator to remove excess water. Eventually, waste will be transferred from the DST to the waste treatment plant to be vitrified.

Tanks are constantly surveyed to verify that none are leaking. Monitoring is conducted both in-tank, i.e., by measuring volume, and in the surrounding soil using moisture and radiation detectors (Ecology, 2019).

Fig. 5.11 shows a detail of waste removal operations being carried out in parallel on two USTs. In August 2012, workers began moving waste from C-112, an SST containing some 4000 m^3 of radioactive and chemical sludge. In addition, workers were soaking the 40 m^3 of concrete-like solids remaining in Tank C-108, so they can be dissolved into a liquid and pumped out.

More detail on the removal operations is given in the following example. For the first time, waste solids were being retrieved in parallel from three USTs: C-104, C-107, and C-109. Crews were working to finalize the removal of hard bottom waste from C-104. They began a month-long recirculation of caustic in C-104 after 60 m^3 of sodium hydroxide was injected the week before to begin dissolution of the remaining hard bottom. During the recirculation period, the caustic reacted with the encrustations to form a soluble salt (Fig. 5.12). Upon completion of the reaction, the remaining material will be pumped to DST AN-101, and C-104 will be finally flushed. As of August 2012, the workers have used modified sluicing and hot water washes to remove more than 98% of the nearly 1000 m^3 of waste initially contained in C-104

FIGURE 5.11 Retrievals continue in underground storage tanks C-112 and C-108. *Credit to DOE, Wiki Commons.*

FIGURE 5.12 Tanker truck injects caustic into tank C-104. *Credit to DOE, Wiki Commons.*

when retrieval began 30 months before. Meanwhile, the Mobile Arm Retrieval System (MARS) moved on to tank C-107, having removed nearly 66% of the tank's contents. MARS' robotic arm employed its high-pressure nozzles to dissolve waste in C-107 and shift it to a pump for removal. The MARS arm flexibility allowed personnel to effectively maneuver around obstacles found in the tank that could have considerably delayed the progress of retrieval. In parallel, a third tank, C-109, was being water-soaked to weaken and break down hard masses into a fluid state. It was estimated that in August 2012, less than 14% of the original 240 m^3 remained in C-109.

Many organic chemicals were used in support of Hanford operations and were discharged to the waste tanks. The chemicals went through thermal, chemical, and radiolytic processes that formed volatile and semivolatile compounds. After initial storage, several activities performed on the tank waste, including removal/retrieval of various materials (e.g., uranium, strontium, and cesium); evaporation; solidification; and settling, tank-to-tank waste transfers, and evaporator batch campaigns mixed and spread many of these constituents through the SST and DST. Organic compounds and their degradation products are the primary sources of gases and vapors (WRPS, 2017). The toxic impacts of chemical vapors and dust are described elsewhere in this book (e.g., mercury in Box 7.28, arsenic, lead, and dusts in Box 5.6).

Chemical vapors from Hanford UST are an actual and potential concern because workers enter the tank farms regularly for waste retrievals, equipment maintenance, and surveillance. During the waste retrieval process, tank farm workers can be exposed to chemical vapors escaping from tank top spaces and other release points. Specific weather conditions, e.g., inversions, also have been increasing the number of vapor incidents.

Concerns about tank vapors at the Tank Farms have been made since July 1987 or even earlier and have been the subject of several evaluations. Changes have been made through the years to better protect workers, including better PPE, alarms activating when chemicals are detected, more industrial hygiene professionals assisting the tank farm activities, and enhanced monitoring for chemicals. Enhanced engineering controls include fans to draw air from the tanks, convey it through HEPA filters, and then discharge it through tall stacks to keep vapors away from workers and favor dispersion: However, not all tanks have stacks.

Symptoms reported have generally included raw throats, coughing, stinging eyes, migraines, dermal irritation, a metallic taste in the mouth, bleeding nose, giddiness, or combinations of these. Independent teams also have reported cases of accelerated pulse and respiratory difficulties. A few workers declared they have gone on long-term disability due to past exposures.

Workers generally have mentioned the smell of ammonia, one of the known chemicals in the top volume of the tanks. In fact the tanks contain more than 1500 different species of chemicals, some of them highly toxic, in addition to radionuclides (Cary, 2018).

BOX 5.6 Some chemicals of concern (NRC, 2002)

Some of the worrisome chemicals found in the course of nuclear decommissioning are structural materials, coatings, bulbs, switches, electrical cables, etc., which can include asbestos, lead, PCBs, and mercury. Besides, quartz and cristobalite silica were identified during concrete demolition. Fumes, especially those containing lead and arsenic, and smoke from flame cutting and welding can cause significant chemical exposure during decommissioning. Copper fumes are also hazardous (see story in the following).

Chemical hazards at one decommissioning plant—unnamed in (NRC 2002)—included lead and arsenic vapors, released from torch cutting and plasma arc, and quartz and cristobalite particles, produced from chipping and hammering. At that plant, air samples indicated a few exposures exceeding OSHA's permissible exposure limit (PEL). Arsenic concentrations (PEL = 0.01 mg/m^3) exceeded the PEL a few times during the sampling campaign. The highest arsenic reading was reached while using the torch and grinder to cut a hole. The same activity caused lead reading at 1.5 mg/m^3 to exceed the PEL (PEL = 0.05 mg/m^3). Quartz (PEL = 0.1 mg/m^3) and cristobalite (PEL = 0.05 mg/m^3) aerosols much exceeded the PELs when using the chipping hammer (817.84 and 1.5 mg/m^3, respectively).

Airborne sampling is essential to monitor exposure to metal fumes (BNL, 2011)

Problem encountered; impacts

At the Brookhaven Graphite Research Reactor (BGRR) Project, exposure levels to copper fumes repeatedly exceeded OSHA (US Occupational Safety and Health Administration) total weighted average (TWA) limit concentrations while workers were torch-cutting low-carbon steel plates inside a ventilated tent. The TWA for copper fumes is 0.10 mg/m^3. At BGRR, copper fume concentrations of 0.15 mg/m^3 were measured. The samples were collected at the lapel location, which is standard practice for exposure monitoring of other airborne contaminants, i.e., within 30 cm of the nose/mouth, but outside the face shield.

During the investigation, BNL found a well-documented sample bias associated with welding and cutting activities.

When conducting these activities, face shields and welding masks may influence the airflow near the breathing zone and can produce biased sample results. To be representative of the worker exposure, samples should instead be collected from the interior of face shields and welding masks. Ideally the collection point should be within 5 cm of the mouth of the worker. This methodology is in compliance with OSHA sample collection methods and ANSI/AWS (American Welding Society) standards.

Lessons learned

Positioning of airborne samplers is a key factor for precise assessment of metal fume exposure during welding and cutting activities.

5.22.2 Analysis; solutions

After a series of reported events in the spring of 2016 and complaints from the trade unions and other stakeholders, the Secretary of Energy directed the independent Office of Enterprise Assessments (EA) to perform an assessment of progress in addressing recommendations issued by previous reports. In January 2017, EA issued its report, which included seven recommendations intended to address tank vapor issues. The seven recommendations in the report addressed technological and administrative aspects of Tank Farm vapor issues and worker concerns and focused on critical areas such as stakeholder communications, engineering controls, industrial hygiene, damage compensation, accident response, and occupational medicine.

Managing Tank Farm vapor issues continues to be a challenge because of longstanding, difficult issues in vapor characterization, the identification and control of fugitive releases (vapor sources that are not related to known emission points, such as ventilation stacks or vent filters), and repeated incidents. To complicate the challenge, the current climate—e.g., multiple lawsuits—remains a barrier to open communications with stakeholders (DOE, 2018): See a recent case at the end of the following paragraph.

For the past 4 years, Stoneturn Consultants have provided independent oversight of safety and health at the Hanford Nuclear Reservation (STC, 2017). This oversight was established through an agreement between the unions and the contractors. Under this program, workers and supervisors have been subjected to several safety culture surveys, which have found that there is a high degree of trust in industrial safety and nuclear safety, but a low level of trust in chemical safety, especially in the tank farms. The level of trust matches the injury and illness reporting, and the fact that facility has not been able to manage unexpected chemical vapor releases. Hopefully, huge investments made in engineering and administrative controls in the past couple of years will correct for this situation: A $13B 10-year contract was awarded by DOE in May 2020 (DOE, 2020a). Unfortunately, a judicial plea to a civil court blocked this significant step only few days after the contract was awarded (ENR, 2020)

5.22.3 Lessons learned

Exposure to chemicals in the workplace is well known to be associated with an increased incidence of incidents and occupational diseases. Nuclear decommissioning makes no exception, although chemical hazards are often given less recognition than radiological hazards. In fact, experience in decommissioning of nuclear fuel cycle facilities (of which Hanford is a well-known example) has shown a number of cases where workers were impacted by chemicals used in former operations. See Section 5.20 for another case in point. Reactors are less vulnerable to chemical hazards, but hazards during chemical decontamination should be taken in due account.

There is a need for employers to develop occupational risk assessments and, consequently, plant-specific preventive arrangements to protect the health of workers exposed to chemical agents in the course of decommissioning. A joint effort between decommissioning organizations, workers, and other stakeholders is critical to the efficient management and control of hazardous chemicals and materials at nation-wide level.

5.23 Occupational risks from asbestos

5.23.1 Decommissioning process

Often, older piping systems were insulated with asbestos. Asbestos has very good mechanical and insulating properties that led to its worldwide use in industrial plants (not excluding nuclear plants) and ordinary dwellings. This practice ceased in the 1970s when the toxic and carcinogenic effects (e.g., on lungs) of asbestos airborne fibers became widely known: In those years, most national legislations proscribed the use of this material (its use continued in several other countries, unfortunately). The generation of airborne particulates is typical of dismantling activities; therefore, special attention should be given to the identification, removal, and separate treatment of asbestos-coated piping and other components in decommissioning projects. Typically, workers wear protective clothing and respiratory apparatus during cutting and handling of asbestos containing materials (ACMs). Waste recyclers or scrap dealers are exposed to serious risks when unconsciously handling ACM. The management and disposal of ACM should be conducted only by certified companies. The following sections illustrate a few cases in which the presence of asbestos has created a serious issue in nuclear decommissioning.

5.23.2 Problem encountered 1; impacts

The demolition of an old chemistry plant at Harwell, UK, was followed by the removal of the building platform and the drains that formerly conveyed radioactive releases to site treatment. Asbestos sealing gaskets between the pipe sections were unexpectedly found. Related drawings showed the gaskets but did not specify their construction materials. Work was deferred until ad hoc permissions were granted for a specialist contractor to remove the asbestos gaskets. The work was successfully completed.

5.23.3 Lessons learned:

Records not necessarily provide all of the details needed when planning for decommissioning. In this case, however, the old age of the drainage system should have led the decommissioning planners to suspect that the gaskets contained asbestos (International Atomic Energy Agency, 2006c).

5.23.4 Problem encountered 2; impacts

When the new licensee ENRESA commenced the dismantling works of the Vandellos 1 NPP turbo blower, it was found that the thermal insulation of the blower and the nearby pipes had an asbestos layer. The presence of asbestos was not documented, and the former plant licensee, who had previously informed ENRESA that the plant contained no asbestos, was unaware of this situation. Some former workers were interviewed and revealed the asbestos layer was installed during a repair work. ENRESA prepared an asbestos removal project including procedures for safe works and requested that the delay in the dismantling schedule was officially endorsed. A new access hatch had to be installed, and workers had to be equipped with the appropriate protection.

5.23.5 Lessons learned

This occurrence highlights that all plant activities should be recorded during the plant's operational lifetime. In respect to decommissioning, it is noteworthy that existing documents do not necessarily match the actual conditions of the plant. Operations staff should be included in the decommissioning team when checking the operational records (International Atomic Energy Agency, 2002).

5.23.6 Problem encountered 3; impacts

The unexpected disclosure of asbestos in piping gaskets highlighted a serious issue during the decommissioning of the Experimental Organic Cooled Reactor (EOCR) at the Idaho National Engineering and Environmental Laboratory, United States. Luckily, the issue surfaced before the building was demolished, and corrective actions were taken. As the demolition was being prepared, a survey and characterization campaign was started to determine the amounts of and localize all ACM within the facility. After identification, the asbestos was removed and disposed of in preparation for the explosive demolition of the building. Explosives were planned to demolish the extensive piping and the building in one operation. During the asbestos removal campaign, however, it was found that a number of other components had asbestos, e.g., in the gaskets, which would have been demolished. This could have produced the release of airborne asbestos fibers and an unacceptable environmental impact. The demolition was postponed until all flanges were broken down and the gasket material was removed.

5.23.7 Lesson learned

ACM can appear in unexpected locations and, to prevent unnecessary delays, much attention must be given to localize all sources of this material before scheduling massive demolitions (International Atomic Energy Agency, 2005).

5.23.8 Problem encountered 4; impacts

Due to the rigor of the relevant auditing process, it was found at Sellafield, United Kingdom, that some metallic waste potentially containing asbestos had been shipped to a facility not authorized to deal with asbestos (incident of July 9, 2018).

All of this waste has been subjected to radiological measurements and found not to contain radioactive material. At the time of reporting, investigations were underway in cooperation with the receiving company to check whether any impacts might have occurred during the handling of this waste (ONR, 2019).

5.23.9 Lesson learned

Workers unaware of dealing with ACM can be exposed to significant occupational hazards.

5.23.10 Problem encountered 5; impacts

Heavy rainfall in July 2005 saturated the insulating layer on an unused steam line on a corner of K-1400 building at ETTP, Tennessee, United States. The saturated material detached from the pipe and its bottom side spilled to the ground. Sampling data verified that the insulating material contained asbestos; a spill of up to 1 kg was estimated, which is greater than the legal reportable quantity for asbestos (DOE, 2020b).

5.23.11 Lessons learned

Many kilometers of insulated steam lines run across ETTP aged installations. The event investigation disclosed that a preventive maintenance program had not be planned to the end that the steam pipes would not degrade over the years. The continual exposure to weather agents caused the degradation of the insulation and the release of asbestos. Responsible organizations should be aware of the need to routinely inspect unused equipment that may have asbestos-containing insulation. Special attention should be given to components/piping that is directly exposed to environmental agents over prolonged periods of time (Box 5.7).

5.24 Vehicle traffic events (DOE, 2020b)

The following vehicle traffic events are a tiny fraction of cases occurring regularly at all nuclear sites. Those quoted in the following have happened at sites undergoing decommissioning activities but were not necessarily linked to such activities. On the other hand, due to vehicle traffic increase (contractors, inspectors, visitors, consumables/spare parts/equipment deliveries, etc.) at

BOX 5.7 Botched demolition, Fernald Centre, Ohio, United States (DOE undated)

Plant 7, the Hexafluoride Reduction Plant, commenced operations in June 1954. The plant converted uranium hexafluoride to green salt, which was used in Plant 5 to produce uranium metal. After just 2 years, the Atomic Energy Commission ordered the shutdown of Plant 7 because a similar processing plant was in operation at Paducah, Kentucky. For the next 13 years, Fernald management reviewed several proposals for the unused facility, but none of them materialized. In 1969, all equipment was removed and sold, and the building was used to store containers of green salt and empty containers. In 1994, Fernald demolished the Plant 7 steel structure to reduce the potential risk for contamination to workers, the public, and environment. Due to an initial failure, demolition required two separate explosions to bring the seven-story structure to the ground.

Failure of a demolition step is potentially hazardous to the demolition workers, who are required to access the structure (made weaker and possibly unstable after the first explosion) and place new charges for the second explosion. Any unexploded charges could inadvertently explode while the workers are there.

nuclear decommissioning sites, an increase in the number of vehicle incidents is quite possible. Additionally, it is well known that any incidents—although minor and unrelated to nuclear activities—that happen at nuclear sites bear a "nuclear stigma" and should be prevented as far as possible at least to safeguard the image of nuclear organizations in charge.

5.24.1 Problem encountered 1; impacts: worker hit by truck suffers fracture to leg and twisted ankle, East Tennessee Technology Park, Oak Ridge, Tennessee, United States, February 26, 2020

Workers were conducting General Material Handling intended to transfer to and store material at K-1316-K, Maintenance and Material Storage Tent. The flatbed truck was parked on the roadway next to K-1316-K to unload the material. Flagmen were placed at the front and back of the truck for traffic control and supervisors checked forklift operations to unload the material.

After all the material was transferred, the truck driver received an "all clear" from the supervisor to leave the area. Before the truck left, the supervisor moved along the passenger side of the truck, and as he approached the front, he noticed a vehicle traveling down the roadway toward the expected direction of the truck. The supervisor stepped out into the blind spot at the front of the truck trying to draw the driver's attention (he did not consider that the flagman was still in place). The flagman, who was concentrated on vehicle traffic, did not notice the supervisor movement. The truck driver, who could not see the supervisor, moved forward. The right front bumper hit the supervisor, throwing him to the ground, and the right front wheel crushed the supervisor's legs. The driver at once stopped the truck, and another worker provided first aid to the injured supervisor.

A worker called the emergency number. The material handling work was halted and placed in a safe condition (truck was secured and flagmen stayed in place to divert traffic from the incident area). The injured worker, who remained conscious, was kept stable. The emergency service transferred the supervisor to local hospital. A medical check disclosed a fracture in the lower left leg and a twisted right ankle. After initial scene investigation, the truck was allowed to go back to its base for end of shift.

5.24.2 Lessons learned

Perform and keep work within controls. Never evade work procedures, take impromptu initiatives, or overdo.

5.24.3 Problem encountered 2; impacts: worker tried to stop a rolling truck, Hanford Site, Washington, United States; March 19, 2019

A heavy equipment operator (HEO) drove a 0.75 t truck up a slightly sloped unpaved road and parked the truck beside an excavator.

The HEO engaged the manual transmission in gear, inserted the parking brake, switched the engine off, exited from the truck cabin, leaving the door open, and then walked some 15 m away. The HEO reports that some 10 min later the truck slowly rolled backward down the slope toward a road shoulder.

The HEO rushed to the truck and tried to hop into the open door of the cab to activate the foot brake. The HEO could not make it into the truck because his clothing was grabbed on the open truck door. The HEO then tried but failed to move the foot brake by hand. The truck was moving faster, and after some 15 m, the HEO hopped off the truck, clearing the truck door and landing on his back on the ground with no injuries. The truck stopped against the road shoulder with no damage.

Federal Engineers and Constructors (FE&C, the contractor providing the truck) Safety staff checked inside the truck cabin and noticed the parking brake pedal was flat on the floor and the ignition key was in the off position. However, they could not remember whether the manual transmission was in gear or when they put it in gear when going into the cabin. After the incident, the truck was secured. The HEO was assessed by a physician and returned to unrestricted work. Notifications were sent to Facility Management and the DOE-RL Representative.

On March 1, 2019, the off-site mechanical assessment confirmed the truck had several problems. Firstly, the parking brake was unable to adapt to the tension applied, and the cable was stretched out and would not retain the truck when engaged. Secondly, oil had leaked and impaired the rear brake system. And thirdly, when parked on a slope, placed in second gear, and without the parking brake engaged, the truck rolled.

5.24.4 Lessons learned

This event was a near-miss to an injury, where something happened that was unexpected and no barrier prevented the event from producing reportable consequences (i.e., luck was the main factor preventing a reportable injury). Causal factors include the following:

- The HEO made an error in judgment.
- Incorrect performance was due to inattention.
- FE&C management did not enforce maintenance requirements for the truck. As the result, preventive maintenance of the truck was inadequate.

FE&C management did not have requirements for daily mechanical inspections or a written vehicle maintenance plan. Lack of required maintenance resulted in the parking brake not having any adaptation ensuring there was enough tension to engage the parking brake. Due to the lack of a daily inspections and inadequate vehicle maintenance, the rear brake system was impaired by oil leaks from the rear axle seals. As confirmed by the mechanical review, the manual transmission would not hold on a slope in second gear.

The terms "equipment" and "other vehicles" are used in the FE&C Equipment Maintenance Program, and in the 100K Waste Sites Remediation and Backfill documents in force. FE&C had determined that the terms did not apply to light duty trucks. Therefore, FE&C did not have written instructions about light duty trucks for documented daily inspections, checklists, and scheduled maintenance. These instructions were available for all other equipment and trucks as specified by the following statement:

> *"Motor vehicles and heavy equipment shall be inspected (including functional checks) at the beginning of each shift to ensure that the equipment is in safe operation condition and free of damage or wear and tear that could cause failure while in use. The inspections shall be documented prior to using the equipment."*

5.25 A high-voltage energized line was severed and fell to the ground, Hanford Site, Washington, United States; August 14, 2018 (DOE, 2020b)

5.25.1 Decommissioning process; problem encountered; impacts

The 2101M Warehouse Shops and Office Building (14,664 m^2) is planned to remain in use until 2057. The design service life of the 2101M Building life is

75 years. As the building was constructed in 1953, it is anticipated to be replaced in 2028. Therefore, the need for a new warehouse should be evaluated urgently (MSA, 2017).

On the day of the incident, a Mission Support Alliance (MSA1) worker was driving a cube truck backward in the parking lot of building 2101M in the Hanford 200 East area. The lift platform on the rear of the truck was in a downward position when it came in contact with a bracing cable to a power pole. The force applied transferred from the cable upward to the power lines cutting one line and dropping a high-voltage live wire onto the ground. There were no injuries or risk of contact with people in the surroundings of where the power line fell. However, it was just luck that impeded this near-miss to result in injury or death.

The truck driver was driving backward between 5 and 15 km/h. The lift platform was lowered adding some 2 m length to the truck. The driver had noticed the cable before, but, as he could not see it in the mirrors while backing, he thought the truck was clear of the cable. When another worker signaled the incident, the driver moved the truck forward some 10 m, lest the power pole should fall.

Surrounding facilities were occupied by many Hanford contractors, not only MSA1. Immediately after the event, stop signs and barricades were installed before these contractors exited from their workplaces.

Response actions resulted in the following:

1. The truck driver at once notified the work supervisor of the incident, whereas another worker began warning individuals nearby to stay away of the hazard.
2. A concrete block was installed in front of the cable to alert drivers to avoid further contact.
3. Electricians shut down power to the affected pole and repaired the power line. Later they restored the power.

5.25.2 Lessons learned

A spotter should have been employed to help the truck driver and allow for better maneuverability in a congested area. In the present case, all MSA1 workers were required to use an assistant during backing operations of service trucks, and anytime, the truck lift platform was not in a secured position.

Demolition activities may result in a considerable growth of on-site movement of heavy vehicles for the transfer of demolition waste, consumables, and equipment, thus increasing the risk of traffic-related incidents and injuries to workers (and local communities, although transport incidents offsite are beyond the scope of this book). The frequency of traffic incidents involving project vehicles during demolition works should be minimized through a combination of education and awareness-raising strategies, and the enforcement of specific on-site procedures.

5.26 Uncontrolled hazardous energy release during removal of temperature sensor, Los Alamos National Laboratory, New Mexico, United States; July 19, 2018 (DOE, 2020b)

5.26.1 Decommissioning project; problem encountered; impacts

An LANL contractor was carrying out work under an ongoing MSS demolition project to take away legacy equipment at the Target Fabrication Facility (TFF). The TFF is a two-story building (TA-35-213) housing activities related to weapons production and laser fusion research. The TFF laboratories and shops specialize in providing precision machining, polymer science, physical and chemical vapor deposition, and target assembly.

The temperature sensors are unused legacy equipment consisting of two parts: a top cap threaded onto the lower part, which is incorporated into a pressurized airline and consists of a well with the temperature sensor. The goal of the task was to take away only the cap since the lower part is incorporated in the pressurized airline. This goal was consistent with the requirement to remove legacy equipment "to the greatest extent possible." The workers were alerted to the airline being pressurized; however, since the cap was not within the hazardous pressure boundary, no release of pressure was expected upon its removal.

The worker carrying out the work (E1) stood below the unit on a 2.5-m ladder, while a second worker assisted with the work. E1 removed the cap from the lower part of the unit with a wrench. However, as the threading connecting the cap to the lower part was unexpectedly damaged, E1 mistakenly loosened the whole unit, releasing pressurized air. The reaction force ejected the whole unit into the air, striking the metal ducting overhead and falling back onto the pressurized line. By instinct, E1 grabbed the unit and reinserted it into the main line while it was still discharging pressurized air. The contractor at once notified the TA-35 Maintenance and Site Services (MSS) Operational Maintenance Coordinator (OMC) who in turn alerted higher-ups. The removal work at TA-35 was later suspended pending investigation. Both workers wore the required PPE. There were no physical injuries or no impact to the plant or the environment.

The event report was based on the fact that the contractor worker did not expect (and took no measures against) the possible release of pressurized air and reinstated the cap while it was still releasing pressurized air, a risky operation indeed.

5.26.2 Lessons learned

This event exemplifies a class of incidents due to failure to implement and/or comply with a hazardous energy control process that causes potential worker exposure to the release of uncontrolled hazardous energy (e.g., live wires, powered mechanical hazards, steam, pressurized gas, etc.).

5.27 Fatality and injuries from release of carbon dioxide at Idaho National Engineering and Environmental Laboratory, Idaho, United States (DOE, 1998)

It is by presence of mind in untried emergencies that the native metal of a man is tested.

James Russell Lowell (1819–1891).

5.27.1 Decommissioning process; problem encountered; impacts

Idaho National Engineering and Environmental Laboratory (INEEL) is located on 2300 km^2 of desert in a rural, unpopulated area of south eastern Idaho. INEEL is a scientific institute whose multifaceted mission is to integrate engineering and applied science to tackle environmental management challenges, waste disposition, nuclear technology and application, and national security.

The Test Reactor Area (TRA) contained an operating test reactor, four unused research reactors, reactor fuel stores, labs, and site support systems. The Engineering Test Reactor (ETR) Facility consisted of many independent buildings that supported the ETR reactor during its operations. ETR was the world's largest and most advanced materials test reactor when it started operations in September 1957. The 175-MWt reactor investigated fuel, coolant, and moderator materials. The ETR was deactivated in 1982 soon after its final shutdown. The reactor was removed and disposed of in September 2007.

Building 648 (Electrical Building) housed electrical equipment for the TRA and the ETR Facility. It consisted of the ground floor containing switching equipment, controls, and power systems and a basement with electrical cables. The accident occurred on the ground floor of the Building. Decommissioning of the Electrical Building was completed in October 2007.

At the time of the accident, contractor activities at INEEL were run by the DOE Idaho Operations Office (ID). The facility where the accident occurred was under the jurisdiction of the Office of Nuclear Energy, Science and Technology (NE). Lockheed Martin Idaho Technologies Company (LMITCO) was the management and operations contractor for INEEL and for the TRA.

The building electrical components were protected from fire by a carbon dioxide (CO_2) fire extinguishing system. When the system is activated, CO_2 is released from several nozzles in the ceiling of the building ground floor. The release of CO_2 is controlled by two electronic control heads situated in a storage building next to Building 648.

The percentage of CO_2 in the building following the accidental activation of the fire extinguishing system was estimated at around 50%. This level is far (and unnecessarily) above the 30% concentration needed for fire protection and is extremely lethal to exposed individuals. This oxygen-deprived atmosphere can cause symptoms of nausea, vomiting, almost total impairment, spasmodic breathing, convulsive movements, and death in minutes.

On July 28, 1998, 13 workers (supervisors, operatives, electricians, and fire protection staff) were deenergizing electrical circuit breakers in preparation for preventive electric maintenance in the Electrical Building.

As the last circuit breaker was opened, the CO_2 from the fire extinguishing system discharged suddenly and without a warning alarm. Within seconds, the workers found themselves fighting to run away from the lethal atmosphere with no visibility and under the confounding effects of CO_2.

In addition to lack of visibility and CO_2 effects, escape was considerably hindered by various pathway obstacles, the failure to designate emergency exit routes, and poor exit direction lighting.

The initial rescue attempts by site personnel—which were urgently needed, given the concentration and toxicity of the CO_2 atmosphere—were hindered by the lack of readily available self-contained breathing apparatus (SCBA) (Fig. 5.13).

This unavailability resulted in several rescuing attempts at considerable personal risk and posed the initial rescuers the agonizing dilemma whether to violate safety and regulatory prohibitions against entry without SCBA or defer the search and rescue of workers stuck in Building 648 until the arrival of the Fire Unit. The rescuers chose to risk their life to rescue coworkers.

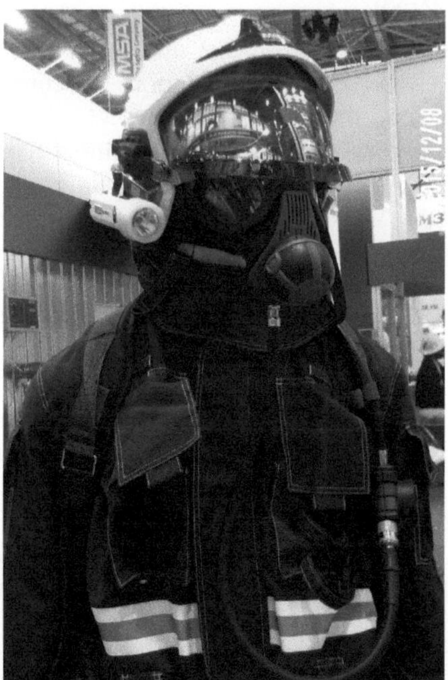

FIGURE 5.13 A model of self-contained breathing apparatus (SCBA) for firefighters. *Credit to Alex Chirkin, Wiki Commons.*

Their determination contributed to three rescues and thus saved the lives of three workers. Without their heroism, the loss of life might have been much greater and could have also included some rescuers.

In total, 15 workers received medical attention or treatment after the accident. One electrician passed away, and several suffered life-threatening injuries and dangerous CO_2 inhalations.

LMITCO had not established a proper, immediate response to lethal levels of CO_2 resulting from an automatic or unexpected discharge and had not considered measures for urgent escape or to launch search and rescue operations. Planning was incomplete and partly erroneous, training insufficient, and PPE was unavailable to assure emergency exit, ease prompt search and rescue, or protect affected workers and responders.

Injuries to the workers and rescuers directly stemmed from the unavailability of SCBA. The failures that caused or worsened the inability of workers to flee, or of rescuers to quickly enter/leave the building, contributed to the severity of the impacts received by the workers, because it extended their CO_2 exposure. While immediate response and evacuation would not have avoided the accidental discharge of CO_2, it would have mitigated the impacts on workers.

5.27.2 Lessons learned

The Investigation Board determined that this accident was avoidable. Since March 1996, INEEL had experienced serious precursor accidents. These events had indicated a need to widely improve work planning and controls, conduct hazard assessments, and assure that specific safety requirements were integrated into work plans and enforced in the field.

The DOE-ID and LMITCO had also not been on time in the implementation of DOE's integrated safety management (ISM) policy to resolve the standing work control issues to improve safety performance. The recurrent accidents and less-than-good level of safety performance confirm the continued tolerance of a loose, expert-based approach to control the work and related hazards.

LMITCO has not been responsive in managing the inflow of requirements and standards regarding CO_2 fire extinguishing systems and worker safety, and incorporating these requirements into manuals, design controls, procedures, and training programs. This failure contributed to inadequate understanding and competencies in dealing with the hazard, erratic and insufficient application to design, work plans and procedures, and inadequate resource allocation.

The physical lockout of the CO_2 system to protect the workers, for example, is a single requirement that would have avoided this accident. The procedure that requires the CO_2 lockout, however, had not been updated and was not used for this work. There are specific training requirements on the emergency response to CO_2 releases, but the CO_2 hazard was not part of LMITCO training programs: Therefore, on the day of the accident, workers were not fully aware of the hazard, emergency measures, or the limitations of the available protection.

Once the CO_2 system discharged, immediately flooding the room and creating invisibility, the workers were not given the means to safely flee, including clear exit routes, respiratory devices, exit lighting, or emergency ventilation. The initial rescue efforts were frustrated by the lethal concentration of CO_2, physical obstacles, low visibility, and lack of SCBA.

The circumstances that eventually culminate in this tragedy began to build up in the past. The CO_2 system design, first commissioned in 1971 and modified in 1997, did not include the monitoring of system status to guarantee at least a 25-second warning alarm regardless of the source of activation. The failure to reevaluate the need for this system, the lack of a corporate policy and procedures to mitigate CO_2 risks, and gradual cost cuts in the site infrastructure that reduced the availability of SCBA and search and rescue training also contributed to the accident. Despite the previous accidents at INEEL, LMITCO and ID failed to implement requirements, ensure corrective actions in work planning and control, or achieve more rigor and discipline.

The Investigation Board concluded that ID has not been determined in monitoring contractor performance or in ensuring that corrective actions and improvements in hazard and work controls were completed and fully enforced. The Board also concluded that LMITCO did not comply with its contractual obligation to protect workers from a potentially lethal hazard by establishing required design, procedures, hazard assessments, work controls, communications, PPE, system lockout, and training. The contractor failed to avoid activation of the CO_2 system in occupied areas or, alternatively, to ensure prompt warning and the needed escape time and means to accomplish search and rescue missions without risking more lives. One of the injured workers said: "It's taken one life. We're lucky it didn't take more."

5.28 Noise hazards in demolition

Noise is the most impertinent of all forms of interruption. It is not only an interruption, but also a disruption of thought.

Arthur Schopenhauer 1788–1860.

5.28.1 Decommissioning process; problem encountered

Jackhammers drilling into concrete, compressors, concrete mixers, ventilation and suction equipment, rivet busters, grinders, saws cutting metals, electrical generators, conveyors belts, dump trucks, and earth-moving bulldozers are just a few examples of the noises that demolition (and construction) workers face in their jobs. It should be considered that a decommissioning process may easily include several simultaneous sources of noise. It is also important to distinguish continuous noise (or with minor variations) from intermittent noise. While noise may not appear as hazardous as other demolition hazards, it can have a heavy impact on worker safety and health. Vibration can be concurrent with noise and is briefly dealt with in Box 5.8. A comprehensive project study focusing on noise and vibration (although a nonnuclear project) is given in Jacobs (2017).

A special case of vibration is due to vibrating hand tools used in demolition, which can cause hand-arm vibration syndrome (HAVS). Exposure to vibration must be managed and minimized. More guidance on this aspect is given by HSE (undated a).

Noise is often ignored because hearing loss is gradual, and the danger is not obvious. This does not mean it is not serious. Most people's hearing worsens as they get older. But normally aging does not impair hearing before the age of 60. People who are not exposed to noise and have no other medical conditions keep their hearing for many years. In contrast, those who are exposed to noise

BOX 5.8 Vibrations during demolition of large structures

Demolition activities can result in different levels of ground vibration, depending on the equipment and techniques used. Ground vibrations spread through the ground and decrease with distance. Buildings in the vicinity of the demolition site react to these vibrations in various ways ranging from no perceptible effects, low rumbling sounds to minor damage in rare cases.

Vibration is a specific risk during demolition. The measures taken to reduce the acoustic impact of these activities will also help reduce the impacts of vibration on local communities and the buildings closer to the demolition site.

Ground vibration induced by the collapse of large structures, e.g., cooling towers in NPPs has recently been identified as a hazard to nearby nuclear facilities, which requires mitigation. The traditional concept of separately designing cooling towers and other installations as isolated components in an NPP site is especially inappropriate in a tight, congested nuclear site. Lin and Zhong (2017) presents a study on the mitigation of ground vibration by means of an innovative use of materials working as cushions underneath the cooling towers.

There are numerous examples of massive NPP cooling towers and reactor buildings that were demolished by explosives and may have caused significant vibrations to nearby installations, for example Trojan, Oregon, United States, 2006; Calderhall, Chapelcross, United Kingdom, 2007 (Fig. 5.14); and Philippsburg, Germany, 2020. At Maine Yankee, Maine, United States, the turbine building and the containment buildings were also demolished with the use of explosives. In all explosive demolitions and resulting structural collapses, the vibrations onto nearby structures were estimated in advance and monitored during the works.

Box 5.8 Vibrations during demolition of large structures—cont'd

FIGURE 5.14 Chapelcross NPP, United Kingdom, demolition of cooling towers. *NPP*, nuclear power plant. *Credit to Lynne Kirton, Wiki Commons.*

and do not protect their hearing start losing their hearing at an early age. To quote one example, the average carpenter at 25 has the same hearing as a 50-year-old person who has worked in a quiet environment.

The average construction/demolition site has a noise level of between 80 and 90 dB. It has been reported that some 51% of construction workers have been exposed to hazardous noise and 31% of those workers were not wearing hearing protection. Moreover, some 14% of all construction workers have hearing difficulty (ASSP, 2020). According to CPWR (undated), demolition work has an average noise level of 99.3 dB, and chipping concrete has an average 102.9 dB.

Typical noise levels depending on specific dismantling/demolition tools are given in Table 5.1 (CPWR undated).

The following cases (1)–(5) are all extracted from HSE (undated b). While not necessarily referring to nuclear decommissioning projects, they are applicable to such projects.

TABLE 5.1 Loudest tools, increasing noise level.

Tools	Average noise level (dB)	Max noise level (dB)
Welding, cutting tools	94.9	122.8
Other portable power tool	95.4	118.3
Handheld power saw	97.2	114.0
Power drill	97.7	123.7
Rotary hammer	97.8	113.5
Chop saw	98.4	117.7
Impact wrench	98.4	131.1
Machine tool	101.8	119.8
Nail gun	103.0	112.8
Chipping hammer	103.0	119.2

5.28.1.1 Metal-cutting guillotine

5.28.1.1.1 The problem

The noise level at the operative's stand of a metal-cutting guillotine was measured at 92 dB.

5.28.1.1.2 The solution

The guillotine was fully maintained, and its hydraulics refurbished.

A collecting tray was equipped with rollers and covered with carpet, to reduce the noise of falling metal pieces.

5.28.1.1.3 The results

The noise level at the operative's stand was lowered to 84 dB.

5.28.1.2 Pedestal grinder

5.28.1.2.1 The problem

An electrical company had a small maintenance workshop in which the pedestal grinder caused a major noise problem.

5.28.1.2.2 The solution

The pedestal grinder was bolted to the concrete floor using rubber mounts, and rubber inserts were used in the paneling.

5.28.1.2.3 The results

The noise levels were lowered from 104 to 102 dB.

5.28.1.3 Cold-cutting circular saw

5.28.1.3.1 The problem

An electrical company had a small maintenance workshop in which the cold-cutting circular saw caused a major noise problem.

5.28.1.3.2 The solution

The saw was remounted using rubber mats.

The feed table was decoupled from the saw table to prevent the transfer of vibration.

5.28.1.3.3 The results

The noise levels were reduced from 94 to 90 dB while segmenting 3-mm-sheet metal.

5.28.1.4 Reducing noise in road vehicles

5.28.1.4.1 The problem

Noise levels in team-carrying road vehicles can cause a significant hazard. In some waste collectors, noise levels have been measured up to 103—105 dB. One local company owning a number of such vehicles assigned a noise reduction study to decide on the practicability of a substantial noise reduction without impairing the vehicle's road effectiveness.

The main source of noise was the engine. However, the cabin was also subjected to noise from the transmission and hydraulic power unit coming through the light steel floor and the rear wall.

5.28.1.4.2 The solution

This noise was lowered by attaching a lead/fiberglass laminate to the undersurfaces of all floors. Noise resounding was also limited by coating the roof and all internal surfaces of the doors and the front and rear walls with an expanded polyurethane foam lined with a porous PVC film.

5.28.1.4.3 The result

A noise reduction of 4—9 dB was obtained. Similar reductions have been achieved on fire and maintenance vehicles.

5.28.1.5 Isolating a hydraulic guillotine

5.28.1.5.1 The problem

The hydraulic power section of a 250-t hydraulic guillotine generated 90—95 dB. Noise was produced by vibration from the motor pump to the machine frame (and through stiff piping to the rest of the machine) and by impacts from the valve block. An acoustic enclosure was ill-suited because its effectiveness would have been limited by the transmission of vibration through the piping.

5.28.1.5.2 The solution

The noise was reduced by isolating the noise sources from the guillotine. This solution was much cheaper than an all-encompassing enclosure and did not affect access to any part of the machinery.

The motor and pump units were mounted on a rigid frame, isolated from the machine. Flexible couplings were interposed within the rigid piping. The valve block was isolated from the machine frame through a cork–rubber combination.

5.28.1.5.3 The result

A noise reduction of some 17 dB was achieved.

5.28.2 Lessons learned

Noise hazards are common at construction and demolition sites including at nuclear decommissioning projects. Basic rules to protect the workers against these hazards include among others (a trade-off may be needed as some recommendations can be inimical to such factors as health and safety, productivity and costs, etc.):

- The first protective step is to understand the level at which workplace noise can be hazardous. According to the US Standard ANSI/ASSP A10.46-2020 "Hearing Loss Prevention for Construction and Demolition Workers," an acceptable noise level is 85 dB over an 8-h shift, with a 3-dB doubling rate (i.e., for every 3-dB increase in noise level, the maximum exposure time is reduced by half, vice versa, a 3-dB decrease doubles the maximum exposure time).
- Noise monitoring at project locations should be conducted regularly, and the frequency of monitoring should be increased as the demolition works approach-sensitive locations or when more noisy operations are planned to take place. The monitored noise levels should be assessed against the noise limits and should contribute to the scheduling of works to make sure that the noise emissions from the various locations are below the specified limits
- Once the noise levels at the demolition site have been determined, engineering controls should be considered to minimize the noise. These could include attenuators or enclosures for older equipment, or siting equipment at a distance from workers. Worker rotation between noisier tasks and quieter tasks is also an option.
- Use rubber coatings in chutes and dampers to decrease impact noise.
- Minimize drop height of waste.
- Minimize the number of vehicles/heavy equipment on-site at any time.
- Ensure that noisy vehicles are parked as distant as possible from noise sensitive areas.

- Ensure that drivers are aware that noise can cause disturbance to workers and the public, especially when entering and leaving the project site (e.g., no unnecessary horn blowing).
- Evaluate alternative types of reversing alarm with reduced noise emissions, such as ambient noise sensing alarms with variable volume or directional modulated alarms—these measures require a case-by-case evaluation taking also account of any conflicting health and safety issues.
- The next protective step is to use PPE (beginning with simple earplugs) as a barrier between the workers and the noise. Technological progress has made it possible for workers to protect their hearing while still being able to communicate with coworkers.

The workers should understand that hearing loss occurs gradually, and they may not be aware of the problem at once. Therefore, it is important for workers to have their hearing tested regularly for hearing loss. Periodic monitoring tends to make the workers more careful about protecting their hearing.

Workers also should be made aware of the fact that workplace noise is not only a health hazard. It can also turn into a safety hazard if noise impairs proper communication or prevents workers from hearing a vehicle or piece of machinery moving close to them. Additional safety-related issues caused by excessive noise may include absent-mindedness, impacts on orientation and focus, ignoring alarms, mental tiredness, irritability, and defensive reactions or overreactions. See also quotation on top of this section.

5.29 Cutting fatality at the K-33 Building, Oak Ridge, Tennessee, United States (DOE, 1997a)

5.29.1 Decommissioning process; problem encountered

K-33, the largest of the gaseous diffusion Oak Ridge plants, covers some 13 ha at the K-25 Site. Until 1985, K-33 was used for uranium enrichment. Considerable quantities of enriched uranium remained in process equipment at various locations in the building. To minimize nuclear criticality risks, these uranium deposits had been safely confined. Hazardous waste was also stored at K-33. There were only a few workers in the building at the time of the accident.

The basic equipment stage in a gaseous diffusion plant has three components: an electric compressor, a converter with a very large surface area of semipermeable material, and the process gas heat exchanger. A group of eight equal stages connected by piping forms a cell, which is encased in sheet metal. K-33 contains 640 stages in 80 cells. The cells are grouped in units, and the units make a cascade (Fig. 5.9).

The accident on February 13, 1997, in Cell 7 of the K-33 Building when the welder (W1), who was using a cutting torch, became trapped in flames. The work involved the removal of six converters from Cell 7 for transport to Portsmouth/Paducah as spare parts. This work started on February 8 and was to be completed by March. The roof of Cell 7, two converters, a side panel on the wall next to Converter 3, and a small side panel on the side next to Converter 5 had already been taken away from Cell 7 prior to the accident.

The removal of equipment from the cells was procedurally classified as routine maintenance, and no specific job hazard assessment was prepared in advance of the work.

The job was behind schedule, and the team worked overtime on Saturday, February 8, and on the evening of the day before the accident. On February 13, 1997, the Service Supervisor (the in-the-field supervisor responsible for the conduct of the work) was not present at the daily safety meeting held by the welders in early morning. Although not attending the meeting, the supervisor signed the work permit but did not appoint a fire watch.

When the accident occurred, W1 was wearing three sets of coveralls, three pairs of gloves, four layers covering safety shoes, a head cap and hood, a personal air monitor, and a full-face respirator with an attached welding mask.

Another welder was in Cell 7 working on Converter 5. More welders and other workers were in the vicinity of, but outside, Cell 7. No workers were in direct sight of W1, who was basically working alone.

W1 was using a cutting torch to dismantle one of the six converters remaining in Cell 7 and, at the time of the accident, was working on a ladder at the end of Converter 4. No accurate description of the accident is available, because there were no eyewitnesses. The Investigation Board reconstructed the most plausible order of events (up to the extinction of the flames by a mechanic), using physical evidence, workers' reports, and the findings of burn tests. The accident began its course when a piece of hot metal shaving or a spark ignited W1's anticontamination clothing somewhere around his left knee; however, he did not realize what was going on for a minute or more when he felt the heat and saw that the clothing on his left leg was burning. W1 tried to extinguish the flames while he was still on the ladder. He climbed down the ladder, walked to the south of Convertor 4, and cried for help while still attempting to extinguish the flames. He took away mask and respirator and cried for help again. Because the respirator had stifled the first scream, only the second scream was identified as a distress call by one of the four maintenance mechanics at work outside Cell 7. Two mechanics requested a radiological control person to be allowed to go into Cell 7. One of them switched off the oxygen and acetylene cylinders of W1's torch, positioned outside Cell 7, while the other went into Cell 7, snaked under the piping, and saw W1 standing on the cell floor near Converter 1 and fully enveloped by flames. While the mechanic was helping W1 take off his burning clothes, the mechanic's gloves also caught fire, so he had to leave W1 to extinguish his own gloves by rubbing one against the other. Then W1 fell on the floor of the cell. The mechanic grabbed a dry-chemical fire extinguisher and put out the flames: It had taken 3 min after the accident began.

W1's screams produced a response including coworkers, fire department, and medical facility staff. Under difficult, restricted access conditions, the responders provided first aid to W1 and removed him on a stretcher from K-33.

Obstacles in the cell and in the building hindered the transport of W1 to the ambulance.

W1 had third-degree burns over 95% of his body. He died at the hospital the next day from acute respiratory distress syndrome.

A review of previous occurrences at K-25 recorded in DOE's Occurrence Reporting and Processing System (ORPS) since 1991 disclosed that the main root causes were poor management systems and human error. The root causes of the November 22, 1992, fatality were that line management did not assign responsibility for safety, effective work controls were not used, and lack of safety culture produced unrecognized hazards.

The following are some of the technical and administrative factors identified by the Investigation Board as having caused the accident:

- Records of six agencies, plus the DOE, indicated only one fatality due to cutting and welding operations resulting in clothing fires.
- The aforementioned lack of information/experience may have led to a situation whereby fire-retardant clothing was not required by regulations or standards or employed. Besides, one national consensus standard on flame-retardant clothing was not enforced by DOE or addressed to by the Occupational Safety and Health Administration (OSHA).
- Although required in the work permit process, no workplace fire hazard assessment was carried out, and no fire watch was appointed.
- The site's safety committee assessed anticontamination clothing from the standpoint of heat stress, not flammability
- The work permit did not consider risks from the use of multiple layers of PPE. As noted earlier, the multilayered PPE worn by W1 delayed recognition of fire burning on his leg.
- Due to unclear in line management responsibility for safety, an inaccurate work package was prepared.
- Several references require a job hazards analysis (JHA) for the type of work carried out at K-33, but nobody was given the task of conducting a JHA.
- The work documents dealt poorly with such safety issues as admin and engineering controls, lighting, and fire watch.
- Classifying the work as "routine" may have led to ignoring some hazards and related controls.
- Lessons learned in previous K-31 work were not applied to K-33. For example, welders' clothing had caught fire in the past at both K-31 and K-33, but these incidents were not recorded.
- Fire watch training for supervisors and planners was not required; therefore, responsibilities were not well recognized.
- The work at K-33 was "work for others." Roles and responsibilities for "work for others" projects were not identified.
- Planning decisions were mostly made by individuals outside of project management and/or not experienced with the type of work.

5.29.2 Lessons learned

In brief, the planning process for the K-33 work did not involve the necessary analysis of hazards, did not provide effective controls, and did not incorporate vital lessons learned. Procedural flaws and failure to observe procedures resulted in unidentified hazards and lack of mitigation measures. Errors are often due to the culture prevailing in the decommissioning team (see Box 5.9, Section 6.1)

Welding/cutting is one of the most common operations that workers from the heavy duty and mechanical sectors regularly carry out during including

BOX 5.9 Decommissioning mind-set and culture (Communication from Grete Rindahl, 2010)

Two representative situations of different cultural approaches toward decommissioning were reported in interviews. "Scenario 1" is based on a description from an early decommissioning project that took a long time to become efficient, due to considerable opposition to performing decommissioning from both workers and management. "Scenario 2" highlights a contractor's approach where work was efficient and the contractor was highly motivated for dismantling and getting the job done, but exhibited poor understanding of safe conduct when working in radiological environment.

Scenario 1: A mechanic is dismantling a pump. He is at work within an industrial environment with radioisotopes and wears protective gloves, a dosimeter and a hard hat in addition to his coveralls. He is familiar with the pump details. He expects there to be some familiar sticky silver residue in the lower left bend—and yes, there is. The reactor circuits were significantly contaminated because of welding corrosion a few years ago, and this item had not been available for manual decontamination before they started plant dismantling. The mechanic keeps his distance and moves the contaminated items swiftly and at arm length into a separate container to minimize radiological exposure. He dismantles the pump carefully, making sure that all parts of the pump are sorted and remain undamaged. Who knows, maybe they will have to reassemble them? The pump is in not really that old, and sometime soon people must realize that shutting down this reactor was not a sensible move...

Scenario 2:

A mechanic contractor is dismantling a pump. He is at work within an industrial environment with radioisotopes and wears protective gloves, a dosimeter and a hard hat in addition to his coveralls. In his view, the pump is just another old pump, and he is chopping it to pieces as fast as he can. He keeps his gloves on most of the time in case any of the dirt being that radioactive stuff about which the protection guys preach on and on. A bend is stuck and hard to dismount, so he absentmindedly sits down on top of the pump to get a better grip. He wipes his brow with the back of his glove. Finally the bend loosens, and he throws it into the container. He dismantles the rest of the pump swiftly, using a hammer and a chisel when needed and throws all the scrap in the container.

nuclear decommissioning. However, the management and the workers themselves do not always give enough consideration to the hazards of these operations. One relevant case is the risk of catching fire.

5.30 Worker fractures leg bone while dismounting equipment, Hanford Site, Washington, United States; February 18, 2020 (DOE, 2020b)

5.30.1 Decommissioning process; problem encountered; impacts

A stationary operating engineer (SOE) was conducting regular surveillance on equipment near the BX Farm in the Hanford 200 East Area. To get readings on a portable light plant generator, and to extend battery life, the engineer needed to access the battery disconnect.

As there were no stairs or ladders, the SOE climbed onto the equipment body by stepping on a projecting arm of the structure, tires, and frame and completed the task. As he was climbing off the structure, the SOE stepped off the projecting arm with a height of some 70 cm to the ground. The SOE twisted his right ankle, heard two explosive sounds in the right knee and fell to the ground with a sharp pain in his right leg. A second SOE transported the injured person to the Site Medical Clinic by car. The pain was so strong that the injured SOE was unable to exit the car.

Subsequently the injured SOE was transported with an ambulance to Kadlec Regional Medical Centre at Richland, Washington. At the hospital, the injured SOE was diagnosed a tibial plateau fracture (a break in the top of tibia [shinbone] at the knee). He was then treated with pain killer, and a splint was placed on his right leg.

5.30.2 Lessons learned

This case belongs to the broad category of slips and trips (see also Sections 5.34 and 5.39). Slips and trips on/from the same level or little above standing position (i.e., where the height of fall is not the major factor) can be related to poor housekeeping, such as the presence of waste debris, abandoned construction materials, liquids (especially lubricating oils or other slippery spills), uncontrolled use of electrical cables and ropes, or inconsiderate movements (such as in the aforementioned case). These events are frequent at construction and decommissioning sites.

Preventive methods to address slips and falls include the following:

- Good house-keeping practices, such as the sorting and moving scattered construction or demolition materials to designated areas far from the workplaces
- Cleaning up waste items and liquid spills periodically
- Positioning electrical cables and ropes in distinct, marked spaces
- Use of antislip footwear
- Training for repetitive or common activities, including use of mockups.

5.31 Two cases of exposure to chemical vapors during D&D activities within the DOE complex

5.31.1 Worker affected by nitric acid vapor, Hanford Site, Washington, United States (DOE, 2020b)

5.31.1.1 Decommissioning process: problem encountered 1; impacts

This event occurred in a room of 234-5Z structure, which was part of Hanford's Plutonium Finishing Plant and was demolished with it (Section 7.1). It is another case of occupational exposure to hazardous chemicals during D&D operations. The event relates to the removal of an overhead pipe that had been supposedly emptied of nitric acid solution. During the pipe removal, some residual nitric acid spilled on both the containment and on a worker's powered air-cleansing respirator. The contact with the acid on the air-cleansing respirator suddenly caused a slight pain to the worker possibly due to his exposure to nitric acid vapor. The investigation following the incident revealed that members of the work team were not aware of preventive measures that would have avoided this incident. In addition, as the work was being prepared, the work team modified the work plan to accommodate a new approach, which caused many hazards being missed or poorly defined. Without rigorous guidance to stop work in case of an unexpected situation, and without a thorough survey of the new conditions encountered, the work was carried out in an unsafe manner eventually leading to unplanned occupational exposure.

5.31.2 Unexpected release of nitric acid in decontamination building; Portsmouth Gaseous Diffusion Plant, Ohio, United States; April 17, 2018 (DOE, 2020b)

5.31.2.1 Decommissioning process; problem encountered 2; impacts

During Portsmouth operating lifetime, the decontamination and recovery facility, X-705 was used for the main purpose of disassembling and decontaminating process equipment that was removed for repair, for uranium recovery, and for plant laundry. X-705 was one of the facilities containing nuclear material that were released under certificate of compliance GDP-2, issued by the US Nuclear Regulatory Commission.

On the day of the event, an odor was detected in the high bay of the X-705 Building by a survey operative. Upon further check, he noticed a slight yellow mist in the surroundings of one of the acid cleaning tables. He started an emergency response. The plant was evacuated. The Emergency Response Organization's immediate investigation found nitric acid (HNO_3) overflowing the drain collection container and spreading onto the floor. In response to the discovery of hazardous material, soda ash was used to control the spill.

Outside air sampling showed less than detectable values for HNO_3, HF, and NO_2. Further investigation determined that ¼ -turn acid feed valve had been inadvertently struck while the table was being placed in a standby mode.

5.31.2.2 Lessons learned

Strong acids are frequently used in chemical decontamination: These acids present inhalation and burning hazards to decontamination workers and others who happen to work in surrounding areas (Boxs 5.10 and 5.11).

5.32 Construction fatality at the Brookhaven National Laboratory Upton, New York; June 20, 1997 (DOE, 1997b)

Experience is the name everyone gives to their mistakes.

Oscar Wilde (1854, 1900) Lady Windermere's Fan (1892).

5.32.1 Decommissioning process; problem encountered; impacts

The Brookhaven National Laboratory (BNL) was established in 1947 as a multifaceted organization conducting research in physical, biomedical, and environmental sciences. BNL's current activities support DOE's scientific and technical role. BNL operates two reactors for medical research and treatment objectives. It also operates particle accelerators for high-energy physics,

BOX 5.10 Uncontrolled chemical spray Hanford Site, Washington, United States; September 6, 2018 (DOE, 2020b)

The scene of the incident was Hanford's 300-FF-5 project. Workers were off-loading a tanker of tetrasodium pyrophosphate (TSPP, formula $Na_4P_2O_7$) to on-site storage tanks. Hose connections were completed, and an air compressor was turned on to pressurize the tanker and start the transfer. The vendor operator (VO), wearing the prescribed PPE, verified the pressure at 180 KPa on the truck gauge and climbed to the top of the tanker for a check. When the VO heard a noise coming from the pressure relief valve (PRV), he shut the isolation valve for the compressed air to the tanker. The PRV lifted, spraying the chemical onto the VO. The VO climbed down from the truck and was subjected to safety shower. Medical evaluation found that he did not require follow-on attention.

The incident reported was fairly benign although it disclosed a potential mechanism for more serious impacts. It has been found that nausea, vomiting, and diarrhea are likely to occur after TSPP ingestion or inhalation. Exposure to this chemical can also produce skin eruption and visual inconveniences.

BOX 5.11 Telehandler mast bumps worker's head while he was moving drums, Idaho Clean-up Project, Idaho, United States; August 7, 2018 (DOE, 2020b)

CPP-1617 (the Radioactive Mixed Waste Staging Facility) was the scene of this event. A telehandler operative and a rigger/watcher were moving 200L drums from a pallet in preparation for radiological surveys (note a telehandler aka teleporter is a vehicle with an extendable boom, which can be coupled with manipulators or lifting devices such as pallet forks). While trying to grab a drum hanging by a sling from the telehandler, using a drum lifter, the telehandler operative mistakenly lowered the boom and the mast hit the rigger/watcher's head. The telehandler worker was qualified and experienced for this work. Upon notification of the event, the supervisor went to the affected worker's home and drove him to the site's medical department: Following medical evaluations on the day of the incident and the following day, no treatment or job restrictions were prescribed. Similar drum activities were temporarily discontinued.

After further investigation, corrective actions included the reviewing of operating procedures, performing expert surveys on fork lift activities, and holding training with other specialist employees. Following another fact-finding assessment on a similar incident involving the same telehandler operative, management concluded the drum-handling process was safe. Events that led to the August 7, 2018, incident were caused by unsafe decisions taken by the telehandler worker including (1) not waiting for or using the watcher properly, (2) using a piece of equipment different from that used previously, (3) rushing to finish their work and not taking time to observe procedures or admin prescriptions.

chemistry, biology, and material research. BNL is located in central Long Island, New York. It has over 3000 staff and hosts about as many visiting scientists and students at any given time.

A large geotechnical program was in progress to upgrade the sewage treatment plant and refurbish its sanitary sewer piping. The pipe project consisted of in situ relining of structurally robust pipes, replacement of weakened pipes, and minor repairs. At the time of the accident, grading was being performed along the sewer pipeline strip of land.

The main tasks accomplished initially were pipeline layout and clearing, including tree removal. The project was on schedule, so there were no undue time pressures.

On the day of the accident, two work items were planned:

- grading of some 100 m of the sewer line strip of land; and
- excavation for and placement of a precast concrete manhole structure at the boundary of the project.

FIGURE 5.15 Volvo front loader similar to the one involved in the accident. *Credit to Math Knight, Wiki Commons.*

Grading was required to limit the trench depth and to allow clearance under an electrical power line.

The construction worker (referred to as "the laborer") was fatally injured when he was run over by a front loader (referred to as "the wheel loader") being driven by "the operative" (Fig. 5.15).

Nobody else was near to the operative and the laborer.

The operative had just dumped the wheel loader's bucket of dirt in the spoil heap. Then, with the wheel loader in reverse gear and driving at less than 10 km/h, the operative dragged the bucket to level the rough surface. Around 16 m from the spoil heap, the operative turned the wheel loader slightly to the left, putting its left side aligned with the laborer. The evidence is not conclusive as to whether the laborer (1) slipped or tripped into the route of the wheel loader while stepping back out of the way and falling on his back or (2) was hit in the back as he was walking away and rotated to land on his back in the route of the wheel loader. The laborer was run over by at least one of the left wheels of the loader. Unaware of the situation, the operative drove further back nearly 25 m and then started moving forward on a new route to the right of the previous path. After moving forward for some 8 m, the operative saw the surveyor's rod on the ground, halted the wheel loader, opened the cab's door, and saw the laborer lying on the ground to the left of the wheel loader.

The first emergency responders arrived within 5 minutes of the operative's call. The first emergency medical technician assessing the medical condition of the laborer observed that he had major trauma to the head and upper body, was not breathing, and had no pulses: It was clear that he was beyond hope.

According to the laborer's general practitioner, the laborer had been diagnosed with adult attention deficit hyperactivity disorder (ADHD), a syndrome causing hyperactivity, short attention span, and easy loss of concentration. Studies suggest that ADHD may be a risk factor for accidents. However, no evidence indicated that ADHD had interfered with the labourer's past job performance. The Investigation Board concluded that ADHD may have been a factor in this accident, but it could not be determined to what extent, if any.

The Board concluded also that the operative did not have any medical or physical conditions that could have contributed to the accident.

The Board found no inadequacies in the overall management of work, safety protocols, hazard controls, or prework identification of tasks. In addition, the wheel loader was compliant with applicable safety standards; the reverse-signal alarm was OK at the time of the accident and was clearly audible during grading activities. The laser leveling instrument did not hinder the laborer's vision. The operative and the laborer were properly qualified and trained for the work.

An analysis was performed to identify the possible lack or malfunctioning of barriers planned for this operation. This analysis addressed admin systems, management systems, and physical systems in place to confine and prevent hazards (barriers). It was proven that most of these systems worked well. One barrier did not, however, the maintaining of a safe working distance between the laborer and the wheel loader. Successful performance of this barrier would have avoided the accident. This barrier may have failed for several reasons. For example, the operative might have placed too much reliance on the reverse-signal alarm to warn others of the route of the wheel loader. The laborer may have become complacent about the reverse-signal alarm and so he did not keep safe distance. The Board concluded that basically it was the lack of visual contact between the operative and the laborer that determined the accident.

5.32.2 Lessons learned

Vehicle traffic and use of lifting equipment in the movement of machinery and materials on a construction/demolition site may pose such interim hazards as physical contact, spills, dust, emissions, and noise. Heavy equipment drivers have limited vision close to their equipment and may not see people close to the vehicle. Articulated vehicles make up a significant crush hazard zone on the outer side of a turn while moving. Techniques for the prevention or mitigation of vehicle impacts include the following (the reader will note that some of these techniques had been adopted in the case of the BNL accident):

- Separating the location of vehicle traffic, machine activities, and walking areas and controlling vehicle traffic through the use of one-way traffic routes, strict speed limits, and trained flagmen wearing high-visibility clothing to avert traffic

- Enhancing the visibility of employees through their use of high visibility clothing when working in or passing through heavy equipment traffic areas, and training of workers to confirm visual contact with equipment operatives before approaching the equipment vehicle
- Ensuring moving equipment is equipped with distinguishable backup alarms
- Using well-maintained and tested lifting devices that are adequate to the load, such as cranes, and securing loads in place when raising them to higher positions.

5.33 Ignition incident, East Tennessee Technology Park, Oak Ridge, Tennessee, United States; February 18, 2003 (DOE, 2003a).

5.33.1 Decommissioning project; problem encountered; impacts

MACTEC is a subcontractor to Bechtel Jacobs Company LLC (BJC), the prime contractor for the site's Environmental Management Program, which addresses work at ETTP (formerly K-25). MACTEC had been subcontracted to decontaminate and demolish buildings at ETTP Project. MACTEC performed characterization, removed contaminated components, decontaminated materials, demolished structures to grade, and disposed of the resulting waste.

At the time of the incident, rubble heaps were being loaded on trucks for shipment to the Environmental Management Waste Management Facility (EMWMF, the Oak Ridge Reservation on-site disposal landfill). The waste transports were halted because the EMWMF was temporarily closed due to heavy rainfall. To keep the workforce busy during this interval, they were directed to remove protruding rebar from the K-1302 (the former Fluorine Storage Building) concrete pad. Since the work area was (wrongly, as it appeared later) considered to be a contamination area, the workers were required to wear anticontamination (anti-c) PPE.

On the day of the accident, a MATEC worker was using a cutoff machine with an abrasive wheel to cut rebar when the sparks ignited the anti-c coveralls and the plastic ankle-length boot on his right leg. The ignition was made evident by the smoldering PPE. The incipient fire was extinguished, and no injuries were reported.

5.33.2 Analysis; solutions

The Investigation Board identified several inadequacies in work planning and implementation. Distinct in categories, inadequacies include, but are not limited to those described under 5.33.3–5.33.6 (the author is responsible for the selection):

5.33.3 Analyze the hazards

- The hazards associated with use of the cutoff machine were not fully identified and analyzed.
- The hazard controls were not fitted to cutting rebar with the cutoff machine.
- The hazard analysis did not consider or raise doubts on the actual state of contamination of the concrete pad.
- The activity hazard analysis (AHA) was too generic to address specifically the cutoff machine operation.
- Tool selection was based on availability, not appropriateness.

5.33.4 Develop and implement controls

- The AHA identified the need for a hot work permit, but the requirement was not enforced.
- Although the site was posted as radiologically contaminated prior to 2001, successive surveys showed only findings below the detection level, which would support reconsideration of the anti-c PPE.
- Equipment maintenance and tests for the cutoff machine were not considered and carried out.
- The BJC management system to control the subcontractors failed in the following elements:
  - ø High turnover of personnel
  - ø No reaction or questioning attitude by the subcontractor
  - ø Acquiescence to a questionable answer rather than acting on an identified risk
  - ø Overruled the MACTEC AHA (a key requirements document) with the BJC hot work procedure (a lesser level document)
  - ø Disregard for the subcontract requirements for written direction from BJC about replacement of key MACTEC staff
- Worker did not have proper training on the cutoff machine and related hazards.
- The MACTEC temporary supervisor did not have adequate hazard and task-specific training to supervise employees.

5.33.5 Perform the work within controls

- The work executed with the cutoff machine did not adhere to the manufacturer's safety manual.
- Inadequate in-the-field controls were exerted on the cutoff machine operation and the rebar cutting job.

5.33.6 Feedback and improvement

- A lesson learned from a previous incident with the cutoff machine and spark damage to anti-c PPE was not circulated beyond the subcontractor environment.
- A 1997 Type A fatality report and the 1998 Defence Program's report *Personal Protective Equipment Use in DOE Programs* have indicated that PPE can per se induce considerable hazards and that overprotection should be minimized; however, these warnings went ignored.
- Continuity of experience and implementation of improvement programs were preempted by frequent turnover of MACTEC and BJC staff.

The following guiding principles should have been followed, but serious deviations occurred in this case:

Clear roles and responsibility for ensuring safety shall be defined and maintained at all levels within the organization in charge and its contractors.

- MACTEC yielded to BJC's determination and did not act on its authority under the AHA to move on with requiring a hot work permit. MACTEC was responsible for conducting its own hazard analysis and following its AHA. In this case, lack of specifics in the AHA and acceptation of BJC's procedural determination removed barriers that would have avoided the incident.

Staff shall have the experience, knowledge, and skills to discharge their responsibilities.

- BJC, MACTEC line managers, and the worker executing the work did not have enough experience with the cutoff machine to realize that the machine could be hazardous, especially when associated with combustible anti-c coveralls.

Sufficient resources shall be allocated to address safety, programmatic, and operational aspects. Protecting the public, the workers, and the environment shall remain a priority at all times and for all parties.

- The single DOE Facility person, who had been given additional commitments, was unable to provide adequate supervision of the plant's D&D activities. No backup DOE person was appointed. (Author's note: This point may reveal an upper management's underestimation of the risk of decommissioning activities.)

Admin and engineering controls to avoid and mitigate hazards shall be adapted to the specific work and related hazards.

- BJC was unable to determine the procedure accurately because the latest modifications of the applicable standards were not known.
- The MACTEC overarching AHA is too vast in scope. Task-specific details need to be included in the AHA; alternatively task-specific AHAs should be drafted.

Worker involvement

- The workers responsible for the cutting activities and the use of the cutoff machine were not involved in the discussion of the hazards.

5.33.7 Lessons learned

Significant changes had occurred in the project organizations for both BJC and MACTEC. Over the previous 11 months, there had been 25 people who had occupied 10 positions relevant to the plant D&D activities. Due to this frequent staff turnover in key critical positions over a short time span, lack of project and work experience/understanding contributed to the failures that caused the incident.

The Investigation Board determined that this accident was avoidable. The observed cause of this incident was the impingement of hot sparks on the leg of the worker's anti-c coveralls. MACTEC selected the cutoff machine based only on its current availability, not its merits. The hazards posed by that machine were not assessed. The worker did not have enough knowledge of the techniques and the hazards of the cutoff machine, and he took no measures to prevent the hot sparks from impacting his leg, which caused the ignition of the coveralls. Besides, the equipment was not properly maintained: The integrity and functions of equipment should be kept at the intent of the original design and as required. The focus should be on preventive maintenance with regular test and overhauls. This incident occurred because the hazard control process missed to assess the hazards posed by the cutoff machine. Although the Activity Hazard Analysis for General Construction Hazards indicates that a hot work permit is required for the job, MACTEC accepted the BJC's position that the BJC procedure for hot work did not require a hot work permit for the planned job. That decision allowed the worker to wear 100% cotton (i.e., combustible) anti-c coveralls. His body position was incorrect when using the cutoff machine.

In summary, the overall lesson learned from this event is that the responsible parties failed to implement adequate work controls. This includes administrative, planning, and technical aspects. This event has many similarities to the one described in Section 5.29.

5.34 Slips, trips, falls, USDOE complex

5.34.1 Bettis Atomic Power Laboratory, Pittsburgh, Pennsylvania; February 17, 2004 (DOE, 2004c)

5.34.1.1 Decommissioning process; problem encountered; impacts

Bettis Atomic Power Laboratory (hereinafter Bettis) is a science and engineering installation tasked with designing, developing, and maintaining nuclear propulsion plants for the US Navy. Bettis is owned by the DOE and operated under DOE contract by Bechtel Bettis, Inc. (BBI) as prime contractor. There are around 2000 contractor staff and 80 DOE personnel based at Bettis. In September 1977, a long-term D&D program was initiated by Bettis to decontaminate and reduce the number of radiologically controlled facilities and areas. As an example, during 2014, 2015, and 2016, the D&D program has accomplished the following:

- Shipped 2400+ t active waste mostly produced by the demolition works.
- Shipped almost 12 t mixed waste.
- Shipped 4600 + t of inactive waste.
- Recovered for reuse around 2400 m^2 of building surfaces, of which the majority was radiologically controlled before.

The incident described in this case study occurred outside (but adjacent to) the plant fence and was not directly linked to decommissioning activities: However, it is representative of a number of slips, trips, and falls that occur during D&D activities, especially when a large number of contractors move in and work onsite. In February 2004, a BBI staff member was going to work. The employee tripped on irregular tarmac in a pedestrian pathway of the staff parking lot (an area which is under full responsibility of the site organization), some 30 m from the main pedestrian entrance to the plant. The employee fell on hands and knees, suffering minor superficial injuries. Probably due to frost, the tarmac where the employee fell had lifted about 1.5 cm along a crack that crossed the whole width of the pathway. First aid assistance was promptly provided, and the employee went to work soon after the incident. It was later learned that the employee's minor knee injury had worsened preexisting medical conditions. These complications resulted in the employee's prolonged hospitalization, intensive treatment, and physical rehabilitation.

5.34.2 Analysis; solutions

The tarmac in the parking area contiguous to the main pedestrian entrance, including the pedestrian pathway, had last been maintained 8 years before the incident. The tarmac had since severely deteriorated in several spots including where the incident occurred.

Since this pathway also serves as the main access to the plant for individuals who park in areas reserved for the disabled and medical services, the pathway falls within the regulations of the Uniform Federal Accessibility Standards. These regulations specify the smoothness requirements for new construction or modifications to accessible routes.

In March 2004, a consulting company was subcontracted to assess the tarmac conditions near the main pedestrian walkway to the plant entrance and nearby parking spaces to determine the cause of the tarmac deterioration.

The company issued a report, which identified several other locations where the tarmac had undergone similar damage and the need for periodical maintenance.

The Accident Investigation Board ascertained that prime contractor managers in charge of the maintenance of the parking areas were not aware of the smoothness requirements for accessible pathways. The Board also ascertained that the prime contractor had not defined maintenance requirements for accessible pathways. The deteriorated walking surfaces had been tolerated based on the typical conditions of paved pedestrian, road, and parking lots in the surrounding population centers. DOE supervisory staff did not check the application of the required maintenance standards.

The Board recognized that the accessibility standards for disabled persons did not apply retroactively to existing parking areas, nor did the standards require immediate repairs of pedestrian pathways any time an irregularity appears on their surface. However, the Board considered that the prime contractor should have a reasonable maintenance program to keep at least the most used pathways in good status.

5.34.3 More cases from (DOE, 2020b): problems encountered; impacts

The following are typical slips, trips, and resulting falls arbitrarily chosen among uncountable cases.

Lawrence Livermore National Laboratory: On May 14, 2018, an electrician was bending a 2.5-cm conduit. The conduit was larger than what the electrician regularly used, and the bending of the conduit was being carried out in a hall on a vinyl floor. The first piece was bent resulting in an irregular ridge. Additional force was applied on the second piece to get a better shape. The electrician had placed one foot on the bender and the other on the floor. As the bender was being pulled toward the worker, the bender slipped out which made the worker lean forward and lose his footing. The foot placed on the floor twisted, and the worker felt acute pain. He had left their mobile phone in his work vehicle, and nobody was around to help. Later on, the plant coordinator found the injured worker sitting on the floor, helped him to a work vehicle, and drove him to the medical service. The medical test determined the worker's ankle was broken.

Hanford site, Tank Farm: On November 20, 2018, a worker slipped on a ramp leading to the rear entrance of a storage tent. Personnel working within the area called emergency and provided first aid until medical help arrived. Preliminary evaluation by the Hanford Fire Department (HFD) indicated the worker had a fractured ankle. The HFD transported him by ambulance to the regional hospital where a fracture to the distal fibula was identified.

The worker was moving materials from pallets outside the tent into the tent through the front entrance. As he had to make a phone call, he departed through the rear entrance of the tent. Upon stepping onto the ramp, the employee felt his left foot slip and turned his left ankle while losing his footing and falling.

The ramp to the front entrance of the tent was treated with ice melt as the tank farms had implemented abnormal operating "Response to Winter Hazards (Snow, Freezing Rain and Freezing Fog)", due to the weather conditions at the time of the event. However, the rear entrance ramp was not treated with ice melt as it was not expected to be walked.

5.34.4 Lessons learned

The site responsible organization needs to provide a safe work environment for the whole plant site, including parking areas and outdoor pedestrian pathways. Special attention must be paid to areas that must be walked by individuals with disabilities. Even minor road or walkway inadequacies can cause significant injury to staff and visitors.

When at work, the personnel should be trained to strictly follow procedures also in terms of correct access to equipment and tools, positioning, and movements. On the other hand, procedures should be tailored to individual operations: "Always" and "never" statements are especially important in this regard.

5.35 Dust events

5.35.1 Decommissioning process; problem encountered; impacts

Decommissioning activities may generate fugitive dust due to such activities as demolition of structures, floors, walls, rubble handling, on-site excavation and movement of soil, contact of machinery with bare soil, and exposure of bare soil and waste heaps to wind. Secondary emissions may include exhaust fumes from diesel engines of earth-moving equipment or from open burning of solid waste on-site.

Sections 5.35, 6.32, etc. are typical events exacerbated by high winds or other decommissioning-related events resulting in the generation of excessive amount of dust. These events may impact the workers nearby and transport radioactive or hazardous contamination off-site, especially for open-air demolition.

The air concentration of contaminants will depend on nature and amounts of the material, the size of generated particles, the meteorological conditions (e.g., the distribution of wind frequencies and intensities), and the duration of the works. Mathematical methods can be used to estimate dispersion and deposition of the contaminated particles.

5.35.2 Lessons learned

Traditional techniques to minimize the generation of dust and/or protect the workers include the following:

- Minimize dust from material generation sources, e.g., by using control equipment (water suppression, baghouse, or cyclone).
- Minimize dust from open area sources, including waste piles, by using controls such as placing enclosures and tarpaulins, increasing the moisture content, or applying water or nontoxic chemicals (e.g., fixatives) to minimize dust resuspension.
- Selectively remove potential air pollutants, such as radioactive substances or asbestos, from existing structures prior to demolition.
- Adopt protective techniques, such as PPE (dust masks).

5.36 Heavy load incidents

The following brief descriptions of incidents have been extracted from (NEA, 2018): Activities pertaining to fuel, refueling, or fuel handling equipment have been omitted, as well as operations that could not be likened to those taking place during decommissioning. This selection has included a measure of subjectivism. The descriptions in the following provide no clue to the plant or country affected.

No.	Event description, causes
1	The load attachment rope ruptured during on-site shipment of a steam generator's leg support. The leg support fell about 25 m onto an open container. The incident was caused by noncompliance of the flame cutting area during the fastening of the rope.
2	During testing of the lifting beam of the reactor pressure vessel (RPV), one of three load attachment ropes got broken. The lifting beam tipped to one side and banged onto two reactor cover slabs of the reactor shaft. The investigation showed that the fastening was not fit for the lifting beam.
3	During transfer with a portal crane of an open concrete cask containing a spent filter, the cask got displaced from its clamp and fell on two open concrete casks in the cask storeroom of the waste treatment building. The filter ended up on the storeroom floor. The two other casks also containing spent filters were damaged. The dose rate in contact of the filter on the storeroom floor was 180 mSv/h. Some weeks later, the filter was hooked up and placed in another cask. The damaged casks were encased and decontaminated after the filters had been removed. It was ascertained that three fingers of the clamp were not properly engaged on the cask. There was no automatic check that the clamp was adequately engaged. Even the visual monitoring, carried out by a camera mounted on the bridge crane, was insufficient. In previous events, the inadequate clamp engagement had been noticed but not properly corrected. Corrective actions were taken after all these incidents. The event was rated INES level 0.

4 For the shipment of bar-shaped contaminated components by crane, a shielding device is employed. The shield is suspended on two synchronized hoists (a master–slave configuration). The lifting of the unloaded shielding device was abruptly stopped, due to the blockage of the brake of the slave hoist. During the investigation, the master hoist slackened, and the shield struck the hatch opening. Consequently, the fastening of the shielding device, which is coupled to the master hoist, was curved. The incident was caused by a malfunctioning brake control of the slave hoist.

5 For the shipment of bar-shaped contaminated components by crane, a shielding device was employed. The component was pulled by a clamp into the shielding device. During this operation, a separation tube could not be entirely pulled into the device. The separation tube was then moved back to its initial point, but the clamp did not set the tube free. It was found that the drive chain between gear and reel had been displaced and the control cable of the clamp had been detached.

6 A container (about 2 t) was being lifted with a 5-t secondary-hoisting device in a shipment hall. The main facets of the incident include the following:

- The container suddenly began to slowly move downward.
- The movement could not be blocked.
- No serious damaged occurred.
- Due to the limited elevation reached above the floor, there were no safety impacts.
- It was disclosed that the spring deactivating the brake was broken.
- Fragments of the broken spring ended up between the anchor and the coil of the engine.
- Brake was blocked completely and could not work as foreseen.
- The issue was not reported to supplier earlier.
- After 40 years of regular operations, a malfunction was quite possible.
- Annual maintenance/inspection activities were not enough to prevent the incident.

7 The decommissioning of the plant was underway at the time of the incident. A rotating lid of a reactor tank was placed in a shipment container to be removed by crane. During the lowering of the load, the hoist failed and the load fell. The fall height was some 1.8 m. The incident was caused by a failure of a weld linking the cable drum and the hub. In turn, the failure of the weld was induced by material and weld defects.

8 Two test devices were employed for the ultrasonic examination of the interiors of RPV. During the test, the support frames of the testing devices were repositioned inside the RPV. The upper support frame was lifted and fixed on the predetermined position. Then the lower supporting frame was repositioned with a tool and fixed inside the RPV. During this process, the upper supporting frame was mistakenly slackened. The support frame fell down into the flooded reactor well and remained on the tool employed for the repositioning of the lower support frame.

9 As part of a routine outage activity, a shielding slab was lifted to access a well enclosure where a moderator system rupture disc was situated. Initially, the hoist engine was activated for the lift of the shielding slab. The load was being then lifted when it halted at 4.5 m height. While the operative was trying to continue the lift, the shielding slab started instead to go down and slowed by the brakes of the hoist engine, striking the wall and the floor until it landed on the moderator system piping: The incident occurred because the lift exceeded the rated capacity of the chain and lifting beam.

The workers responsible for this task were blamed for this occurrence: They assumed that the shielding slab was made of concrete with a weight of 1600 kg. It was later found the slab was made of steel with an actual weight of 5100 kg, which much exceeded the maximum load capacity of the hoist.

This incident had no direct impact on nuclear safety, the workers, or the public; however, this near-miss highlights the need for complete knowledge of a load before it is lifted.

10 During the moving of an empty HLW container from one room to another, the container lid fell to the floor when one of the slings slid off the pivot and the container inclined sharply to one side. The lid hit the toes of one of two workers who happened to be in the operations area of the bridge crane. These workers were not involved in the container handling and were not allowed to be there. After the event, the container handling activities were suspended. The injured worker was attended at the NPP's first aid station and later dispatched to the medical service.

Immediate causes of the event include unreliable coupling of sling to the container; failure to secure the container lid to the casing by the designated fastenings; human errors.

Root causes include shortcomings in the planning and management of HLW container handling activities; poor training of workers; container design flaws. This incident was classified at INES level 0.

11 A basket was used for the transfer of dismantled activated components. A hoist with a shielding device was employed to move this basket into a steel container. This fully automated process was monitored by camera. When the basket was lowered into the steel container, the clamp unexpectedly got detached from the basket. The basket fell into the container. The incident was caused by the wrong centring of the shipping device.

12 A container filled with control rod parts was being repositioned in the fuel pool. For the container's transfer, the refuelling platform was used. During the transfer, an operational error made the container strike the adjacent container's top. The container was lifted again. This led to its angling. It remained in a sloped position above three adjacent containers. Parts of the container's inventory fell between and into the adjacent containers. The event was caused by an erroneous judgment of the crane operator.

13 A container was transferred from the fuel pool to a drier tank in the conditioning area. The reactor building crane and a shielded cask were employed to this purpose. During the descent of the container into the drier tank, the workers did not see that the container was in a sloped position. Due to this position, one of the two shut-off valves of the shielded cask was blocked open and could not be closed after the unlocking of the container. The loading of the drier tank was suspended. The sloped position of the container was not observed because of the missed check of the height display.

14 This incident involved incorrect actions on the part of reactor staff and subcontractors while dismantling, using a lifting crane, a heavy suspension structure employed for hanging equipment in the reactor central hall. This event was caused by staff missing to comply with work safety rules, especially with the labor protection standards for load handling and lifting operations. Due to this inattention, the load straps of the bridge crane sheared off, and the metal suspension structure dropped into the transport passageway. Ensuring the safety of heavy load handling operations is critical when this work is carried out in the reactor areas. The observed cause of this event was the breaking of the portal crane load straps. Root causes included inadequacies in the planning and implementation of work involving the dismantling and transfer of heavy loads, especially the incorrect strapping of a suspension structure which was being dismantled; use of untested and uncertified load straps, poor worker training, insufficient prejob briefings; violation of work authorization process, execution of unplanned work, failure to ensure work safety. This event had no direct impact on plant safety and workers.

15 During an unloading operation of the reactor building crane four (out of 15) screws were torn. The screws were employed to attach the driving pulley's support to the bridge's support. The cause of the event could be attributed to a fatigue rupture of the screws.

16 During the unloading of a container with several drums of conditioned NPP MAW, a 400-L drum fell from the clamp of the crane. The drop (from a height of some 4 m) caused a 2 mm opening of the drum lid, along 40 cm of its border. A small quantity of concrete powder spread from the drum. A β-contamination of 30 Bq/cm_2 was found on the drum surface (near the lid opening).
Neither α-contamination on the drum nor airborne contamination in the room was measured. The clamp was checked, but no technical defect was noticed. The clamp was then refitted using a new 400-L drum as reference (matching the drums handled in the incident). It was discovered that during the event, the clamp was still fitted to an old drum design reference, which was 1 cm larger in diameter than the new design. The occurrence was then attributed to this anomaly. The event has been rated as INES level 0.

17 During the unloading of a MOSAIK cask, the cask's shock adsorbing upper part had to be repositioned. A special crane was employed. The cask's shock adsorbing top was being lifted when it fell some 5 m. Two guide rods of the cask and a 200-L drum, which was filled with radioactive waste, were damaged. The fall of the cask was due to the rupture of the lifting cable. The cable defects could be traced back to former shipping operations where the cable rubbed the driving pulley's edge.

18 During an unloading operation of the crane, a bolt of a holding brake's pad got fractured. This led to a partial failure of the crane's brake. Damage to other components of the brake could be observed. The incident could be due to fatigue fracture of the screws. The fatigue effect in turn was due to a change of the brake housing.

19 For survey of a filter vessel in the reactor water cleanup system, a mobile hoist was employed to lift and remove two concrete pads. After lifting the first pad into the upper shipping position, the hoist trolley slid out of the crane rail and fell some 5 m taking with it the attached concrete pad. The follow-on survey showed that during the fall of the load, components, holding devices, and support structures sustained damages. The integrity of the piping was not compromised. Minor surface damages occurred on the biological shield. Regardless of these damages, the incident did not cause any safety-related impacts. The licensee stated that the event cause was a combination of influencing factors and assembly defects.

20 The transfer of a shielded measurement chamber was underway. While being raised, the chamber tripped. A worker got slightly injured. The position of the attachment riggings was chosen without consideration of the center of gravity and especially the weight of the lead shielding. The event was attributed to an operational error.

21 While performing the transfer of the lower half reception tube from the pond maintenance room to the reception tube position in the pond, the following problem was identified. When using the north auxiliary hoist, which is coupled to the pond crane, the hoist brake was sliding each time a slow steady movement was sought.

5.37 Manual lifting, overexertion (DOE, 2016b)

Enough work to do; and strength enough to do the work.

Rudyard Kipling (1865–1936)

5.37.1 Decommissioning process

Incidents leading to musculoskeletal disorders (MSDs) and injuries are frequent in a number of industries: nuclear decommissioning and demolition are no exceptions. Acute or chronic MSDs affect the muscles, nerves, blood vessels, and sinews. Workers can be subject to risk factors, such as lifting, pushing and pulling heavy items, bending, stretching overhead, taking uncomfortable body positions, and continuously repeating the same movements.

The reader will note that the three case studies presented in the following all refer to typical activities in decommissioning processes.

5.37.2 Problem encountered 1; impacts

On December 23, 2015, a hazardous waste handler (WH) at Sandia National Laboratories (SNL), Albuquerque, New Mexico, United States, suffered a hernia while loading drums. First, the WH tried to position a 200-L drum of diesel fuel (weight almost 200 kg). As there was space for only one person in the storage area and the drums were pressed against each other, the WH had to squeeze in and reach to grasp the far edge of the drum. As he moved the drum a few centimeters to be able to position a wheeled platform under it, he felt a sharp pain twinge in his lower belly. He went on working. After the WH and a coworker slipped the drums onto the shipping vehicle, they went back to the storage place. There, the WH moved a 120-L drum containing lab trash (weight 15 kg) off a bulk container onto the ground. At this moment, the WH heard a light explosive sound in his belly and felt a swelling. He was taken to the site Medical Service and then to the hospital, and diagnosed with a hernia. In January, he had surgery to correct the condition.

5.37.3 Analysis; solutions

At SNL, communicating the expected strategy for positioning of drums had to be improved to obtain the optimal position and facilitate pickup. Corrective actions include seeking advice from an industrial ergonomics specialist to observe work conditions and train workers on handling and positioning drums.

5.37.4 Problem encountered 2; impacts

Production of enriched uranium at Oak Ridge, Tennessee, United States, ended in 1964, and gaseous diffusion was finally shut down in 1985. Demolition of

K-27, the last of the five gaseous diffusion facilities at Oak Ridge, began in February 2016 and was completed in late 2017.

On September 16, 2015, a K-27 worker helping to load 12-kg cases of water onto a truck felt a sharp sound in his upper arm. Since he had no pain, he went on working, but 15 min later, he felt pain and was escorted to the Health Department for a check. He was diagnosed with a strained bicep but went back to work the next day with limitations and had surgery to correct the injury.

5.37.5 Analysis; solutions

As K-27 work had been carried out as planned with no changes, the moving/lifting practices were assessed against the ergonomics handbook to determine any need for changes. It was determined to decide on a case-by-case basis if some of the handling could be performed with a forklift instead of manually. Reviewers determined that all coolers could be dealt with by moving the whole pallet from cold storage with a forklift, filling the coolers, and returning the rest of the pallet to cold storage. With this approach, some manual handling was still needed but much less than before.

5.37.6 Problem encountered 3; impacts

On January 8, 2015, a worker at the National Renewable Energy Laboratory (NREL), Colorado, United States, injured his shoulder as he pulled down on a crowbar while cleaning the inner components of a wind turbine transmission. This work had been performed many times and was described in a safe operating procedure. Because they found it difficult to get metal particles removed, they had to turn the shaft by placing a 1.8-m crowbar into one of the bolt holes at around head height and pull down. Finding a point of strong resistance, while the worker tried to pull down on the crowbar, he felt a sharp sound in his left shoulder. He managed to complete his shift, but while driving home he felt pain and reduced strength and pricking in his left arm. On his next work day, he felt increasing pain and reduced ability to lift and rotate his arm. He went to the NREL Occupational Health Clinic and was referred to an off-site orthopedic doctor who decided that shoulder surgery was required.

5.37.7 Analysis; solutions

It is essential to know when to stop and reassess work when facing unexpected conditions. For instance, should material handling work activity include a choice between damaging expensive equipment or potential injury, the decision requires further review by higher-ups; in this case, workers should stop work. A certified ergonomist assessed material handling operations concerning the transmission assembly, disassembly, and cleaning for any alternative strategies. That assessment indicated that the higher one's hands are when applying strength, the greater the loss in strength, to the extent that when pushing or pulling at shoulder height, one's strength is only 40% of that actually applied at waist height. The project also upgraded procedures so as to incorporate ergonomic ameliorations.

5.37.8 Lessons learned

Physiological and psychological conditions associated with decommissioning activities produce ergonomic hazards. Discomfort and fatigue are two signs of ergonomic stress resulting in reduced productivity, unsafe operations, and heightened likelihood of injury. Typical sources of ergonomic stress during D&D include vibrations, manual lifting of heavy loads, and static work.

Overexertion, and ergonomic injuries and illnesses in decommissioning projects can be prevented and controlled by any or all of the following means:

- Training of workers in lifting and materials handling methods, including the specifications of weight limits above which mechanical support or two-person lifts are mandatory.
- Planning workplace layout to minimize the need for manual transfer of heavy loads.
- Selecting tools and designing work stations that reduce strength requirements and work times, which encourage better body positions, including custom-adjustable work stations as needed.
- Enforcing administrative controls such as job rotations, shifts, and breaks.

5.38 Beryllium-related events

5.38.1 Decommissioning process; problem encountered; impacts

Beryllium is one of several toxic chemicals to be found in nuclear facilities or used during operations or decommissioning. The list also includes, among others: sodium (Sections 5.21 and 5.44), mercury (Section 7.28), lead and arsenic (Box 5.6), decontamination chemicals (Sections 5.31, 5.44, 6.19, etc.), asbestos (Section 5.23), and polychlorinated biphenyls (PCBs, Section 7.12).

In addition to being used in several nonnuclear sectors, beryllium metal and beryllium oxide are used as moderators and as reflectors to augment the thermal neutron fluxes in the cores of many research reactors. Beryllium has been also used as an important component in the production of nuclear weapons.

Soluble beryllium is toxic if absorbed through the skin. Beryllium remains permanently in tissues as no self-elimination mechanism is known. The main concern is the adverse health effects following the prolonged inhalation of airborne beryllium particulates (scarring of lung tissue, beryllium sensitization, and a condition known as chronic beryllium disease). There is also evidence that inhaled beryllium is carcinogenic.

Beryllium is normally not radioactive. However, after exposure in a reactor core, it may contain considerable amounts of tritium. Activation of beryllium impurities can also produce other radionuclides, especially ^{60}Co. This is a serious concern when dismantling and handling beryllium components during decommissioning (International Atomic Energy Agency, 2006c).

In Canada and other countries, an occupational exposure limit (OEL) of $0.05 \mu g/m^3$ has been adopted for airborne beryllium particulates for the average exposure of 8 h, above which health impacts may occur.

For nonroutine work such as cleaning, maintenance, and leakage handling (activities that can be assimilated to decommissioning for the purposes of this book), the approach to safe work includes identification of hazards present, adequate training to beryllium operations, and the use of PPE like powered air-purifying respirators (PAPRs) with beryllium-specific filters.

On August 22, 2017, BWXT (a company manufacturing CANDU fuel using beryllium as a process component) found out that improper respirator filters were being used with PAPRs while conducting nonroutine work. BWXT determined that incorrect filters were procured in 2013. Findings based on general air sampling from the affected areas indicated the following average and maximum air concentrations: $0.12 \mu g/m^3$ and $0.29 \mu g/m^3$, respectively. Later personal air samples from the affected areas indicated average concentration $0.39 \mu g/m^3$ and maximum concentration $4.63 \mu g/m^3$. A total of 15 instances emerged where two workers had been exposed to airborne beryllium well above the OEL. Both workers were addressed to the plant occupational health service for follow-up monitoring including tests for beryllium sensitivity. No health effects have appeared since the incident came to light.

The licensee's investigation identified several causal factors that led to the use of inadequate filters, e.g., insufficient verification upon delivery of the filters and purchase orders missing a complete description of the item. Finally, BWXT did not realize that improper filters were taken from the plant store to be used for PAPRs.

BWXT launched several corrective actions to prevent recurrence of a similar event (NEA, 2019).

The following events involve legacy beryllium contamination at Lawrence Livermore National Laboratory (LLNL), Livermore, California, United States. A 2003 case study is given in Box 5.12. Two more occurrences at LLNL are given in the following.

In November 2007, at LLNL, workers removed ventilation ducting to nonberyllium work areas without checking the risk for transfer of beryllium contamination. Contamination values above legal levels were discovered on the external surfaces of nearby ducting that were formerly linked with the removed ducting in one piece, as well as in areas of the same room. As beryllium contamination was not checked prior to the ducting removal, the workers were exposed to beryllium.

BOX 5.12 Significant airborne beryllium concentrations found during decontamination, LLNL, California, United States; May 2003 (DOE, 2003b)

Problem encountered; impacts

Workers wearing respiratory protection were decontaminating the firing chamber of the LLNL Contained Firing Facility (CFF) after a detonation experiment containing beryllium. The Hazards Assessment and Control (HAC) procedure required the floor to be wetted to minimize dust production, followed by wet sweeping to remove particulate contamination. While surveying the chamber prior to starting work, the workers noticed the chamber had puddles of water with few dry areas. Assuming the chamber to be wet enough for decontamination, the workers started work.

When the exposure monitoring results were reviewed, it was discovered that three of the four workers had largely exceeded the Chronic Beryllium Disease Prevention Program permissible exposure limit (PEL) of 2 μg Be/m^3 averaged over 8 h, without considering their respiratory protection factors. Shortly after, the chamber was meticulously wetted before the decontamination was resumed. Subsequent monitoring results indicated a considerable reduction in the airborne concentrations, although still exceeding the action level. However, because all the workers were wearing PPE, including respirators, none of them exceeded the PEL, and only one reached the action level. The action level for beryllium is 1/10 of the PEL or 0.2 μg/m^3. If the concentration is at or above the action level, worker protection provisions are legally required.

Following further wetting of the chamber prior to more decontamination work, the exposure monitoring results went down below the action level.

Analysis; solutions: When employing any beryllium decontamination technique (e.g., sweeping) with a potential for even light abrasion and generation of airborne particulates, water must be applied in advance to avoid elevated airborne concentrations. It is not enough to count on the spread of water from adjacent puddles. Fine particles of beryllium tend to remain airborne for a significant time regardless of water being applied to surfaces or even misting water in air. When conducting area beryllium decontamination, the following actions are recommended:

1. Wet all surfaces thoroughly, leaving no dry spots, prior to using any (even slightly) abrasive decontamination process.
2. Wet surfaces even beyond the affected area to avoid the spread of Be to adjacent areas.
3. Be wary of changing conditions such as accelerated drying due by ventilation and exposure of formerly inaccessible surfaces during cleanup. Reapply water abundantly if dry surfaces appear.
4. Utilize cleaning techniques that minimize abrasion.
5. Spot cracks, crevices, and collection points associated with moving areas (such as door seals) that can generate airborne particulates and specify procedural controls.

Box 5.12 Significant airborne beryllium concentrations found during decontamination, LLNL, California, United States; May 2003 (DOE, 2003b)—cont'd

6. Evacuate the area if airborne contamination is believed possible. Time, ventilation, and wet surfaces are all necessary conditions to minimize airborne beryllium concentrations.
7. Use wetting processes during the removal of dust and debris containing other hazardous materials (e.g., asbestos and lead). Dry cleanup processes (e.g., dry sweeping and shoveling) may generate the airborne release of particulates.

Lessons learned: Prior to using any beryllium decontamination technique (such as sweeping) that has even minor abrasive action, accurately and abundantly wet all surfaces and leave no dry spots to avoid hazardous airborne concentrations.

In April 2008, LLNL worker's monitoring found he had been exposed to beryllium levels above legal limits while conducting work in a beryllium-contaminated installation. The worker did not wear respiratory protection while handling cables in a normally inaccessible below grade area. The hazard evaluation prescribed respiratory protection for all dust-producing operations and when intruding the inside of components.

On October 3, 2008, a subcontractor removed a beryllium-contaminated ventilation grating and installed in its place a prefabricated vent blocking panel with no authorization. Although the worker wore protective gloves to prevent exposure from the beryllium-contaminated component, he was not trained for this specific operation. The prescribed prejob personal and air monitoring were not carried out. See more LLNL cases of beryllium contamination in East Bay Times (2016).

The aforementioned cases disclosed inadequacies in LLNL Chronic Beryllium Disease Prevention Program concerning identification, evaluation, and control of legacy beryllium hazard (DOE, 2010a).

5.38.2 Lessons learned

Typical deficiencies in dealing with beryllium contamination relate to the following:

- Prejob characterization and identification of beryllium-contaminated parts
- Specific training of workers
- In-the-field air monitoring
- Regular medical tests of affected workers
- Disposal of beryllium-contaminated items

5.39 Recurrent incidents in forklift, telehandler and heavy equipment use in material handling. Idaho Cleanup Project, Idaho, United States; 2018 (DOE, 2020b)

All experience is an arch, to build upon.

Henry Brooks Adams (1838–1918)

5.39.1 Decommissioning process; problem encountered; impacts

In 2018, a number of forklift and heavy equipment-related events, which have created a negative performance trend in material handling and the operation of heavy equipment, prompted Fluor Idaho LLC, the managing contractor for the Idaho Cleanup Project to appoint a working party of experienced personnel to review all applicable events and to identify causes.

More events occurred after the working party had been appointed and were subsequently reviewed. Although no serious injuries or SSC damage occurred, there was a potential for more serious impacts. Fluor Idaho LLC recognized weaknesses existed in material handling and in inadequate communication between spotter and equipment operative. A selection of events is briefly described in the following.

Case 1, September 03, 2018. An operative maneuvering a forklift inadvertently impacted a worker who was shifting a rope barrier for the forklift. Upon being impacted by the forklift, the worker alerted the forklift operative (FO) who immediately stopped the maneuver. The forklift was transferring a waste box to a flat truck when the forklift's left rear wheel touched the worker's foot. The worker was soon transported to the Site Medical Department facility and was examined with no injuries found. Then the worker went back to unrestricted work.

The fact-finding review of the incident showed the worker was replacing a rope barrier and was moving away from the forklift holding the rope. He erroneously assumed he was out of the swing radius of the forklift. The FO had a look at the worker walking away and also assumed he was safely out of the forklift path.

Case 2, April 11, 2018. At the Sludge Repackaging Project (SRP), workers were offloading 40-L overpacks from a flat truck. As the FO grappled an overpack and began retreating from the truck with a drum grabbed to the drum handler, the hooked drum contacted and stuck to a near drum, and was then pulled away from the truck's bed and fell to the ground.

Case 3, August 7, 2018. A worker handling a drum lifter with a forklift was hit in the head with the forklift. Inside the Radioactive Mixed Waste Storage Facility, a telehandler operative (TO) and a rigger-spotter (RS) were moving 200-L drums from a pallet to allow radiological surveys be carried out. While trying to pick a drum with a lifter hung by a telehandler sling, the RS was holding the lifter on the drum to be lifted; at the same time, the TO lowered the boom mistakenly and the mast hit the head of the RS. It was reported this event occurred while the third drum of the day was being moved. TO asserted that this was his first time using this telehandler with the "jib" configuration for moving drums with the specific type of drum lifter. TO was qualified to operate the equipment and had used it previously. RS used both hands to hold the drum lifter in place, so he could not use hand signals to direct the TO. The supervisor was notified of the injury first. After contacting upper management and the project health and safety supervisor, the supervisor was requested to go to the home of the RS, where he picked up the RS and drove him to the site medical service. Evaluation resulted in no treatment or work restrictions.

Case 4, August 13, 2018. During forklift operation at the Radioactive Mixed Waste Storage Facility, a waste box that a forklift operator was attempting to move slid forward as the forklift prongs were thrusted into their pockets. Since other workers were in the proximity to where the waste box slid, a step back was taken, and a team meeting was called. A little later, it was found out one of the workers was slightly hit by the waste box during the incident. Heavy equipment operations were temporarily suspended. A fact-finding meeting highlighted the potential chance of an injury, e.g., a twisted or broken foot, anklebone, or knee.

Case 5, September 20, 2018. Two construction workers were inside a perimeter fence carrying out cleanup from previous tar paving activities: Their work consisted of loading spades and other materials into the loader bucket of an excavator in view of later removal.

Once the items were placed in the loader bucket, the driver of the excavator started slowly to withdraw the equipment from the area. While moving back, the excavator contacted a light pole, causing the light pole to collapse to the ground.

Nobody was injured during this incident. Moreover, nobody was near the fallen light pole. The only individuals in the work area were the two construction workers, one driving the excavator and one around 3 m away in front of the excavator.

The worker driving the excavator, while in the cab and before moving the vehicle, ensured nobody was in the proximity by turning and looking back and using the vehicle mirrors. Besides, although not an appointed spotter and standing in front of the excavator, the second worker looked to ensure nobody was nearby.

Following this event, all construction operations were suspended pending evaluation of the facts. The excavator remained undamaged.

Case 6, November 23, 2016. SRP drums were being offloaded from an open body trailer when a lid retention bolt from an adjacent drum hooked the lid ring of the drum being lifted. As the FO lowered the drum in the drum handler to the ground, the second drum, which was mistakenly attached to the first, came free and fell around 30 cm to the ground. Personnel left the area.

Radiation protection staff wearing PPE surveyed the drum and checked its integrity. No contamination was found either on workers or the drum.

Case 7, December 18, 2017. Workers were moving drums onto pallets in preparation for transportation. This activity required workers to single out selected drums from cargo containers and group them onto shipping pallets. The pallets would then be loaded onto the truck for final arrangement. During the process, a drum was removed from an oil containment pallet and positioned onto a receiving shipping pallet. The FO then tried to put this pallet with the remaining drum back into storage. When the oil containment pallet was raised from the ground with the fork lift, the entire pallet and drum tilted and the drum came to lie sideways. Removal of the "shipment drum" caused the oil containment pallet with the remaining drum to get unbalanced. When the pallet was raised, the unbalanced load tilted making the drum fall sideways. Work was halted. Workers left the area. Higher-ups were notified. RCT carried out a radiological survey of drum and checked its integrity. No contamination was detected, and the drum was not damaged. The tilted drum was then safely put back into the cargo container.

Case 8, February 15, 2018. While lowering a port cover into a storage area, a crane operative used the wrong control to execute a downward motion, which made the overhead lifting equipment contact the roof of the contamination tent. The crane operative was following the applicable procedure to move the port cover off the port hole and put it in the storage area. When lowering the port cover, the crane operative used the "crane down" control in lieu of "hoist down" control. This lowered the whole equipment onto the contamination tent.

Case 9, February 20, 2018. SRP drums were being offloaded from a truck. As the FO moved the load from a graveled area to a pavement area, the forklift crossed a rough area at the dirt-to-pavement connection. The asperity made the forklift rebound and the drum to detach from its handler, fall, and topple.

Case 10, June 15, 2018. During a waste handling operation, a 200-L drum was being picked by a forklift with an attachment from the top level of a three-layer stack of drums. Both the spotter and the FO noticed the container was slanted in the attachment, as the forklift began to raise the drum from the stack. The FO at once stopped the lifting and started to take the drum back on the stack. As the FO lowered the drum to the stack, the container angled forward and slid out of the attachment, and fell 2−3 m to the floor. Damage to the drum was limited to only a small notch on the drum bottom. Another small notch was seen on the drum lid ring (which may have been there from an existing defect). The waste handling operations were paused, and top managers were notified.

Case 11, September 13, 2018. Workers were repositioning a surface cleaner (around 450 kg) when the cleaner toppled from the forklift prongs and fell to the tarmac. The cleaner toppled soon after the initial lift as the FO turned toward the building. The forklift prongs were some 15 cm off the ground, and nobody was close to the toppled cleaner. Around 2 L of diesel fuel leaked from the cleaner onto the tarmac after the fall. The management directed the workers to straighten up the cleaner and clean up the leakage. Forklift use at the plant was restricted pending assessment of the case.

Case 12 and 13, February 3, 2019. An external support operative (ESO) escorted a fuel truck to refill fuel for a ground heater. The fuel truck driver, a D&D skilled tradesman successfully maneuvered the fuel truck between a light pole and a fire hydrant and filled the ground heater tank. In late evening, the driver was to exit this place and return to the vehicle parking in accordance with ESO's instructions. A large snow pile placed in front of the fuel truck forced the driver to move the truck forward and in reverse two or three times to do the turn. Although the ESO escort was helping the driver as a spotter, the back bumper hooked the light pole and was torn off the fuel truck as the driver tried to go away.

Hearing a sharp sound, the driver halted, left the vehicle to survey any damage, notified the shift supervisor, and secured the area.

Case 13, November 30, 2018. An event occurred when a mechanical sweeper used by the snow removal team contacted a barrier pole. There were no injuries. A minor quantity of hydraulic fluid leaked from a broken hydraulic line.

Case 14, January 17, 2019. An equipment operative (EO) hit a fence with an excavator's bucket. The EO was clearing snow when the backhoe's bucket caught the ground and the backhoe slid to the side. The EO lifted the bucket off the ground and contacted a security fence. No security violation occurred.

Case 15, January 17, 2019. A maintenance worker using a skid loader with a blade attachment was removing snow from behind a building, when he hit the building with the blade; he was temporarily distracted by two individuals walking nearby. He notified his supervisor, and follow-up notifications were made to higher-ups later. The supervisor escorted the maintenance worker to the Medical Department for mandatory drug screening.

Case 16, October 02, 2019. While driving a rented bus, an operative struck a waste skip, due to an ongoing snowstorm. The operative watched another truck coming down the road. The operative then moved to allow more room for the semitruck. The bus contacted the waste skip: The side of the bus was scratched, and a window was broken. The incident was due to the bus driver assuming he had enough room around the skip. There were no injuries.

Case 17, February 11, 2019. An operative was moving a semitruck backward into a door airlock. To enter the airlock, the truck must move at an angle. The truck's driver had performed this maneuver many times.

While trying to take the right angle, the truck ran over an orange traffic drum, which made a fiberglass piece be torn off the truck's fender. As the driver was watching his spotters, he did not see the orange drum. Abundant snow was falling, and visibility was poor. The drum was partly covered with snow. No injuries occurred.

Case 18, February 13, 2019. A worker was removing snow using a snow sweeper. The snow sweeper inadvertently impacted the snow boundary rope. As the snow rope got stuck to the sweeper's arm, it pulled the building electrical conduit the rope was attached to, the conduit in turn was torn off the building, the electrical panel was damaged, and the building power was disconnected.

The sweeper operative did not notice the rope due to abundant snow. The rope barrier was slack and hidden in the snow. As the sweeper's arm contacted the rope, it caught on the rope. The other end of the rope was tied to an electrical conduit. The operative did not realize the sweeper got stuck to the rope. When the rope became tightly stretched, the conduit was pulled off the electrical box, which sparked when it was torn off the building. Electricians cordoned off the area at once and placed an operations lock to the building panel. Nobody was injured.

Case 19, February 13, 2019. A worker was moving drums between areas. To this end, he was using a propane forklift. While moving in reverse through the door between the two areas, the forklift mast hit the door. The worker halted the process and placed drum and forklift in a safe mode.

The worker and spotter had moved three drums previously between those two areas using an electric forklift. Another area was using the electric forklift for their job. As the worker still needed a forklift to move drums, a propane forklift was employed. However, the mast on the electric forklift is shorter than on propane forklift. The worker did not realize the mast would not pass through the door. While moving backward, the forklift mast bumped into the top of the door. Nobody was hurt.

5.39.2 Lessons learned

The main common factors of these incidents were as follows: (1) worker experience was not equal to their self-confidence and (2) complacency caused inadequate situational awareness, when the workers failed to identify and recognize the meaning of new conditions. In detail, these factors include the following:

- Inadequate proficiency about specific equipment configuration and/or tasks. Moreover, the lack of proficiency in operating rarely used equipment was exacerbated by the presence of snow and ice.
- Failure to maintain good situational awareness under unexpected conditions was apparent in many of these cases.
- Prejob and daily communications occasionally lack rigor. Individual tasks (spotter, operative, etc.) are often discussed only in general terms. Often, these discussions do not address unexpected conditions/hazards.
- Although not contributing causes to any aforementioned event, workers and supervisors voiced frustration regarding (1) consistency in holding

individuals accountable for inadequate performance and (2) lack of clear communications about accountability-related actions taken in consequence of wheeled vehicle events.

Fluor Idaho LLC took the initiative to improve forklift operative and spotter proficiency by

(1) reducing the identification of "qualified" forklift operatives only to those who do carry out this work and
(2) reducing the identification of "qualified" spotters only to those who do carry out this work. By reducing the number of qualified individuals, it is expected that proficiency will increase due to the following:
  (1) "Seat time" on forklift equipment will increase
  (2) Spotters will acquire more in-field time and familiarity with equipment operatives

5.40 Laser incidents

5.40.1 Decommissioning process; problem encountered; impacts

There is a lot of experience in nonnuclear laser cutting applications: On the other hand, nuclear applications of this technique, while being quoted in the technical literature as generally successful, are not considered to have reached a fully mature status. Therefore, most experience on laser hazards—including incidents—is drawn from nonnuclear applications.

In laser cutting, a laser beam is delivered by mirrors and a fiber to the cutting point, and focused and irradiated by a lens on the worksheet. The worksheet surface is rapidly heated to a temperature far exceeding ignition. A molten puddle is formed. Injection of the assist gas (oxygen) into the molten puddle forms a hole. As the laser beam is moved, the kinetic energy of the injected gas and the oxidation reaction with oxygen forcibly remove the material successively in the depth direction and form a continuous kerf. The assist gas generates the heat of oxidation and removes the molten metal from the kerf. The safety of laser operatives is provided by shields made up of windows absorbing laser wavelengths, safeguarding devices to ensuring safe distance, and goggles to protect the eyes.

Based on experience, the most common laser hazards are the following:

- Damage to the eye is common in laser cutting; depending on the specific type of laser, different parts of the eye can be damaged.
- Hazards of toxic gases: This depends on the gases released by the material being cut.
- Fire and dust explosion hazards are related to combustible materials (either the material being cut or combustible materials in the immediate vicinity).

- Dust from the laser cutting of titanium, aluminum, zinc, and their alloys triggers thermite reaction when mixed with dust of iron oxides or copper.
- Hazards of machine's moving parts to human body can be prevented by enclosures, fences, and other safeguarding devices (AMADA undated).

A brief selection of laser incidents is given as follows (OSHA, 2020):

Problem encountered 1: Employee's arm is fractured by laser table, August 15, 2018.

Employee was working alone inside of metal fabrication facility, cutting parts with a laser cutting table.

He began cleaning metal dust from the lower section between two material tables.

As he stuck his arm between the two tables, the machine cycled, moving the two tables and catching his arm.

The electronic barrier system around the loading table had been bypassed, letting the machine cycle with the employee inside the banned area. He suffered a fractured arm and was transported by ambulance to hospital.

Problem encountered 2: Employee dies in laser cutter accident, June 7, 2010.

Employee was using an automated laser cutter within the production area. The front guard was raised, and the employee was making adjustments. He became caught in a pinch point when the structure holding the laser cutter moved to the right front area near the metal structure of the machine. Employee's head got taken in the small space between the metal structures, some 8 cm wide. A coworker heard a noise and found the employee fatally injured.

Problem encountered 3: Employee's arm is amputated by laser cutter, June 1, 2006.

Employee was removing a metal piece from a laser cutting table. He reached over the table as the cutter ran its cycle. The laser cutting head amputated the employee's arm.

Problem encountered 4: Employee's eye injured when exposed to laser beam, March 14, 2003.

An employee was working with an open beam laser. He wore laser eyewear during beam alignment, but then he took off the eyewear. While taking a meter measurement for output energy, his eye caught a flash from a hidden optic in the beam's path. Exposure exceeded current limits for lasers, damaging the macula in his retina, causing incurable retinal damage and eyesight loss.

Problem encountered 5: Employee's legs crushed in laser cutting pit, January 7, 2003.

A machine operative was attempting to reattach a laser-cutting head to its connection point on the laser cutting machine.

The machine had a table on which a sheet metal was laid and moved along a vertical axis.

A pit under the table collected the cut out pieces falling from the sheet. When the laser head was moving over a bumpy point of the sheet metal, it detached from the connection. Just before the incident, the laser head detached four times at the same point on the sheet while the sheet was being cut. Each time, the operative moved the connecting point to far end and reattached the head from the side of the machine. As the problem continued, he climbed into the pit and reattached the head so the machine could skip the difficult spot. After reattaching the head the fourth time, he told the next shift operative to restart cutting. Being unaware of the operative being in the pit, the new operative pressed the "start" button, which made the table of the machine move toward the coworker in the pit.

The operative's lower legs were entrapped between the frame of the table and a pit pipe and were fractured.

Problem encountered 6: Two employees suffer eye injuries from indirect laser light, March 18, 1996.

An assistant project scientist was aligning a laser system for an experiment. As he was viewing the laser focus through a chamber window, he was exposed to indirect laser light. During the short period of alignment, he had removed his protective eyewear. He sustained serious injury to both eyes with a partial vision loss.

5.40.2 Lessons learned

The following lists typical causes of laser incidents (Imperial undated; Industrial Electronics undated):

1. Misaligned optics, upwardly directed beams, beams crossing gangways, and beams at eye level
2. Fatigue, leading to carelessness or ill-considered shortcuts; rough, boisterous play
3. Available eye protection not worn or wrong/damaged eyewear worn
4. Complacency
5. Device malfunction
6. Operative unfamiliar or insufficiently trained with specific laser process or model
7. Improper postservice resetting of device
8. Failure to follow procedures due to haste, poor coordination of responsibilities and communications, and other factors
9. Manufacturer-installed safety features (protective enclosures, interlocks, beam stop, etc.) removed or ignored

5.41 Dislodging of concrete block causes serious injury, K-31 Building, East Tennessee Technology Park, Tennessee, United States; December 17, 2003 (DOE, 2004d)

The chapter of knowledge is a very short, but the chapter of accidents is a very long one.

Lord Chesterfield (1694–1773)

5.41.1 Decommissioning process; problem encountered; impacts

In August 1997, British Nuclear Fuels Limited (BNFL) was awarded by Oak Ridge Operations Office (DOE ORO) a contract for the ETTP Three-Building (K-29, K-31, and K-33) D&D and Recycle Project. More details of this project are given elsewhere in this book (Section 6.35). In mid-December 2003, work was underway to demolish the K-31 Control Room.

A BNFL Pipefitter (PF) was injured while trying to remove a concrete block from within a wide-flange, steel column (a structural beam having am H-shaped cross section; has wider flanges than an I-beam). On November 27, 2003, demolition of the K-31 Control Room had been authorized. The workers were using a power shovel with a standard bucket attachment to take down the block walls. The excavator worker removed the block flush with the wide-flange column; however, the excavator could not be used to remove the rest of the blocks within the wide-flange columns due to accessibility problems. Several columns were left with blocks remaining in the webs. Hand removal of blocks was not addressed in the Enhanced Work Planning (EWP) nor was it debated during safety meetings. The team took own decision on how to best carry out the work. A team member used a scissor lift near one column to reach the top of the block stack and manually removed the blocks using manual tools. A scissor lift could not be used near some columns due to the big heaps of debris. The PF used a 2-kg manual sledgehammer to break the cement and a pry bar to loosen and remove the block at chest height. As the block was removed, the remaining upper blocks slid down onto the lower stack of blocks. The PF reiterated this approach on the lowest block in the upper stack until all blocks in the top stack were removed. The lower stack of blocks was then removed by prying out a block at a time. The injured PF had carried out hand removal of concrete blocks for some $2^1/_2$ days before the accident (Fig. 5.16).

During this work, a loosened 22-kg top block fell about 3.25 m hitting the PF on his left foot. This was the observed cause of the accident. To gain better visibility, the PF had his cutting shield in the up position because he did not need it for that work. The PF did not report to the BNFL Clinic and remained at work for several hours although he was visibly limping. The PF demonstrated for the Building Manager the approach used for removing the blocks. During the demonstration, a couple of the remaining blocks fell out of the wide-flange column. After this demonstration, the Building Manager did not

FIGURE 5.16 Oregon army national guardsmen use pry bars to lift a section of cement off a simulated victim during training exercise. *Photo by Spc. Matthew Burnett, Wiki Commons.*

realize the event was a near miss and did not issue a stop work on block removal using hand tools. After leaving work, the Pipefitter drove home; then as the conditions of his foot worsened, his spouse transported him to the Oak Ridge Medical Centre. The PF was hospitalized for 4 days during which period he underwent surgery. The surgery involved insertion of pins in the fourth and fifth toes and lancing of the bottom of the foot to relieve pressure. Complications arose over the end-of-year holidays, and the PF had to go back to the hospital on January 7, 2004. Due to advancing necrosis, he had his fifth toe amputated. Besides, full-thickness skin from the abdomen was removed and grafted to the top of the left foot due to hematoma and necrosis. The PF remained in the hospital for 8 days and returned to restricted work on February 3, 2004.

Based on the height of the fall and the weight of the block, the impact would release about 6650 m-kg of kinetic energy. The work boot worn by the PF was rated to offer a protection of about 1460 m-kg of kinetic energy. Neither this work boot nor the use of a metatarsal guard would withstand the energy produced by the falling block.

5.41.2 Analysis; solutions

The Investigation Board highlighted one root cause and several contributing causes.

The root cause was that BNFL failed to establish a valid work planning process. This root cause can be articulated in the following aspects:

- Work plans for K-33 were considered to be transferable to K-31 without prior walk-downs or specific assessments.
- Management and work team failed to stop work when conditions were different from EWP (and they had not been trained to the stop work process).
- EWP was unclear and open to different approaches that could be used to conduct work.
- Hazards were poorly addressed.
- Integrated safety management core functions were not implemented.

Twelve contributing causes were identified by the Investigation Board:

1. Unauthorized and inappropriate work approaches were used to remove blocks from wide-flange columns.
2. The EWP and Field Change Notice (FCN) planning was inadequate to provide the job steps necessary to safely carry out Control Room demolition.
3. Implementation of the EWP was inadequate.
4. Incorporation of lessons learned was inadequate.
5. Identification and assessment of the hazards related to the manual removal of concrete block from the column were poorly conducted.
6. Supervision of the task was inadequate.
7. Working conditions did not comply with the EWP-specified hazard controls or were insufficient during task execution.
8. Communications were insufficient to ensure safe conduct of work and obtain timely medical assistance for the injured PF.
9. Training of managers and workers engaged in work planning and execution was insufficient to ensure safe conduct of work and proper response to the accident.
10. BNFL staff failed to follow internal policies, procedures, and EWP requirements.
11. BNFL responded inadequately to the accident.
12. DOE ORO's response was also inadequate.

5.41.3 Lessons learned

The Investigation Board determined that the work control planning was insufficient to ensure worker safety. In addition to the poor work control process, BNFL did not address revisions and FCNs to ensure that work steps and control of hazards were applied to the new work scope. Experience shows that insufficient management of enterprises has caused accidents. A system should be in place to ensure that changes do not augment risk or make up additional hazards. A change may include modification in equipment, tools, software, staff, administrative provisions, or hierarchical adjustments.

Moreover, the Board determined that BNFL continued to miss comprehensive incorporation of lessons learned from previous incidents to avoid recurrence of similar inadequacies in all areas of the demolition project. DOE ORO management had not ensured that corrective action plans consider negative trends, provide full solutions to avoid recurrence, and promote higher performance.

5.42 Injury due to flying piece, K-33 Building, East Tennessee Technology Park, Oak Ridge, Tennessee, United States; March 26, 1999 (DOE, 1999b)

5.42.1 Decommissioning process; problem encountered; impacts

The K- 33 Switchyard is located at the ETTP Three-Building (K-29, K-31, and K-33) D&D and Recycle Project. More details of this project are given elsewhere in this book (Section 6.35).

The Project site was being managed by BNFL under contract to the US DOE.

An ironworker apprentice employed by IDM Environmental Corporation (IDM) was injured by a flying piece of wire netting fence. At the time of the event, he and a coworker were leaving a work area next to the K-33 Switchyard where they had been removing electrical conductors from air circuit breaker (ACB) cabinets. When not carrying out the removal of electrical conductors, an excavator/shear was demolishing concrete and removing metal from the roof slab on a side of the ACB cabinets.

Just before the incident occurred, the excavator/shear was trying to roll up a section of dismantled wire netting fence with metal railing attached on each end. After the fence and railing were amassed into a roll, the excavator/shear rotated anticlockwise to remove the roll to a scrap metal heap. The trailing end of the fence got caught on an obstacle, and some 10 m of the fence and attached railing were dragged from the jaws of the shear. The rest of fence and attached railing were still grasped of the shear jaws. As the excavator/shear went on rotating, the fence was stretched, and the excavator/shear operative felt the fence was being tensioned. The operative tried to halt the shear but failed to halt the rotation before the fence disentangled from the obstacle. Consequently the railing end flew over the boom and shear, and hit the exiting ironworker on the back of the head. The flying piece knocked the ironworker apprentice's hard hat and goggles from his head, and he underwent scalp, ear, and facial injuries. He passed out and fell facedown. Following first aid by coworkers, the injured ironworker was stabilized by the emergency medical service responders and transported to the Oak Ridge hospital. The injury resulted in a 6-day hospitalization and follow-on outpatient treatment.

5.42.2 Lessons learned

The Investigation Board identified as a root cause for the accident the failure of IDM management to implement a proper an adequate Enhanced Work Planning (EWP) that includes task hazard analysis (THA), as mandated by the BNFL contract with DOE and IDM subcontract with BNFL. Contributing to this, BNFL failed to verify that corrective actions were taken by IDM further to previously identified deficiencies. Besides, DOE failed to identify inadequacies in the BNFL and IDM process caused by insufficient supervision program (audits, reviews, and in-field monitoring).

The Board also identified several contributing causes that may have made the accident more probable without individually causing it. These contributing causes are as follows:

- Lack of safety barriers and alerts (e.g., barricades, flagging) and a clearly marked, safe exit path at the workplace
- Lack of recognition of the hazards posed by the concurrent operations of demolition and removal of electrical conductors
- Lack of worker training in the specific procedures
- Lack of timely and effective communications between different teams of workers at the work site

5.43 Personal injury during table saw use, Heyrend Way Facility, Idaho National Laboratory, Idaho Falls, Idaho, United States; February 10, 2006 (DOE, 2006c)

5.43.1 Decommissioning process; problem encountered; impacts

The Heyrend Way Facility is a leased facility a few kilometers south of Idaho Falls, Idaho. The facility is part of the INL's Science and Technology Campus, where the INL accomplishes work in basic science and engineering resulting in new products and processes.

Contractor activities at the INL are managed by the DOE Idaho Operations Office (DOE-ID). Battelle Energy Alliance (BEA) was DOE's prime contractor since February 1, 2005.

Among other activities, work at the Heyrend Way Facility included the fabrication of customized storage containers: The accident occurred during this activity.

A newly recruited technician suffered a serious injury to his right hand while using a table saw. On the day of the accident, the technician moved to rip cut ETHAFOAM 900 sheets (this material is semirigid polyethylene foam procured for use in the Heyrend Way Facility in 5-cm-thick sheet form). The blade guard assembly was not reinserted prior to this operation. The Laboratory Custodian observed the first cut of the ETHAFOAM 900 sheet before leaving the technician but did not see that the blade guard assembly was not installed. The technician noted that the material was hard to control while cutting with the table saw. While cutting the final piece of the sheet, the material became increasingly harder to control, and the technician stopped using the push stick before completing the cut. He then put both hands on the scrap portion of the material, leaving the product piece between the saw blade and rip fence unconstrained. Within several centimeters of completing the cut, due to a sudden movement of the material, the technician's right hand contacted the rotating saw blade. The wound went across the palm of his hand between the base of his little finger to the web between his thumb and index finger. Other workers provided emergency first aid assistance. As no first aid kit was available, cleaning rags were pressed on the wound with pressure to stop hemorrhage. The technician was laid on the floor and treated for shock.

The injury caused the traumatic amputation of the little finger, and considerable reconstructive surgery was needed to fix other fingers. The technician remained at the hospital following surgery and was discharged a few days later.

5.43.2 Analysis; solutions

Based on physical evidence, the Investigation Board determined that the sudden forceful recoil of the ETHAFOAM 900 caused the final accident sequence.

The following recaps the main conclusions of the investigation:

- The total time between the emergency call and the arrival of the ambulance was 13 min. When the response time is greater than 3–4 min for work areas where severe bleeding or life threatening or permanently disabling injuries are possible, OSHA prescribes that first aid kits and personnel trained in first aid be available. Neither first aid kit nor personnel trained in first aid were available in this case.
- Regardless all the efforts spent for preventing accidents, there must be a preparedness to deal with them to reduce impacts should they occur. This is a responsibility for the operating organization management, individual workers and supervisors, medical services, and other emergency responders including off-site organizations. However, the operating organization has the essential role to plan and facilitate interparty coordination. A crucial aspect of emergency planning is regular training of individuals in the discharge of their emergency-related responsibilities.
- Scheduled self-assessments and surveys were regularly held at the Heyrend Way Facility. However, the assessments were mostly conducted by facility personnel, who were not trained to conduct safety and health assessment; besides, checklists cannot replace thorough reviews by subject matter experts.

- Although the National Security Team had management responsibility for safety in their programs (including activities at the Heyrend Way Facility), there was no documented safety supervision carried out in the period June 2003–October 2005.
- BEA did not conduct a prejob briefing or ensure workers were adequately trained and hazards were identified and controlled. Instructions about the use of the blade guard assembly were not clearly forwarded to and understood by the technician. The technician was allowed to carry out work without proper training.
- The procedure the technician was required to comply with did not clearly define the extent of the work, control performance, or provide for worker feedback and constant improvement, as prescribed by ISM. Table saw manufacturer's safety warnings were not included into the job safety analysis (JSA).
- A formal review of project activities was not performed against the JSA used as the work control document to make sure that the project did not present any unreviewed hazards. Besides, using a JSA instead of a formal work control document overrelied on the assumption that training was sufficient to mitigate hazards. As said earlier, training was inadequate to this project.
- DOE-ID and its contractors did not ensure corrective actions from precursor events were established and formalized to avoid similar events.
- Conflicting instructions about the use of gloves and the handbook's prohibition against the use of gloves were not noted and were not reconciled.

5.43.3 Lessons learned

The Investigation Board identified three root causes, one local and two systemic, whose elimination would have avoided the injury:

- Local root cause—Failure to verify the blade guard assembly was in place as required for the specific cutting operation.
- Contractor systemic root cause—BEA ignored a number of inadequacies in work control processes and allowed informal behavior to implement the ISM program, including competence proportionate to responsibility. In particular, to ensure proper competence, all workers should have full understanding of their job tasks. Besides, the workers should be given feedback on safety elements of their jobs.
- DOE systemic root cause—DOE-ID did not ensure supervision activities were able to identify inadequacies in the contractor's implementation of ISM.

5.44 The Rapsodie accident, Cadarache Nuclear Centre, France; March 31, 1994

Accidents will occur in the best-regulated families.

Charles Dickens (1812–1870), David Copperfield (1850)

5.44.1 Decommissioning process; problem encountered; impacts

Rapsodie was an experimental nuclear reactor sited at the Cadarache Nuclear Centre, France, operated by the French Atomic Energy Commission (Commissariat à l'Énergie Atomique; CEA). It was France's first sodium-cooled fast neutron reactor and went critical in 1967. It was a reactor with a thermal energy of 20–40 MW. It eventually closed down in 1983.

On March 31, 1994, an explosion occurred during the cleaning of the residual sodium stored in a tank located in a hall near the reactor. At that time, Rapsodie was being decommissioned, and the sodium treatment was part of the decommissioning activities operations (Marmonier & Del Negro). Following the accident and related investigations, the decommissioning process went on. ASN (2019) describes the current status of Rapsodie as follows. The damaged building was reconstructed 1998–2003. Since 2004, work has been underway to restore facility conditions, to dismantle all redundant equipment, and to collect and take away waste. The reactor core has been fully downloaded, the fuel removed from the facility, the reactor vessel sealed; the reactor pool has been emptied, partly decontaminated, and dismantled. The last sodium residues were transferred in late 2018 to the Phénix installation at Marcoule for treatment. The postirradiation fuel assembly sectioning and inspection laboratory (LDAC) was shut down in 1997 and partly decommissioned: Its decommissioning is part of the broader Rapsodie decommissioning project. To complete the project, the following activities remain:

- Decree to allow final dismantling: expected at the time of writing
- Removal of the remaining sodium present in the vessel. Estimated time: 5.5 y
- Decontamination of the buildings after the removal of the sodium contained in the reactor vessel (9 years)
- Dismantling of the block reactor and completion of decontamination (15 years)
- Buildings released for free use but not demolished

The shock wave produced by the explosion made a 300 m^2 concrete slab fall, destroying the circular passageway located nearby. A dead worker (a member of the operating organization) was found under the slab. Four workers with multiple fractures were hastily hospitalized.

This explosion was due to unexpected chemical reactions and gas released following alcohol treatment of the slightly contaminated sodium.

This accident did not cause contamination of wounded workers, other staff, or the environment. Slight contamination around 10 Bq/cm^2 was measured close to the explosion pit. This contamination was due to the cesium present in the sodium. The reactor vessel remained undamaged. The ventilation system in the contiguous laboratory (LDAC) was stopped for 3 hours as windows and a door were shattered. Although this ventilation system works as dynamic containment for LDAC, the accident did not enlarge the level of contamination in the laboratory.

The accident was graded INES level 2. Summary data (relevant to the INIS classification) for this event include the following:

	Yes	No
Release beyond authorized limits	☐	☒
Overexposure of a member of the public	☐	☒
Overexposure of a worker	☐	☒
Contamination spread within the facility	☐	☒
Damage to radiological barriers	☐	☒
Degradation of defense-in-depth	☒	☐
Person injured physically or casualty	☒	☐
Is there a continuing problem	☐	☒
Was a press release issued	☒	☐

5.44.2 Analysis; solutions

The accident background and circumstances can be described as follows. After the permanent shutdown of the reactor, all 37 t of primary coolant sodium were drained in 1985 and filtered to remove most of its radioactive contamination; ^{137}Cs was reduced to 1/100 of its initial concentration. Following decontamination, the sodium was stored for 8 years in a 55 m^3 tank sited in a basement next to the reactor containment building; in 1993, the sodium began to be drained out of the tank to be converted into sodium hydroxide at the Desora plant. Desora had been specifically designed for that objective. After drainage, still some sodium remained at the bottom of the tank.

Before the tank could be dismantled, the remaining sodium had to be removed. The process selected to this end, already carried out several times, consisted of gradually injecting in the tank ethyl-carbitol (2-(2-ethoxyethoxy) ethanol) in 50-L batches, while monitoring the reaction through measurements of such parameters as temperature, pressure, and the concentrations of hydrogen and oxygen.

The primary cause of the accident was the formation of a complex environment, consisting of three main components (alcohol, alcohol crystalline salt, sodium). This environment was formed through the following subsequent process phases:

- Cooling of several days due to the alcohol injection being stopped
- External electrical heating in addition to the local heat produced by the chemical reactions
- Alcohol injection for the past 2 days before the accident

This environment was especially conducive to the progression of thermal decomposition and catalytic exothermal reactions. Large amounts of gases (hydrogen and light hydrocarbons) were then generated. Soon after the last alcohol injection on the day of the accident, the process became uncontrolled, causing a pressure surge, the abrupt burst of the tank, and the explosion of the gaseous mixture in the hall. The energy released was 200 MegaJoules (the equivalent of about 50 kg of trinitrotoluene, TNT).

5.44.3 Lessons learned

The state of knowledge at the time the accident occurred did not make possible to foresee the nature of the chemical reactions nor the loss of control. Therefore, the Commission of Inquiry established soon after the accident proposed a large-scale program of investigations including the following:

- Extensive investigations in situ
- Studies and analyses on all possible chemical reactions
- Chemical investigations of the reaction residues
- Metallurgical, mechanical, and thermal analyses
- Impact of human factors

CEA laboratories, the chemical industry, the universities, and other parties were involved in this program. Pending completion of the above investigations washing of sodium pools with alcohol was forbidden. In following years, the use of alcohols to dissolve the residual sodium has been fully discontinued. Currently, carbonation (a humidified carbon dioxide process) and water wash are the preferred routes for sodium treatment before dismantling (CEA, 2016).

A generally applicable requirement regarded the availability after the final shutdown of a facility of a team fully familiar with that facility.

Several sodium-related incidents can be found in the literature (e.g., Sections 5.21 and 5.44). Two cases not described in detail in this book also refer to nuclear decommissioning sites (DOE, 2000a, 2000b, 2004e).

For comparison, see also a relatively minor incident caused by chemicals in Box 5.13.

BOX 5.13 Chemical event at Jaslovské Bohunice, Slovakia (Slovak Spectator, 2014)

The NPPs A1 and V1 at Jaslovské Bohunice, Slovakia, are being decommissioned. The decommissioning organization is Nuclear and Decommissioning Company JAVIS, a joint-stock company whose 100% owner is the Slovak state, which also enforces the shareholder's rights through the Ministry of Economics.

While chemicals were handled, there was an accident in August 2014 in a former NPP turbine hall at Jaslovské Bohunice (Trnava Region), currently converted to decommissioning-related activities, e.g., decontamination. A casing was damaged, and two workers in the area were injured. This was an industrial accident, and no radiation was involved.

The injured men were hospitalized in the Teaching Hospital at Trnava; one was treated for acid burns of eyes, whereas the other was in stable conditions, in no life danger, and was held at the hospital for potential respiratory problems.

5.45 Hot cell fire due to size reduction activities, Oak Ridge National Laboratory, Tennessee, United States (DOE, 2011d)

A great flame follows a little spark Dante (1265–1321).
Paradise, I, 34

5.45.1 Decommissioning process; problem encountered; impacts

The Transuranic (TRU) Waste Processing Center (TWPC) is a multipurpose nonreactor nuclear facility constructed by the Department of Energy (DOE) to house treatment processes for high-α TRU waste. On January 7, 2011, a fire erupted during waste processing activities in the hot cell of the TWPC next to the ORNL. While performing approved size reduction activities of a waste carbon steel box (using an industrial saw with an abrasive blade), sparks made their way under or around fire blankets covering spent HEPA filters and other combustibles staged adjacent to the hot work operations. The sparks ignited filtration loading (dust and debris) on the spent HEPA filters. The spent filters had recently been removed from the in-cell filter bank of the TWPC hot cell and were awaiting size reduction operations. The reader should note that many models of HEPA filters (Fig. 5.17) are vulnerable to fire.

TWPC workers identified smoke coming from underneath a fire blanket covering one of the filters. When the fire blanket was lifted to gain direct observation, a small flame was noticed. The fire blankets were quickly

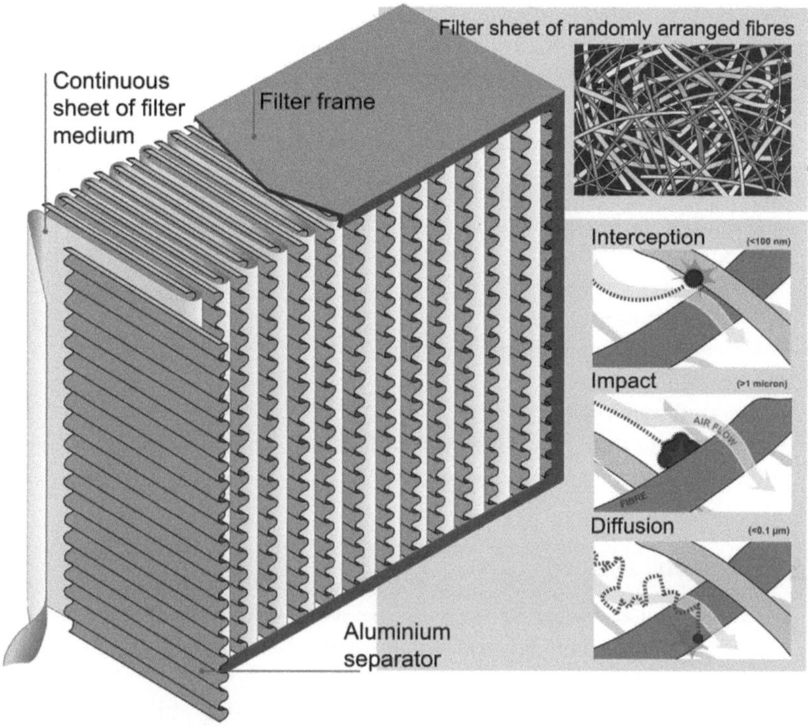

FIGURE 5.17 Schematic HEPA filter (Philip Greenspun illustration project, Wiki Commons). *HEPA*, high-efficiency particulate air.

replaced, and Operations personnel segregated the waste involved in the fire from other items on the hot cell work table. Concurrently with notification to the Duty Waste Operations Lead, the fire alarm was immediately activated promptly notifying the ORNL Lab Shift Superintendent and the ORNL Fire Department, and the process building was evacuated.

The ORNL Fire Department responded in a timely manner, assumed on-scene command and control of the fire event, and worked with TWPC on-shift management to extinguish the fire.

5.45.2 Analysis; solutions

There was no release of radioactive materials or increased levels of radiation as a result of the fire; no injuries occurred. The equipment damage was minimal, and no other plant systems were impacted.

The primary cause leading up to the fire event was determined to be less than adequate implementation of the hot work procedure controls, which resulted in failed or compromised barriers required to prevent a fire during hot work operation.

Contributing causes were identified as follows:

- Prejob briefing was inadequate.
- Segmenting tools and techniques generated a large rain of sparks.
- Elements of conduct of operations were inadequate including the following:
  - Work planning
  - Process sequencing/housekeeping
  - Communications
  - On-shift supervision
- Self-imposed time pressure to complete visual examination before midnight of second shift operations.

5.45.3 Lessons learned

- Personnel must always be alert of the work environment and changing conditions. Personnel did not recognize the hazards and conditions that resulted in the fire.
- Personnel in the field must always have a questioning attitude.
- Personnel are to perform work in accordance with approved procedures.
- All work and procedures to be used for the work need to be discussed in a prejob briefing, even if work appears to be routine.

See also Boxes 5.14 and 5.15.

BOX 5.14 Fire at Ignalina Nuclear Power Plant (NPP), Lithuania

This sketchy piece of information is reported only to reinforce the point that small fire events are commonplace at decommissioning plants (in fact at any industrial plant).

In Lithuania, Unit 1 of the Ignalina NPP, where the incident occurred, was shut down in December 2004. Unit 2 was shut down in 2009. The NPP is supposed to be totally decommissioned by 2029.

On August 1, 2016, smoke was noticed at the decommissioned deaerator room 719. The incident was at once notified to the local fire brigade: The firefighters arrived at Ignalina in some 15 min and found small wooden items burning in room 719. All measures were taken to extinguish the fire and ensure safety. The incident did not interrupt decommissioning activities ("The Lithuania Tribune" 2016).

BOX 5.15 Sparks setting fire during cutting operations; November 3, 1997 (DOE, 1997c)

Problem encountered; impacts

Workers at a commercial nuclear hot-cell installation reported a small fire in some flexible exhaust duct. It is assumed the fire started when a fragment of hot slag produced by the cutting of steel plates removed from a decommissioned hot cell fell on the duct. The fire watch put out the fire, and workers evacuated the installation. The fire watch was later hospitalized overnight for smoke inhalation. There was no radioactive release to the environment and no plant damage. All smoke was filtered through the HEPA filter.

Lessons learned

This event illustrates the hazards associated with welding, cutting, and grinding activities. These activities pose special unique hazards to workers carrying out nuclear D&D activities. Fire prevention is an important factor in planning for these activities. Open flames, electric arcs, hot metal, sparks, and spatter present serious hazards ignition.

This incident also stresses the importance of the fire watch in identifying the fire and alerting the welder. It is also important that the fire department be notified prior to fighting a fire with a fire extinguisher. If the fire cannot be contained, the area should be evacuated and turned over to the fire fighters. In this case study, timely identification by the fire watch of a fire started by hot slag avoided serious injury to the welders.

5.46 Heavy load drop at decommissioning division, Karlsruhe Research Centre, Germany, June 25, 2003 (Pfeifer, 2006)

Safety is understanding, it is an attitude of mind, it is not necessarily a simply following of rules or directions.

Garry Richards

5.46.1 Decommissioning process; problem encountered; impacts

A contaminated steel slab of around 2.5 t fell while it was removed by fork truck to a container in the controlled area. The event was rated 0 on the INES and N (normal) in the German notification priority order.

5.46.2 Analysis; solutions

Investigations proved that this incident was caused by inadequacies in safety culture as follows:

- Several organizational groups had determined an unauthorized transport path. The selected path was authorized for transport of contaminated items in drums, but not for steel slabs packed in foil.

- The fork truck employed was not certified for this operation. Again the decision had been taken by several organizational groups.
- Inadequacies of the quality assurance regime were disclosed. The method chosen for the identification and labeling of loads and authorized by managers in charge is unfortunately vulnerable to human errors (e.g., confusion of labels).

The following actions were taken after the event:

- The plant manager issued a work order clarifying the determination of weight and the labeling of the waste materials.
- The amended procedure was forwarded to the dismantling, radiation protection, and waste teams: Additional training was provided accordingly.

5.46.3 Lessons learned

At the time of the accident, some 15% of the reportable events at Karlsruhe were attributed to human errors. It is likely that, however, the human errors and organizational inadequacies (e.g., overlapping, ambiguous or blurred responsibilities) play a much greater role in reported events attributed to other causes, e.g., operation mode, planning, poor regulations/procedures.

5.47 Chemical explosion, Hanford Site, Washington, United States; May 14, 1997 (DOE, 1997d)

That's why it's time for a change

Thomas Edmund Dewey (1902–1971)

5.47.1 Decommissioning process; problem encountered; impacts

The incident occurred in the Chemical Preparation Room of the Plutonium Reclamation Facility (PRF) at Hanford's Plutonium Finishing Plant (PFP). The facility had been unused for many years and was being deactivated in view of decontamination and decommissioning. The explosion occurred in a room where miscellaneous nonradioactive chemicals had been accumulated in the past in support of plutonium production.

The cause of the explosion was a spontaneous reaction of hydroxylamine–nitrate (HAN) and nitric acid mixture. The HAN/nitric acid tank is a 1.5 m^3 stainless steel tank. The tank originally housed a dilute solution that was arranged for a training session in 1993. Over several years, concentrations of the reactants grew due to evaporation from this vented tank until they reached the critical mixture. It appears that the volume of the reactants had shrunk from the initial ca. 0.75 m^3 to less than 0.15 m^3.

The explosion forcibly ejected the bolted lid of the tank and a 3.75 cm fire protection water line broke up, resulting in water flowing across the plutonium plant and to the parking lot. Structural damage involved the distortion of a wall and doors, and two minor holes in the roof. Air samples inside the facility and air monitors in the nearby areas detected no radiological releases. No personnel were injured, although several (8−10) were addressed to hospital with some feeling a metallic taste in their mouths. An investigation of the event was launched, and its findings and conclusions are reported in (DOE 1997d). Apparently, there was evidence from shift surveillance data since 1993 that a trend of weight loss was in progress, indicating a concentration increase in the tank's inventory. Unfortunately the meaning of that data was not recognized.

A spontaneous reaction in a piping system was indicated as the cause for a similar, previous occurrence at the PUREX Plant (1989). A solution of HAN and nitric acid with hydrazine remained enclosed between two valves. This solution probably concentrated over time due to evaporation, and the resulting reaction eventually broke a gasket. Laboratory testing concluded that a spontaneous reaction would occur at a sufficient concentration. Hydrazine was not involved in the aforementioned PFP−PRF explosion.

In another event involving the same chemicals at Savannah River (1996), a sump high-level alarm activated. Investigation concluded that the origin of the level increase was a canyon sump receipt vessel. Evidence revealed that the vessel overflowed resulting in spillage of liquids to the sump. Chemicals (HAN and nitric acid) were routinely added to the receipt vessel in preparation of sump receipts to prevent criticality. Overflow of the vessel was caused by autocatalytic decomposition of HAN in presence of nitric acid. Over time, several sump receipts were added to the vessel. The acidity and HAN concentrations in the sump receipt vessel increased through the chemical additions. At these conditions (concurrently with high temperatures), the reaction commenced and went on until completion, resulting in up to 1 m^3 being erupted to the sump (Fullwood, 2000).

5.47.2 Lessons learned

Nuclear fuel cycle facilities in a shutdown, operation-to-decommissioning transition, or deactivation phase may be subject to hazards that do not exist in operating facilities. Chemicals remaining in unused tanks, piping, drums, or other stores may be in the long term subject to deterioration or concentration changes. The safety of the evolving condition of chemicals should be assessed, even though the safety of the formerly operating process systems had been ensured. These longer-term changes could cause spontaneous reactions or adverse effects such as follows:

- Corrosion product catalyzed reactions
- Gradual chemical deterioration (i.e., loss of chemical properties)

- Concentration via evaporation
- Unexpected cross-contamination due to leaks, incorrect routings, and so on
- Some chemicals decompose to form peroxides that are extremely flammable and possibly explosive

The DOE Chemical Vulnerability Assessment identified weaknesses in the storage of chemicals within the DOE Complex. Corrective actions were established (DOE, 2000b).

5.48 Electrical incidents at Oak Ridge National Laboratory, Tennessee, United States

Both these ORNL events belong to the vast category of electric incidents during D&D: Further examples are given in Sections 5.11, 5.25, etc.

5.48.1 Problem encountered 1; impacts: work scope exceeded during demolition results in worker contacting live wire (DOE, 2013b)

This 2011 incident occurred at the Y-12 plant: Workers contacted live wires when they went beyond the scope of the work package to dismantle and remove a conveyor. This occurrence involves inadequate management aims and insufficient communication between work teams. The root cause of this event was the failure to include all parties in the utility survey of the conveyor and its surrounding structures. As a result of this failure, when the workers exceeded the boundaries of the work order, they did not understand the implications or the underlying risks. Worsening the situation was the transfer of oversight from one supervisor to another prior to work onset, but after the utility survey. The lack of communications left the workers and the new supervisor without the needed knowledge to decide if the work they were about to carry out was really covered by the work package.

These factors indicate inadequacies with the direct supervision of work and the general contractor's supervision of operations. In the lack of a specified procedure for supervisor succession, the work in question was inevitably spoilt since the very beginning. The management guidance exhibited its inadequacy when the new supervisor went on with the work package without recognizing the hazards. Moreover, nobody enquired whether having a new supervisor required to resurvey the job.

5.48.2 Problem encountered 2; impacts: electrical deactivation (DOE, 2010b)

While carrying out activities preparatory to demolition, contractors unexpectedly found out an electrical energy source (105 V, A/C) in an instrument panel. The building had formerly been declared as "cold and " (the USDOE term for a passive safe enclosure—see Glossary) by a previous DOE contractor; it had in recent times been subject to a Utilities Isolation Verification walk-down by the incumbent contractor and was confirmed as electrically deactivated the day before the incident by an electrical subcontractor.

5.48.3 Analysis; solutions

Further to this occurrence, the former contractor's approach for defining a building as "cold and dark" was considered inadequate. The current contractor did not review their approach but just notified the occurrence to former DOE contractor. Therefore, even the present contractor's approach for utility isolation and electrical deactivation was considered inadequate. It was established that although all electrical wires were verified entering the building, wires pulled through a floor penetration were not air gapped or tested. The present contractor initiated an effort to improve the situation.

5.48.4 Lessons learned

Even after a building planned for demolition has been defined as cold and dark, it should be understood that this definition does not guarantee that all hazards have been actually eliminated. Chemical, radiological, electrical, structural, and free liquid hazards may remain: Therefore, hazard analyses should continue as work goes on. In this event, deenergizing sources only at the distribution panels are inadequate. Unknown sources may go undetected through the slab or from ground penetrations.

5.49 Employee fall injury, 336 Building, Hanford Site, Washington, United States; July 1, 2009 (DOE, 2009)

5.49.1 Decommissioning process, problem encountered; impacts

The High Bay Testing Facility (336 Building) was constructed in 1969 and housed experimental apparatuses in support of the Fast Flux Test Facility. The building is 15 m × 15 m by 20 m high. The building is not radiologically contaminated as no radioactive materials were ever handled in it. The redundant building was eventually transferred to Washington Closure Hanford, LLC (WCH) for decommissioning and demolition purposes in 2008.

During preparation work for demolition on July 1, 2009, a mechanic fell 15 m from an open bridge and sustained severe injuries. The mechanic was a member of a WCH team arranging a bridge crane for removal from the building. The mechanic was walking toward the end of the bridge when he stepped through an opening. The mechanic struck an intermediate platform 7.5 m below the bridge and then fell another 7.5 m to the concrete floor. The mechanic survived but had two vertebrae broken in his back, bones broken in both legs, and his left knee hurt. There was no internal damage.

In planning the demolition of the 336 Building, the safest method was assumed to remove the building bridge crane prior to dismantling the building structural beams. Once the crane was pulled out, high reach equipment would be used to dismantle the building. The original plan to prepare the crane for removal was to access the crane bridge, remove the rail stops, detach the travel wheels to allow free movement, and cut the crane's lifting cable, so it could be used to pull the crane out. Three craftsmen were involved in this work. They identified some more tasks needed to prepare the crane for removal and discussed these plan modifications with the work supervisor. These changes included the support of two mechanics to carry out the task of achieving free movement of the crane's travel wheels. On the day of the accident, two craftsmen and two mechanics climbed the ladder to the bridge and started work. The mechanics soon realized that the gear boxes attached to the travel wheels had to be removed first. In the meantime, one of the craftsmen climbed down the bridge to replace the ground support craftsman who had left the scene for a related commitment. The remaining three workers believed they were almost done with the job. Mechanic 2 began climbing down the ladder. The bridge opening was left open, with the assumption that Mechanic 1 would follow. During this time, the ground support craftsman had come back and called up to direct one more task that required two workers: secure the crane trolley to the bridge with chains and tensioners. The remaining team members on the bridge, a craftsman and a mechanic, decided the safest method to anchor the chains would be to bring the crane closer by pushing the crane wheels from both ends of the crane bridge. As Mechanic 1 moved toward the end of the bridge, he did not see the opening, stepped into it, and fell 15 m.

After the fall, his coworkers called emergency center and assisted the injured mechanic by keeping him calm and still. Emergency personnel arrived at the scene in few minutes. The mechanic remained conscious until he was transported to a Richland, Washington, hospital. After 2 weeks in the hospital, the mechanic was discharged and continued recovery over the next several months.

After the accident, WCH management secured the scene, issued required notifications, and discontinued all elevated work. An Accident Investigation Board was appointed by, and reported its findings to, the DOE.

5.49.2 Analysis; solutions

Following the Board investigation, the DOE identified a number of violations regarding (1) fall protection and ladder safety and (2) general requirements and construction safety management, as follows (DOE, 2010c).

I. Fall protection and ladder safety

Contrary to legal prescriptions, WCH failed to protect employees from fall hazards and to implement fall protection and ladder safety requirements. Specific points are given in the following:

A.
- WCH staff climbed a 15 m ladder and crossed elevated platforms at heights of 7.5 and 15 m without the protection of a guardrail, safety net, or personal fall arrest. WCH staff on a 7.5 m platform disengaged from a lower retractable lanyard and attached to an upper retractable lanyard before climbing to the bridge above. When the workers were not hooked to the lanyard, they were exposed to a 7.5 m fall hazard, as the ladder access opening did not have a closure system. Moreover, workers at the 15-m height disengaged from the self-retracting lanyard (SRL) that was attached to their harness to make the SRL useable by a coworker at a lower level. Since the opening was not covered at the time, workers were exposed to a 15-m fall hazard.
- On July 1, 2009, two workers at the 15-m level had no fall protection, a guard rail, or a safety net near an unprotected opening.
- Workers did not employ any fall protection during and after cutting a bridge guardrail. The guardrail was cut in two places and did not provide any real protection to the two workers on the bridge. Workers had instructions to fasten to a guardrail one section back from the cut guardrail, contrary to requirements that forbid hooking personal fall arrest systems to guardrails.

B.
- WCH did not ensure the cover was in place when the opening was unused. In lieu of this, WCH employed an informal administrative control, nicknamed as the "team up team down" scheme, as the only means to prevent access to the bridge opening when it was not covered. This method proved to be ineffective to control the hazard.

C.
- WCH did not comply with the structural engineering recommendations for anchorage. The anchors were placed on flat iron supporting the fixed ladder and structural cross-bridging for the whole building. The anchorage points used were not compliant with the available structural analysis. WCH did not have a structural engineer or anyone with comparable qualifications to determine if the selected anchorage spots were strong enough to support workers safely (according to regulations, they should be capable of supporting at least 2.5 t).

D.
- The three craftsmen and two mechanics working in the 336 Building when the accident occurred were not retrained on updated regulations on Fall Protection. These updates included requirements for controlled access, safety monitors, procedures, and Project Safety Representative (PSR) duties.

E.
- A craftsman climbed a 15-m fixed ladder without fall protection to install two SRLs—one at the 7.5 m height and the other at the 15-m height, to allow access to the bridge to get the crane ready for removal. Although the fixed ladder was intended to be used with a fall

protection device, WCH had no provisions to ensure that workers climbing up and down the ladder were actually using this device.

F. • WCH did not provide needed training for personnel using stationary ladders during demolition work in the 336 Building to ensure they had information and skills to minimize the hazards. The WCH documents available to the project team had no relevant information for the safe use of the ladder installed in the building.

G. • The ladder shaft access to the intermediate platform had a chain that was too short to extend across the entry and serve as guardrail protection. The access opening to the ladder was not equipped with a swing gate nor was the passage through the opening offset to safeguard employees from a fall hazard.

II. General Requirements and Construction Safety Management

A. • WCH did not set up and enforce a worker safety and health program that included measures against fall hazards related to stationary ladders. There was a WCH procedure stating requirements for portable ladders, but there were no similar requirements for fixed ladders. Furthermore, the WCH procedure on fall protection did not include OSHA requirements for fall protection on fixed ladders.

- Another WCH procedure identified requirements that allowed personnel to carry out work without fall protection when they are 2 m away from such fall hazards as a floor opening or open-sided floor. This procedure is in direct conflict with a legal requirement that does not define a minimum distance at which fall protection must be employed.

B. • The JHA applicable to the work being carried out on the day of the accident did not identify the bridge opening as a hazard. As a result, the JHA's Fall Protection Checklist only addressed the hazards of the stationary ladder and did not address measures for the fall hazards related to work on the bridge.

- Hazard analyses were not performed for individual tasks such as cutting the guard rails on the bridge, chaining the cable drum to the crane structure, installing alternate fall protection, and draining gear box oil. According to WCH procedure, these are major changes in work scope. Major changes of this kind required the work supervisor, the responsible manager, and subject matter experts to approve the work details and redo the JHA.
- WCH did not include a lighting survey in the 336 Building hazard assessment. Consequently, WCH did not consider the interactions between the lack of lighting and multiple safety hazards related to working at heights while getting the bridge crane ready for removal (e.g., trips and falls, uncovered openings). A lighting survey conducted a week after the accident indicated that the work area would have been lit to less than 10 lux, which is less than the 50 lux legally required for construction workplaces (note to the reader the US units—foot candles—have been converted here into the SI units-lux).

5.49.3 Lessons learned

The Investigation Board concluded that this accident was clearly avoidable. The worker was very fortunate to survive the fall. The Board determined that the root causes of the accident were the following (DOE, 2009):

- The recognized understaffing of critical resources such as work supervisors and PSRs, the time pressure, and the inadequacies of work planning have been accepted by contractor management and the workforce. No measures were taken to balance work priorities and the discipline of work planning, control, and supervision. Perceived schedule pressure was a significant contributor in failing to adequately plan the work.

Multiple assignments did not allow human resources to be sufficient to provide adequate supervision. For example, the work supervisor initially assigned to the building demolition had several concurrent assignments and was not familiar with the work plan, which led the area superintendent to assume the duties of the direct work supervisor.

- The work control process was not sufficient to identify the work scope, hazards, and related controls to safely carry out the bridge crane removal. For example, the two walk-downs performed during the 336 Building work package and JHA planning phase covered the floor level only, omitting a complete examination of the workplace, which resulted in insufficient work scope and hazard identification. Coordination and communications were also inadequate. Additional hazards and controls resulting from work scope changes were poorly assessed.
- The fall protection procedure did not fully comply with OSHA standards and did not ensure protection to the workers. This procedure was inadequate to effectively convey the applicable regulatory requirements to the workforce. Had the relevant OSHA requirements been applied, the injured mechanic would have been fastened when the bridge opening was not covered or an engineering control would have been installed between him and the opening.

The unsafe configuration shown in Fig. 5.18 highlights the hazards posed by work at height.

5.50 A review of electrical incidents at D&D workplaces (DOE, 2005a,b)

5.50.1 Problem encountered 1; impacts

On June 16, 2005, at the SRS F-Canyon, a D&D team pierced into a duct with a movable band saw, touched the live electrical wiring inside it and caused a ground fault circuit interrupter to trip. The team was mandated with air-gapping electrical ducts for a chemical store and assumed that all circuits had been deactivated. This event caused no injuries.

FIGURE 5.18 Unsafe scaffolding configuration. *Photo by M. Laraia.*

The engineering team had considered hazardous energy sources in the workplace, scrutinized the electrical drawings, and carried out a walk-down. They found that no hazardous energy existed and determined the air gap spots. Workers had also conducted voltage checks and measured no voltage.

Investigators concluded that several factors caused this event, including the following:

1. Electrical circuits in the building were interlinked with electrical control panels in other facilities; most of these were deserted and deactivated. However, one panel was not fully deactivated when it was taken out of service, leaving one circuit active.
2. Based on unclear, ambiguous drawings, engineering team wrongly assumed that the power source had been fully removed from the building.
3. The electrical source was located in a contaminated area that was not covered in the walk-down on radiation protection grounds.
4. The field condition (e.g., rust on duct and wire ends that were hardly accessible) made it difficult to conduct precise voltage checks.

5.50.2 Problem encountered 2; impacts

On May 23, 2005, at the Savannah River Facilities Disposition Project, a crew of workers air-gapping conduit noticed sparks when they cut a duct that went into a junction box. While securing the area short time after the event, the workers found that the 110-volt electrical control wiring for the junction box was live. The workers were wearing certified electrical gloves and PPE, and nobody was injured.

Investigators concluded that an isolation engineer missed to follow the duct up to the end after he arrived at a "T." The electrical drawings indicated that the duct deriving from the "T" went to buttons that had been employed to action a pump. The voltage stemmed from a control circuit that was fed from a lighting panel in another building. Changes made during the installation of the new building switched the power source from a button that had been removed in the older building to one in the new building, but only the drawings for the new building were updated to show the change.

5.50.3 Problem encountered 3; impacts

On June 2, 2005, at the Lawrence Berkeley Laboratory, California, United States, a mechanic was cutting off electrical ducts with a movable band saw: When the saw cut into a live 120-volt wire, he heard an explosive sound. The ducts were residues of a transformer substation that had been demolished many years before. The mechanic supposed that all parts of the high-voltage distribution pad were deactivated because the transformer had been taken away. He was wrong, but luckily nobody was injured.

5.50.4 Problem encountered 4; impacts

On July 19, 2004, at the Hanford PFP, an electrician cut an electric cable to a piece of glovebox equipment and noticed an electrical spark. The glovebox had been isolated, but some circuits remained in service to power D&D work. Because of poor documentation, the equipment removal crew was not aware of which electrical circuits remained active. As the building was classified "cold and dark," the electricians did not feel it necessary to carry out a zero-energy check before cutting the cable.

5.50.5 Problem encountered 5; impacts

On August 12, 2005, at the Fernald Environmental Management Project, Ohio, United States, electricians noticed an arc and heard an explosive sound. The circuit breakers tripped while the workers were removing 110-volt electrical leads from two solenoid valves in a mixer room. The electricians were replacing valves and assumed, following a discussion with the shift engineer, that the solenoid control devices were isolated. The shift engineer had assumed all system parts were isolated, when in actuality only the mechanical components and mixer motors were isolated, not the valve controls. Based on their assumptions, the electricians missed to carry out a zero-energy check of the solenoids.

5.50.6 Lessons learned

Confirming a zero-energy status is an essential element of any hazardous energy control plan. Failure to do so can cause occupational injuries. The DOE's Office of Environment, Safety, and Health assessed occurrences in the period 2000–05 in which workers assumed a safe state without checking the lack of hazardous energy in the field. Most of these events (80%) were related to electrical energy. The term "cold and dark" is used in the DOE environment to describe the state of a deactivated installation or building just before it is delivered to demolition. In this strategy, the cold and dark state is mainly intended to ensure worker safety. For facilities subject to a long period of safe enclosure, the meaning of cold and dark state is to place the plant in a passive configuration that will minimize care and maintenance costs. However, a potential safety issue stems from the possible variants of the "cold and dark" strategy. When workers carry out work on—wrongly supposed—deactivated systems, near-misses or injuries can occur.

The aforementioned events highlight the need of a complete hazard assessment before electrical work commences, especially during D&D activities. The following actions are highly advisable:

- Conduct comprehensive surveys in all areas where work is foreseen.
- Ensure that all drawings are precise and match current state.
- Confirm that all circuits have been deactivated.
- Clearly identify temporary power sources in "cold and dark" installations to ensure that workers cannot confuse temporary power sources with deactivated sources.

References

American Society of Safety Professionals, March 17, 2020. Preventing Hearing Loss in Construction and Demolition. https://www.assp.org/news-and-articles/2020/03/17/preventing-hearing-loss-in-construction-and-demolition.

Autorité de Sûreté Nucléaire, 2019. Rapsodie/LDAC Summary (In French), ASN. https://www.asn.fr/L-ASN/L-ASN-en-region/Provence-Alpes-Cote-d-Azur/Installations-nucleaires/Rapsodie-LDAC.

BBC News, 20 January, 2015. Japan Worker Killed in Fukushima Nuclear Plant Accident. https://www.bbc.com/news/world-asia-30892730.

Brookhaven National Laboratory, 2011. Placement of Airborne Sampling Device Is Important when Assessing Exposure to Metal Fumes. Lesson ID: 2011-BNL-ERP-0005, 9/8/2011 (Available upon DOE Authorization).

Cary, A., June 23, 2018. Harmful Chemical Vapours Again Suspected at Hanford. https://www.tri-cityherald.com/news/local/hanford/article213706614.html.

Center for Construction Research and Training (CPWR), Hazard Analysis —Noise. http://www.cpwrconstructionsolutions.org/hazard/919/noise.html.

Commissariat À L'énergie Atomique Et Aux Énergies Alternatives, 2016. Monograph, Sodium-Cooled Nuclear Reactors. CEA, Paris. http://www.materials.cea.fr/en/PDF/Sodium-Cooled%20Nuclear%20Reactor_CEA-en.pdf.

Communication from Grete Rindahl, Heimdal, J., Hildebrandt, M., Øyvind Braarud, P., November 11, 2010. IFE, Norway to Michele Laraia. IAEA.

Department of Ecology, November 2019. Washington State, Tank Waste Monitoring and Closure. Last Update. https://ecology.wa.gov/Waste-Toxics/Nuclear-waste/Hanford-cleanup/Tank-waste-management/Tank-monitoring-closure.

East Bay Times. 2016. Feds to Investigate Livermore Lab's Handling of Toxic Metal Beryllium, Updated August 15, 2016. https://www.eastbaytimes.com/2008/07/06/feds-to-investigate-livermore-labs-handling-of-toxic-metal-beryllium/.

Engineering News-Record, May 29, 2020. Jacobs, Ex-AECOM Teams Protest Losing $13B DOE Nuke Award. https://www.enr.com/articles/49460-jacobs-ex-aecom-teams-protest-losing-13b-doe-nuke-award.

Fellhauer, C.R., 1998. In: Reactor D&D at Argonne National Laboratory - Lessons Learned, ANL/TD/CP-95532, Spectrum '98, Denver, CO (US), 13–18 September 1998. https://www.osti.gov/servlets/purl/10621.

Fullwood, R.R., 2000. Probabilistic Safety Assessment in the Chemical and Nuclear Industries. Butterworth-Heinemann.

Health and Safety Executive, UK, Hand-Arm Vibration at Work. https://www.hse.gov.uk/vibration/hav/index.htm.

Health and Safety Executive. https://www.hse.gov.uk/noise/casestudies/fullindex.htm.

Imperial College London, Common Causes of Laser Accidents https://www.imperial.ac.uk/safety/safety-by-topic/laboratory-safety/laser-safety/common-causes-of-laser-accidents/.

Institute for Nuclear Power Operations (INPO), December 2005. Radiological Protection at Nuclear Power Stations, INPO 05-008. Atlanta, GA (and subsequent revisions).

Institute For Radiological Protection and Nuclear Safety, 2015. (in French, Institut de Radioprotection et Sûreté Nucléaire, ISRN), Safety at Civil Basic Nuclear Installations Other Than Nuclear Power Plants in France-Lessons Learned by IRSN from Significant Events Reported in 2013 and 2014, Report. https://www.irsn.fr/EN/publications/technical-publications/Documents/IRSN-Report_Safety-Nuclear-Installations-France-2013-2014.pdf.

International Atomic Energy Agency, 2002. Record Keeping for the Decommissioning of Nuclear Facilities: Guidelines and Experience. IAEA, Vienna. Technical Reports Series No. 411.

International Atomic Energy Agency, 2005. Dismantling of Contaminated Stacks at Nuclear Facilities. IAEA, Vienna. Technical Reports Series No. 440.

International Atomic Energy Agency, 2006a. Decommissioning of Research Reactors: Evolution, State of the Art, Open Issues. IAEA, Vienna. Technical Reports Series No. 446.

International Atomic Energy Agency, 2006b. Decommissioning of Underground Structures, Systems and Components. IAEA, Vienna. Technical Reports Series No. 439.

International Atomic Energy Agency, 2006c. Management of Problematic Waste and Material Generated during the Decommissioning of Nuclear Facilities. IAEA, Vienna. Technical Reports Series No. 441.

International Atomic Energy Agency, 2018. Lessons Learned from Deferred Decommissioning of Nuclear Facilities, Nuclear Energy Series NW-T-2.11. IAEA, Vienna.

International Finance Corporation (IFC), April 30, 2007. Environmental, Health, and Safety (EHS) Guidelines - General EHS Guidelines: Construction and Decommissioning. https://www.ifc.org/wps/wcm/connect/7d708218-2a9e-4fcc-879d-9d5051746e7d/4%2BConstruction%2Band%2BDecommissioning.pdf?MOD=AJPERES&CVID=ls62NKq.

Jacobs, 1999. Environmental Impact Assessment Report (EIAR) Volume 2 of 4: Main Report. Bryan St, Suite 1200, Dallas, TX, US. http://www.epa.ie/licences/lic_eDMS/090151b28064b659.pdf.

Laser Accidents (part 1). https://www.industrial-electronics.com/Laser_Safety_09.html.

Laser Cutting Machine Safety Guide, AMADA Holdings Co., Ltd. 200, Ishida, Isehara-shi, Kanagawa 259-1196, Japan. https://www.amada.co.jp/english/products/pdf/saftyguide_laser_e.pdf.

Kerr, D., 2008. A Decommissioning Project manager's Perspective. IBC's 24th Annual residential Summer School on Decommissioning and Radioactive Waste Management, 30 June–4 July. Cambridge.

Knoxblogs, 2011. Atomic City Underground, The low point for Bechtel Jacobs. http://knoxblogs.com/atomiccity/2011/02/02/the_low_point_for_bechtel_jaco/.

KRIPO, 2019. Bavarian Police, Accident at Work during Dismantling Work - Fatally Injured Workers in the Nuclear Power Plant (German original: Die Bayerische Polizei, Arbeitsunfall bei Rückbauarbeiten – Arbeiter im Kernkraftwerk tödlich verletzt). https://www.polizei.bayern.de/news/presse/aktuell/index.html/298056.

Lin, F., Zhong, Q., 2017. Mitigation of ground vibration due to collapse of a large-scale cooling tower with novel application of materials as cushions. Hindawi Shock Vibrat. 2017, 6809246. http://downloads.hindawi.com/journals/sv/2017/6809246.pdf.

Marmonier, P., Del Negro, R., 1996. Information About the Accident Occurred Near Rapsodie. https://inis.iaea.org/collection/NCLCollectionStore/_Public/31/040/31040829.pdf?r=1&r=1.

McWhirter, A.F., 2017. In: Laraia, M. (Ed.), Lessons Learned from Decommissioning: What Went Wrong? Chapter 7 of "Advances and Innovations in Nuclear Decommissioning", ISBN 9780081011225. Woodhead Publishing.

Mission Support Alliance (MSA), September 2017. Hanford Site 2017 General Purpose Facilities Master Plan, HNF-61280, vol. 1 (Available upon DOE authorization).

OECD Nuclear Energy Agency, 2018. Heavy Load Accidents in Nuclear Installations, NEA/CNRA/R (2017), vol. 4. NEA, Paris. http://www.oecd.org/officialdocuments/publicdisplaydocumentpdf/?cote=NEA/CNRA/R(2017)4&docLanguage=En.

OECD Nuclear Energy Agency, May 2019. Proceedings of the Nuclear Energy Agency International Workshop on Chemical Hazards in Fuel Cycle Facilities Nuclear Processing, Appendix C, NEA/CSNI/R (2019)9/ADD1. https://www.oecd-nea.org/nsd/docs/2019/csni-r2019-9-add1.pdf.

Office For Nuclear Regulation, 2019. Quarterly Statements of Nuclear Incidents at Nuclear Installations- Updated 5 March, 2020. http://www.onr.org.uk/quarterly-stat/index.htm.

Pentek Inc., September 24, 2014. Respirator Reduction Aids ALARA Programs at DOE and Utility Nuclear Facilities. https://www.pentekusa.com/doe-alara/.

Pfeiffer, W., Goerinch, H., 2006. Minor incidents during the decommissioning of prototype operations and research facilities of the Karlsruhe research centre. In: Francis Lambert and Yuri Volkov. Safety Improvements through Lessons Learned from Operational Experience in Nuclear Research Facilities. Springer, pp. 67–78, 10 1-4020-3887-9.

Roberson, J.L., July 9, 2015. TA-53 Arc-Flash Accident Joint Accident Investigation Team (JAIT) Report. https://www.scribd.com/document/271487620/TA-53-Arc-Flash-Accident-JAIT-Report.

RPR1. 2016. Fatal Accident in the Mülheim-Kärlich Nuclear Power Plant (German original: Tödlicher Unfall im AKW Mülheim-Kärlich). https://www.rpr1.de/nachrichten/regional/ein-toter-bei-unfall-im-akw-m%C3%BChlheim-k%C3%A4rlich.

Serrato, M. G., John C. Musall, J. C., Bergren, C. L., In Situ Decommissioning (ISD) Concepts and Approaches for Excess Nuclear Facilities Decommissioning End State, WM2013 Conference, February 24-28, 2013, Phoenix, Arizona, US, 2013. https://xcdsystem.com/wmsym/archives//2013/papers/13367.pdf.

Stoneturn Consultants (STC), January 30, 2017. Third Party Qualified Independent Review: Hanford Tank Farm Respirator Cartridge Testing Report No 2: Review of: Use of Full Face Air Purifying Respirators in AP Tank Farms for Low-Hazard Tasks. TOC-IH-58345 Rev. 0. https://hanfordvapors.com/wp-content/uploads/2017/03/STC-Report-No-2-_-APR-Use-in-AP-10-Tank-Farm.pdf.

The Lithuania Tribune, Fire incident reported at decommissioned. Ignalina Nuclear Plant, August 3, 2016. https://lithuaniatribune.com/fire-incident-reported-at-decommissioned-ingalina-nuclear-plant/.

The Slovak Spectator, August 20, 2014. JAVYS: Accident in Jaslovské Bohunice Nuclear Plant Not Explosion. http://spectator.sme.sk/articles/view/54999/10/javys_accident_in_jaslovske_bohunice_nuclear_plant_not_explosion.html.

US Department of Energy, April 1997a. Type A Accident Investigation Board Report on the February 13, 1997, Welding/Cutting Fatality at the K-33 Building, K-25 Site, Oak Ridge, Tennessee, Final Report. https://www.energy.gov/sites/prod/files/2014/04/f14/9702k25_full.pdf.

US Department of Energy, July 1997b. Type A Accident Investigation Board Report on the June 20, 1997, Construction Fatality at the Brookhaven National Laboratory Upton, New York, Final Report. https://www.energy.gov/sites/prod/files/2014/04/f14/9706bnl.pdf.

US Department of Energy, 1997c. Hot Slag Ignites Flexible Exhaust Duct at NRC Facility. Lessons Learned Data Base, ID No: L-1997-OEWS-45-02 (Available upon DOE Authorization).

US Department of Energy, July 26, 1997d. Accident Investigation Board Report on the May 14, 1997 Chemical Explosion at the Plutonium Reclamation Facility, Hanford Site, Richland, Washington. DOE/RL-97-63. https://www.osti.gov/servlets/purl/331557.

US Department of Energy, September 1998. Type A Accident Investigation Board Report of the July 28, 1998 Fatality and Multiple Injuries Resulting from Release of Carbon Dioxide at Building 648, Test Reactor Area. Idaho National Engineering and Environmental Laboratory. https://www.energy.gov/sites/prod/files/2014/04/f14/9807ineel.pdf.

US Department of Energy, 1999a. Office of Health, Safety and Security, Lessons Learned Database, Fork-trucks in Confined Areas Need Spotters. Lesson ID: 1999-NVO-BN-002, 3/1/1999 (Available upon DOE Authorization).

US Department of Energy, May 1999b. Type B Accident Investigation Board Report on the Worker Injury at the BNFL Inc. East Tennessee Technology Park Three-Building Decontamination and Decommissioning and Recycle Project Site. DOE/ORO-2083. https://www.energy.gov/sites/prod/files/2014/04/f15/9905ettp.pdf.

US Department of Energy, October 2000a. Type B Accident Investigation Injury Resulting from Violent Exothermic Chemical Reaction at X-701B Site Portsmouth Gaseous Diffusion Plant. DOE/ORO-2103. https://www.energy.gov/sites/prod/files/2014/04/f15/PORTS_TypeB_082200.pdf.

US Department of Energy, 2000b. Assessment of Chemical and Radiological Vulnerabilities, DOE-0285-FP. https://www.osti.gov/servlets/purl/803622-tHABXr/webviewable/.

US Department of Energy, July 25, 2001. Office of Environment, Safety and Health, Operating Experience Summary 2001-01. https://www.dnfsb.gov/sites/default/files/document/1851/ltr_2001725_9991.pdf.

US Department of Energy, March 2003a. Type B Accident Investigation Board Report of the Subcontractor Employee Personal Protective Equipment Ignition Incident on February 18, 2003 Bechtel Jacobs Company LLC, East Tennessee Technology Park, Oak Ridge, TN. DOE/ORO-2151. https://www.energy.gov/sites/prod/files/2014/04/f15/TypeB_BJC.pdf.

US Department of Energy, 2003b. Insufficient Water Leads to Elevated Airborne Beryllium Concentrations during Decontamination. Lesson ID: LL-2003-LLNL-24 (Available upon DOE Authorization).

US Department of Energy, September 2004a. Office of Health, Safety and Security, Type A Accident Investigation, Subcontractor Fatality at the Pond B Dam Upgrade Project on July 26, 2004 at the Savannah River Site, Aiken, South Carolina. https://www.energy.gov/sites/prod/files/2014/04/f14/Type_A_Investigation_SavannahRiverSite_PondBDamUpgradeProject_2004_07_26_0.pdf.

US Department of Energy, August 2004b. Type A Accident Investigation of the July 15, 2004 Hanford 200 East Area Fall Fatality. DOE/RL-2004-63. https://www.energy.gov/sites/prod/files/2014/04/f14/Hanford200EastArea_Fall_Fatality_Report.pdf.

US Department of Energy, 2004c. Type B Accident Investigation Board Report on a Personnel Injury at Bettis Atomic Power Laboratory. https://www.energy.gov/sites/prod/files/2014/04/f15/0402bapl_0.pdf.

US Department of Energy, February 2004d. Type B Accident Investigation Board Report BNFL Inc. Employee Foot Injury on December 17, 2003, at the East Tennessee Technology Park, Building K-31. DOE/ORO-2175. https://www.energy.gov/sites/prod/files/2014/04/f15/bnfl0402.pdf.

US Department of Energy, June 2004e. Type B Accident Investigation Board Report May 8, 2004, Exothermic Metal Reaction Event during Sodium Transfer Activities East Tennessee Technology Park Oak Ridge, Tennessee, DOE/ORO-2182. https://www.energy.gov/sites/prod/files/2014/04/f15/200405ettp.pdf.

US Department of Energy, 2005a. Operating Experience Summary, Issue Number 2005-15, Published 12/20/05, Article 3: Be Alert for "Hidden" Electrical Hazards When Performing D&D Activities. https://ehss.energy.gov/oesummary/oesummary2005/OES2005-15screen.pdf.

US Department of Energy, 2005b. Operating Experience Summary, Issue Number 2005-16, Published 12/30/05, Article 2: Never Assume a Zero-Energy Condition — Verify and Work Safely. https://ehss.energy.gov/oesummary/oesummary2005/OES2005-16screen.pdf.

US Department of Energy, 2006a. Lessons Learned Data Base, ID No: RCCC-06-0011(available upon DOE Authorization).

US Department of Energy, February 2006b. Type B Accident Investigation, Board Report, Bechtel Jacobs Company LLC. Employee Fall Injury on January 3, 2006, at the K-25 Building East Tennessee Technology Park, Oak Ridge, Tennessee, DOE/ORO-2212. https://www.energy.gov/sites/prod/files/2014/04/f15/ETTP_Type_B.pdf.

US Department of Energy, February 10, 2006c. Type B Accident Investigation Board Report for the January 11, 2006, Personal Injury during Table Saw Use at the Heyrend Way Facility Idaho Falls, Idaho. https://www.energy.gov/sites/prod/files/2014/04/f15/INL_Type_B.pdf.

US Department of Energy, March 14, 2006d. East Tennessee Technology Park (ETTP) Site Inactive Steam Line Asbestos Spill. B-2006-OR-BJCEPC-0302 (Available upon DOE authorization).

US Department of Energy, July 2009. Type B Accident Investigation Report Washington Closure Hanford, LLC Employee Fall Injury on July 11, 2009, at the 336 Building, Hanford Site, Washington. DOE/RL-2009-83. https://www.energy.gov/sites/prod/files/2014/04/f14/board_report_final.pdf.

US Department of Energy, 2010a. Consent Order Incorporating Agreement among the US Department of Energy, the National Nuclear Security Administration, and Lawrence Livermore National Security, LLC, Consent Order 2010-01. https://www.energy.gov/sites/prod/files/hss/Enforcement%20and%20Oversight/Enforcement/docs/consentorders/WCO-2010-01_Consent_Order.pdf.

US Department of Energy, 2010b. Office of Health, Safety and Security, Lessons Learned Database. Lesson ID: 2010-ORNL-SEC-0001 (Available upon DOE Authorization).

US Department of Energy, August 19, 2010c. Letter to Mr. Neil Brosee, President Washington Closure Hanford, LLC 2620 Fermi Avenue Richland, Washington 99354, WEA-2010-02. https://violationtracker.goodjobsfirst.org/archive/245.pdf.

US Department of Energy, 2011. Richland Operations Office, Failure of Filter on Tiger-Vac 4D-22 MR NUC HEPA Vacuum. Lesson ID: 2011-RL-HNF-0046, 11/4/2011 (Available upon DOE Authorization).

US Department of Energy, 2011a. Office of Health, Safety and Security, Lessons Learned Database. Lesson ID: 2011.SR-SRNS-0001, 2/28/2011 (Available upon DOE Authorization).

US Department of Energy, August 8, 2011b. Office of Environmental Management, Accident Investigation Report, Fall Injury Accident at the Savannah River Site on July 1, 2011. https://www.energy.gov/sites/prod/files/2014/04/f14/SRS_Fall_Injury_Report.pdf.

US Department of Energy, 2011c. Grinder Fell from Scaffolding Nearly Striking Employee below, Idaho Cleanup Project EM-ID–CWI-IWTU-2011-0003 (Available upon DOE Authorization).

US Department of Energy, 2011d. Office of Health, Safety and Security, Lessons Learned Database, Hot Cell Fire Due to Size Reducing Activities (ARRA). Lesson ID: TWPC-LLI-2011-04-002, 4/19/2011 (Available upon DOE Authorization).

US Department of Energy, Office of Health, Safety and Security, August 1, 2013. Analysis of integrated safety management at the activity level: work planning and control. In: Work Scope Exceeded during Demolition Resulting in Worker Contacting Energized Electrical Conductor (Available upon DOE Authorization).

US Department of Energy, December 28, 2015. Severe Finger Laceration during Machine Shop Operations (OPEXSHARE available upon DOE Registration).

US Department of Energy, 2016a. Preliminary Notice of Violation (WEA-2016-04). https://www.energy.gov/sites/prod/files/2016/11/f34/UChicago%20Argonne%20PNOV%20%28WEA-2016-04%29_0.pdf.

US Department of Energy, March 2016b. Injuries: Recent Trend in Musculoskeletal Disorders, OE-3: 2016-03. https://www.energy.gov/sites/prod/files/2016/03/f30/OE-3-2016-03.pdf.

US Department of Energy, February 2018. Office of Enterprise Assessments, Follow-Up Assessment of Progress on Actions Taken to Address Tank Vapour Concerns at the Hanford Site. https://www.energy.gov/sites/prod/files/2018/02/f49/EA%20Follow-up%20Assessment%20of%20Progress%20on%20Actions%20Taken%20to%20Address%20Tank%20Vapor%20Concerns%20at%20Hanford.pdf.

US Department of Energy, Office of Health, Safety and Security, June 2019. OPEX Awareness, Dangers in Loading and Unloading Heavy Equipment. https://www.energy.gov/sites/prod/files/2020/12/f81/OPEX_Awareness_Dangers_in_Loading_and_Unloading_Heavy_Equipment.pdf.

US Department of Energy, May 14, 2020a. DOE Awards Hanford Tank Closure Contract. www.energy.gov/em/articles/doe-awards-hanford-tank-closure-contract#:~:text=The%20total%20estimated%20value%20of,%2C%20past%20performance%2C%20and%20cost.

US Department of Energy, April 30, 2020b. Occurrence Reporting and Processing System (ORPS) updated. https://data.doe.gov/MS/asp/Main.aspx.

US Department of Energy, Fernald Closure Project. https://www.lm.doe.gov/land/sites/oh/fernald_orig/50th/p7.htm.

US Nuclear Regulatory Commission, September 2001. Hanford Tank Waste Remediation System Pre-treatment Chemistry and Technology, NUREG/CR-6714. https://www.nrc.gov/docs/ML0126/ML012690529.pdf.

US Nuclear Regulatory Commission, November 2002. Generic environmental impact statement on decommissioning of nuclear facilities, Vol. 1. NUREG-0586, Suppl. 1. https://www.nrc.gov/docs/ML0235/ML023500395.pdf.

US Occupational Safety and Health Administration, 2020. https://www.osha.gov/pls/imis/AccidentSearch.search?acc_keyword=%22Laser%22&keyword_list=on.

Washington River Protection Solutions (WRPS), 2017. Radioactive & Chemical Waste Storage, 2017 and updates. https://hanfordvapors.com/what-is-hanford/radioactive-chemical-waste-storage/.

Further reading

Germany's Federal Office for the Safety of Nuclear Waste Management, Monthly reports on reportable events. In: German, Bundesamt for die Sichereit der nuklearen Entsorgung, Monatsberichte zu meldepflichtigen Ereignissen, 2020. https://www.base.bund.de/EN/ns/events/reports/monthly/monthly_node.html. www.base.bund.de/DE/themen/kt/stoerfallmeldestelle/berichte/monatsberichte/monatsberichte_node.html. (Please note that the monthly reports are only in German).

Chapter 6

Radiological events impacting the workers

6.1 Plutonium exposure incident at Idaho National Laboratory, Idaho, United States

6.1.1 Decommissioning process; problem encountered; impacts (DOE, 2010a)

The Zero Power Physics Reactor (ZPPR) at US DOE's Idaho National Laboratory (INL) was a very-low-power research reactor. Used for testing nuclear reactor designs, it was operated from 1969 to 1992. Following decommissioning, activities at the ZPPR installation include storage of materials in the Vault (mainly in the form of ZPPR fuel plates), handling for surveillance and inspection, and packaging for transport. Workers access the ZPPR installation to inspect and register nuclear materials and to arrange old reactor fuel plates for shipment to outside installations.

Sixteen workers were exposed to plutonium (Pu) in an incident at ZPPR on November 8, 2011. The workers were packaging Pu reactor fuel plates. Two of the fuel storage containers had nonstandard markers pointing out at irregularities with the fuel plates inside. When opening one of the storage containers—a routine, scheduled work—the workers noticed a Pu fuel plate packaged in plastic and bands. When the workers tried to remove the packaging, a sudden release of radioactive substances unexpectedly occurred in the ZPPR Workroom, which contaminated the workers and the surfaces nearby.

Based on external surveying, low-level contamination was detected on six workers, two of them fairly extensively, and all of them underwent whole body scans to evaluate internal exposure. Additionally, the workers were offered treatments to accelerate their bodies' elimination of plutonium. As no contamination was detected outside of the ZPPR installation, there was no risk to the public or the environment.

Early investigations indicate intakes 0—5 times the yearly limit. Intakes of this level are equivalent to doses 0–2.5 Sv committed equivalent dose to the bone (roughly 0–0.08 Sv committed effective dose to the total body). Later evaluations led to the best estimate of the bounding committed equivalent dose

Nuclear Decommissioning Case Studies. https://doi.org/10.1016/B978-0-12-823700-7.00006-0
Copyright © 2021 Elsevier Inc. All rights reserved.

(to bone) being less than 0.71 Sv, with two workers likely to have exceeded the 0.50 Sv yearly limit to a single organ.

This event is representative of situations occurring in a decommissioned plant, which is being reused for handling and storage of nuclear facilities: In fact, such situations would be typical of a decommissioning project proper. See similar cases in Sections 6.4, 6.19 etc.

6.1.2 Analysis; solutions

In the months that followed the accident, it was determined that the fuel plate that caused the accident was already damaged prior to being stored over 30 years before. This damage caused a rupture of the steel sheath, which allowed air and moisture to penetrate the fuel plate and react with the transuranics. Some of the compounds that resulted from the reactions were gaseous and were released in an uncontrolled manner during the fuel handling activities in the workroom.

"The Accident Investigation Board determined that the local root causes were as follows:

- The DOE's management and operating contractor (Battelle Energy Alliance, BEA) did not accurately analyze the Pu hazard in the safety basis and establish commensurate controls.
- The management system lacked requirements intended to influence the decision making of the nuclear facility manager and shift supervisor, resulting in a single-point decision to cut the wrapping.

The Board determined the following systemic root causes:

- DOE—Idaho accepted the risk of known safety basis deficiencies and allowed continued operation of the ZPPR Facility within the framework of a multiyear safety basis upgrade plan without putting effective interim controls in place.
- BEA continued operation of the ZPPR Facility with known safety basis deficiencies and without adequately analyzing the hazard to the worker or establishing effective work control processes.

Contributing causes are events or conditions that collectively with other causes increased the likelihood of the accident but that individually did not cause the accident. The Board identified three contributing causes to this accident:

1. The organizational transition resulted in a loss of knowledge and past practices and records that indicated the conditions associated with the fuel plates.

2. Senior Materials and Fuels Complex (MFC) management did not recognize the significance of information provided by the history of Pu fuel plate failures and by the MFC Independent Safety Review Committee Chairman's white paper.
3. The process worksheet used to conduct the work did not contain directions governed by any of the referenced operating instructions, leading to the creation of work steps without an appropriate hazard analysis or accompanying means of mitigation" (DOE, 2011).

6.1.3 Lessons learned

Important lessons learned of generic relevance include the following:

The supervision systems of the responsible organizations should be managed to the effect that they could identify and fix legacy inadequacies in the technical bases supporting the safety basis of the activities being carried out.
The work planning process should include a comprehensive review of available historical data and identify the need for more data or extra contingencies.
The work team should realize when operations are outside the boundaries of the applicable radiation work permit or are procedurally undefined: If so, they should stop work. The training provided to the workers should warn them to stop work when they face unexpected circumstances.
Self-assessments of the operating organization should be conducted in a way to identify those issues whose resolution would greatly improve safety and performance.

As long as work goes on in some areas of shutdown installations, hazards are there and workers' safety should be accurately checked and evaluated in all areas, including those areas contiguous to where active work is underway.

6.2 Radiological work permit suspension guide exceeded, Savannah River Site, US, August 5, 2010 (DOE, 2010a)

6.2.1 Decommissioning process; problem encountered; impacts

The event in question has produced no significant impacts to the workers, the public, or the environment but had the potential to cause more damage. Area Completion Projects (ACP) Deactivation and Decommissioning (D&D) workers at the Savannah River Site (SRS) identified a small leak coming from a basin water line while they were in the process of removing the lead shielding jacket from the line. The work was being performed in a contamination area/radiation area (CA/RA) in accordance with work package "105-R: Disposition Disassembly Decon Room Equipment/Piping" and associated radiological work permit (RWP). The workers were dressed for the CA/RA work and were wearing respiratory equipment for the lead work and also personal air samplers (PAS).

The lead was being removed by a D&D work crew trained and experienced in lead removal. The workers were cutting the lead shielding from the piping and equipment using a reciprocating saw with a wood/nail blade. This crew has used this same method for removing lead in various areas of the site without any issues for several years and was using it for the fourth day for this particular job. This particular blade was thought to be durable enough to cut the soft lead metal, but the spacing of teeth width was wide enough apart so that it was not expected to significantly damage the underlying carbon steel piping.

The lead shielding in this area varies in thickness from approximately 1.8 cm on some piping to 15 cm around valve bodies and is formed to the pipe and has a sealed seam, so that it is not easily removed. The lead thickness was approximately 3.0 cm thick for the specific piping that leaked. Due to the thickness of the lead, it was being removed in small sections and size reduced as it was removed from the piping. The workers had made cuts around the circumference of the piping at each end of the section and had just started the longitudinal cut to separate the lead from the pipe when the trickle of liquid appeared. The workers had noticed some rusty areas on some of the piping that had previously been exposed but had not seen any evidence of liquid before this event.

As soon as the trickle of liquid appeared, the worker called out for one of the radiological protection (RP) inspectors and moved an aluminum trash can under the leak, where most of the liquid was caught. The trash can had some absorbent pads in it. The inspector took a wet swipe of the liquid and determined that it read <0.03 Bq/cm^2 α and 50 Bq/cm^2 β-γ. This exceeded the suspension guide of 15 Bq/cm^2 for this specific task on the RWP. A timeout was called, and the workers exited from the area. Supervision (D&D and RP) was notified, and the personnel doffed their protective clothing and were surveyed out of the area. The affected workers were also escorted to undergo additional monitoring. As the workers were in anticontamination clothing and already had a respirator for lead dust, no personnel contamination was found. RP reposted the area as a high contamination area (HCA) and relocated the motor air pump sampling at the high-efficiency particulate air (HEPA) filter exhaust to the room where the leak occurred.

Corrective actions implemented by ACP D&D to correct the situation and prevent recurrence included the following:

- Change the work package to address draining of all lines prior to continuing lead removal activities and to include leak collection controls for additional lead removal to ensure residual liquids were captured if a similar breach occurred.

- Engineering evaluation has recommended the following: (1) use of a different blade (wood blade with fewer teeth per centimeters) that will result in the dulling of the blade prior to breaching the carbon steel piping or (2) use of the existing blade with a change in method to use the toe end of the blade until the actual depth of the lead can be determined.
- Prepare and submit Lessons Learned to the site Operating Experience Coordinator.

6.2.2 Analysis; solutions

Engineering review indicates that the piping in question displayed some surface corrosion but was sound and that the wood/nail blade in use did breach the piping. The employee inadvertently cut the line with the reciprocating saw while making the cuts to remove the lead from the line. This happened because of the work crew's previous successful lead removal operations and their perception that the blade being used would cut through the lead but not the underlying pipe.

6.2.3 Lessons learned

Personnel should not allow previous successful operations with a particular method or tool to decrease their awareness or focus on the task at hand.

When performing decommissioning or similar work, personnel should review differences in the type of work being performed from previous experiences to ensure that those differences are understood and any hazards associated with them properly analyzed for implementation of appropriate controls. See also Box 6.1 for a similar accident.

6.3 Radioactive liquid unexpectedly found during dismantling (IAEA, 2016)

You will find it a very good practice always to verify your references, sir.

Martin Joseph Routh (1755–1854)

6.3.1 Decommissioning process; problem encountered; impacts

During breaking down a valve, a liquid spill unexpectedly occurred, which contaminated the worker and the premises.

6.3.2 Analysis; solutions

The facility records did not indicate any residual liquid in the system being dismantled.

BOX 6.1 Multiple radioactive waste spills at Savannah River Site (Knoxblogs, 2010)

Another incident similar to the one described earlier happened at Savannah River Site (SRS) in July 2010, when a waste drum containing ^{238}Pu was being moved and some 50 L of contaminated liquid spurted from the bottom. The drum was packed in 1974 and inserted in a carbon steel "overpack" in 2008.

However, the major concern was not about the break of the container, but the slow response of SRS workers.

The area was evacuated at once: Monitors later measured contamination levels so high that cleanup teams had to suspend work. It took 3 days for workers wearing full personal protective equipment to continue cleanup work. The accident report concluded that it took too much time to gather the resources to tackle the accident.

In this regard, the report stated "While no workers have been contaminated, they were not dressed adequately during the initial two entries for the contamination present. This prevented the facility from performing meaningful mitigation while the contamination was still wet and less dispersible."

Instead of immediately cleaning up the contaminated area, 3 days were spent identifying needed equipment and trained workers for the decontamination task. Apparently, recovery work did not begin until it was directly solicited by the top management.

To obviate these problems, several options should be pursued. Firstly, the history of the system should be reviewed accurately. Secondly the layout of the system should be checked to ascertain whether there are any drainage lines. And lastly dismantling procedures should specify how to ensure that the system is empty before it is opened.

6.3.3 Lessons learned

In many incidents, the lack/insufficiency of design, construction, or operations records was the root cause or contributed to the occurrence reported in this book.

Regardless of case studies specifically mentioning missing or incomplete records, it is good practice in decommissioning to verify the availability and reliability of relevant records in drafting procedures for decommissioning tasks. In lack of the detail, plans should be developed to produce or reconstruct missing information. See also Box 6.2.

BOX 6.2 Some personal experiences: poor historical records of material used, stored, or removed

"We had a major project to decommission a redundant facility with a particularly challenging inventory of nuclear material. Whilst significant precautions were in place ensuring the safety of the personnel and the environment, this plan was based on dry solid contamination. During a routine entry the operative noticed the sight glass on a redundant safe shape liquid storage tank appeared to contain liquid" (Kerr, 2008).

6.4 Workers handling unidentified materials could be exposed, Lawrence Livermore National Laboratory, California, United States (IAEA, 2016)

6.4.1 Decommissioning process; problem encountered; impacts

Workers demolishing a wall at Lawrence Livermore National Laboratory (LLNL) suddenly found some half a kilogram of an unspecified white powder inside a wall hole. The shift supervisor sought advice from the environmental specialist and the waste workman but did not contact the industrial hygienist. The supervisor requested the waste workman to brush away the powder and disposition it into a waste container. The workman doing this operation did not wear personal protective equipment (PPE) and did not know the origin of the powder. The day after, the supervisor requested the waste worker to vacuum up what remained of the previously cleaned-up substance. The worker utilized a vacuum cleaner from a hardware store. Again the hygienist was not contacted. As the material swept was dry and the vacuum cleaner was not equipped with an HEPA filter, some of the white powder was released into the air.

Two contractors were near the airborne substance. Management ordered all workers involved in this event to consult a doctor. No sanitary consequences were reported.

A similar event occurred at LLNL on an on-site parking lot. A worker found a small plastic bag with some hundred grams of white powder inside. The bag had no information printed. The worker took the bag, looked inside, smelled the substance, put it back in the parking lot, and went to the office. Later, on second thought, the worker reported the event, as he was worried that the substance might have been noxious. The substance actually was analyzed and found innocuous; however, had it not be the case, the worker and possibly others could have been subject to unplanned risks.

6.4.2 Analysis; solutions

In both cases, the workers handled unidentified materials for which no prior hazard assessment had been conducted, they did not wear PPE and used improper tools and procedures. Consequently, they posed undue risks to themselves and others from exposure to unknown substances, which could have been noxious. LLNL has trained personnel and procedures to deal with unidentified substances: Unfortunately, the workers involved in these events did not take advantage of these opportunities.

6.4.3 Lessons learned

Not all unexpected events in decommissioning result in damage to people or the environment; nonetheless, they offer lessons that should be considered in planning and implementing decommissioning activities.

When coming across an unidentified substance, workers should not handle it prior to seeking advice from health and safety staff; any further operations on the substance will be based on a hazard assessment and the controls needed to protect the workers and others.

6.5 Americium uptake during the decommissioning of the experimental boiling water reactor, Argonne National Laboratory, United States

6.5.1 Decommissioning process; problem encountered; impacts (Fellhauer, 1998)

The 100-MWt experimental boiling water reactor (EBWR) went first critical in 1957 and ceased operations in 1967 (Fig. 6.1). The decommissioning project duration was 118 months. The project was initiated in April 1986, and the final decommissioning report was released in February 1996. The total collective exposure to the project workforce was 208.7 person-mSv, nobody exceeded the individual project limit of 15 mSv. After the final survey, the EBWR Facility was released as a "radiologically controlled area" (RCA), as residual elevated activity remained in inaccessible areas. Instead, radiation levels in accessible areas are background levels. Therefore, the personnel working in accessible areas do not need any radiological controls.

Three radiological characterizations of the installation were performed in the period 1986–92. ^{60}Co was the only isotope detected.

In September 1994, preliminary bioassay results indicated that seven workers had detectable uptakes of ^{241}Am and minor uptakes of ^{137}Cs and ^{3}H. Following confirmation of bioassay results, decommissioning work inside the EBWR shell was stopped. Investigations and health physics surveys were performed to identify the source of ^{241}Am. Analysis of air samples taken in July 1994 near the fuel pool during control rod cutting activities indicated ^{137}Cs and transuranics (^{241}Am, ^{238}Pu, and ^{239}Pu). Previous characterizations had not indicated these radionuclides.

FIGURE 6.1 Experimental boiling water reactor (EBWR) at Argonne National LABORATORY. *(Courtesy of USNRC, Wiki Commons).*

6.5.2 Analysis; solutions

The use of bioassays, daily air samples, and administrative control sheets tracking personnel movements allowed an accurate reconstruction of the events. Six of the seven affected workers were involved with plasma arc operations in or near the fuel pool. The uptake occurred over a period of 4 days, with the largest uptake in individuals working longest with cutting activities. The americium source is still uncertain. It may have been a product of the ^{241}Pu foil lost during EBWR experiments held in 1967: The ^{241}Pu would have decayed to ^{241}Am; however, no trace of ^{241}Pu was found. Another scenario is that although no fuel element failure was reported at EBWR, microscopic cracks may have allowed the release of transuranics.

Two contractors received some 3 mSv from the americium uptake. The other workers are estimated to have received 500–600 μSv from this event.

Based on these investigations, decommissioning activities were allowed in November 1994 to continue.

6.5.3 Lessons learned

Perform monitoring aimed at radionuclides that can be expected based on plant history, even if they are not found during initial characterization.

Accurate knowledge of historical operations is essential in characterization, especially at research and prototype installations (see Box 6.3).

Be conservative in adopting PPE.

BOX 6.3 Some personal experiences: changes resulting from decommissioning

"During a facility clear out, higher than average levels of activity were detected on the personal air sampler of an operative working in the corridor just outside a room where waste processing activities were being carried out. Investigation revealed that the ventilation flowed from the laboratory out to the corridor rather than the other way. The poor facility ventilation combined with minor changes to extracts that had taken place during decommissioning and slightly increased airborne levels due to waste processing resulted in a mechanism for an unplanned exposure route (Kerr, 2008)."

Increase the number of bioassays.

Make sure bioassay findings are given to the responsible managers in a timely manner.

Good-quality air monitoring and dosimetry equipment is advisable.

Investigative committees should be preselected, trained, and fully dedicated to the requested investigation.

Spare analytic capability is necessary for sample analysis in an emergency.

Consensus is needed on trading off internal exposure against other health and cost parameters.

Entry and exit bioassay data are very important.

Keeping air samples is vital to dose and event reconstruction.

Accurate records help understand the cause and evolution of events.

6.6 Conflicting work documents result in workers entering contaminated area, Hanford Site, United States (DOE, 2012a)

Be not careless in deeds, nor confused in words, nor rambling in thought.

Marcus Aurelius Antoninus (121–180), Meditations

6.6.1 Decommissioning process; problem encountered; impacts

At Hanford, the 100N area covers some 9 km^2. As part of the overall Hanford decommissioning and cleanup program, 148 facilities were demolished in this area by 2012. An inactive facility, the 105N/109N Safe Storage Enclosure, is also located in this area. N-Reactor (the 105N Reactor and the 109N Heat Exchanger Building) reached the Interim Safe Storage (ISS) state in 2012. ISS comprises the Fuel Storage Basin (FSB), auxiliary support facilities, and most parts of the shield wall encasing the 105N Reactor. The 109N Building was

FIGURE 6.2 Aerial Photo of 105-N109-N Reactor Interim Safe Storage (ISS) at completion of decommissioning, Jan 2013. Note the excavations where the facilities formerly surrounding N reactor have been removed: now only ISS remains *(T. T. Yamamoto and J. P. Delaney—WCH-567, Washington Closure Hanford, Wiki Commons).*

dismantled up to the steam generator cells. A new steel cover was placed over the residual structures (Fig. 6.2).

On April 27, 2012, Washington Closure Hanford (WCH) work personnel were involved in demolition activities at the 100N area. While installing a rope and radiological postings around an area of soil that had been placed over the FSB HCA, two decommissioning (D&D) workers entered the FSB area without lapel samplers as required by the FSB RWP.

A heavy equipment operative (HEO) and two D&D workers attended a prejob brief for work in the FSB. The purpose of the prejob brief was to ensure that the responsible person and the workers assigned to the job understand the scope of the work through a comprehensive discussion. To this end, workers needed to fully understand what was expected of them and what was to be avoided. Part of the brief was to instruct the workers on the posting of new airborne radiation area/high contamination area/high radiation area (ARA/HCA/HRA) boundaries. The workers then left the FSB prejob brief and attended a brief at the 1303N for work involving pushing material from the 1303N CA into the FSB HCA. Lapel sampling is required for all HCA entries.

Two radiological control technicians (RCTs) entered the FSB with the two D&D workers to install T-posts, rope, and signs. As the D&D workers were exiting the area, one of the RCTs remaining in the area requested assistance from a D&D worker. One of the D&D workers had doffed their PPE; the second was still in PPE but had removed the lapel sampler. An RCT was at the step off pad. The D&D worker asked the RCT if a lapel was required to go back into the area. After the RCT consulted the lead radiological technician (RadTech), the D&D worker was informed that a lapel was not required. The D&D worker entered the area to change some radiological postings that were incorrect and exited. It is important to note that the RWP for 1303N does *not* require a lapel for nonintrusive work.

The HEOs proceeded to push dirt from the CA into the HCA. The HEOs were wearing lapel samplers in accordance with the RWP. After the HEOs completed their work, the D&D workers prepared to enter the FSB to install T-posts around the area covered by the HEOs. The D&D workers asked the 1303N RCT if lapel samplers were required. The 1303N RCT called a lead RCT. The lead RCT, without reviewing the RWP, stated that lapel samplers were not required. The D&D workers then entered the area and completed the radiological postings.

On April 28, 2012, pre-job briefs were held for the FSB and for 1303N. The scope of the work was similar to the work performed the day before. During the 1303N brief, it was asked if lapel samplers were required. It was confirmed that lapel sampling was required for all entries into the CA and HCA under a different RWP. Work was halted, and a fact-finding task was scheduled. Key points regarding prejob briefings are given in LANL (2010).

6.6.2 Analysis; solutions

Multiple work crews working to different work documents with conflicting requirements resulted in workers entering a contaminated area without proper access controls. The event described here could have resulted in inadvertent contamination of the workers.

Under these circumstances, prejob briefs need to be sufficiently detailed to ensure all personnel in attendance know the scope of work, required PPE, and the work documents associated with the activities. Supporting work documents should be periodically reviewed to ensure there are no conflicts or ambiguities.

RCTs and industrial hygiene technicians (IHTs) are an important resource in the field and should be able to answer the questions asked of them by the workers or know when to forward the questions to supervisors. The supervisors and technicians should have sufficient knowledge of the work area, work scope, and supporting documentation and make necessary inquiries prior to answering questions. When looking to supervisors for direction, all reference information needs to be passed to the supervisors before making a decision.

When performing work during overtime, it is important to ensure that personnel who are not familiar with the worksite completely understand the scope of work, the location, boundaries, PPE, and supporting documentation. A questioning attitude is encouraged: Personnel should also confirm they completely understand the information presented to them.

6.6.3 Lessons learned

During decommissioning work, especially when it nears completion, it is inevitable that work groups with different work scope and goals will be working in close proximity, possibly leading to overlapping. Care should be taken so that there are no conflicting approaches or requirements.

6.7 Work control must address latent field conditions, Oak Ridge Site, United States (IAEA, 2006)

> *God, give us grace to accept with serenity the things that cannot be changed, courage to change the things which should be changed, and the wisdom to distinguish the one from the other.*
>
> Reinhold Niebuhr (1892–1971)

6.7.1 Decommissioning project; problem encountered; impacts

In July 1995, at the initial start of a construction project, work activities involving the removal of five underground waste and process lines to facilitate installation of a new tank vault were initiated. Initial site characterization identified the pipe line contents to be typical waste/sludge mixture of α and β-γ isotopes. Additionally, the surrounding soil was identified to be potentially category 2 soil (contaminated but below 50 μSv/hr dose rate). Field surveys by the construction manager (CM) health physics (HP) technicians, during the initial excavation, identified the soil surrounding the pipe lines to have higher dose rates, which prompted recategorizing the soil as category 3 (i.e., contaminated and greater than 50 μSv/hr dose rate). The presence of CM HP technicians is a requirement during excavation activities until such time as initial characterization results can be confirmed or as otherwise might be required by radiological conditions. The discovery of the category 3 soil required the CM to accelerate the development of its as low as reasonably achievable (ALARA) plan to address excavation activities in addition to the plan's previous scope of addressing pipe cutting operations. Also, during the initial excavation, a drain line was found in a location different than that shown on the design drawings. The CM work control process requires the cessation of relevant activities when underground items not adequately identified by as-built drawings are encountered. The construction workers recognized the hazards and stopped work immediately. Plant personnel were contacted for additional information, prior to work restart.

6.7.2 Analysis; solutions

In this circumstance, a greater potential existed for personnel contamination from the category 3 soil than was originally thought to have existed. The CM had a work control process in place to address and account for changes in field conditions such as inadequate soil characterizations and the legacy problem with as-built drawings not being updated. This work control process also allowed the CM to stop field activities until the unidentified underground items could be properly identified and characterized.

6.7.3 Lessons learned

Potential accidents such as human injury or fatality, equipment damage, and/or costly project delays are avoided when a work control process is established to compensate for field work conditions, which often differ from the initial site characterization, as-built drawings, etc.

6.8 Incident in the ATPu nuclear facility, Cadarache CEA site (ASN, 2009)

6.8.1 Decommissioning process; problem encountered; impacts

The French Alternative Energies and Atomic Energy Commission (CEA) began construction of the Plutonium Technology Workshops (ATPu) at Cadarache in 1959, and production of plutonium fuel for fast neutron reactors was initiated in 1962. The Chemical Purification Laboratory (LPC) began operation in 1966. Production of plutonium fuel for the Phenix and Super-Phenix fast breeder reactors started in 1978. The manufacture of MOX (mixed uranium and plutonium oxide) fuel for light water reactors started at Cadarache in 1989. MOX production was taken over by Areva in 1991, but in 2000, Areva moved MOX production to its Melox plant at Marcoule.

Cadarache produced the last batch of MOX fuel rods in July 2003. The ATPu was shut down in June 2008. Over its 40 year operation, it had recycled more than 50 t of plutonium. CEA holds the ATPu's license, and AREVA runs the plant.

The Decree dated March 6, 2009, authorized the decommissioning of the ATPu. Dismantling operations were launched. Operational waste generated by ATPu's past activities had been removed prior to the start of dismantling activities. The waste removal ended in June 2008. Up to March 2017, more than 450 glove boxes, 30 tanks, and 4 km of piping (the equivalent of 30,000 drums) have been segmented and conditioned for shipment to storage or disposal. Around 60 cleared process areas have been returned to the CEA, the owner of the Cadarache center (WNN, 2017). The ATPu is currently nearing completion of the decommissioning process.

The incident that occurred at the ATPu was announced by the CEA on October 6, 2009. CEA notified the French Nuclear Safety Authority (ASN) of the underestimation of plutonium quantities in the plant glove boxes. While plutonium quantities were estimated at around 8 kg during plant operation, they were later estimated at around 22 kg. CEA also warned at the time the issue was disclosed that the total plutonium amounts could be as high as 39 kg.

Immediately the ASN Commission suspended the ATPu dismantling activities accountability. Dismantling activities were discontinued until prior approval was granted by ASN. An inspection conducted by ASN stated that CEA has been aware of this situation since June 2009, several months before it was publicly announced.

Investigations disclosed that the origin of the event was the accumulation between 1966 and 2004 of fine plutonium dust in the workshop glove boxes. Later inspections disclosed that another Cadarache facility, STAR, also housed plutonium exceeding the licensed quantities. The incident was classified INES level 1 by the French regulator. The incident has had no consequences.

6.8.2 Analysis; solutions

The underestimation of plutonium amounts has produced a considerable reduction in the security margins allowed by the plant design to prevent a criticality accident, whose potential consequences could be dangerous to workers.

ASN considered that the unawareness of this underestimation during the plant operation, as well as the late declaration of the incident to ASN, indicates a significant lack in safety culture from the plant licensee and operative. This led the ASN to rerating this incident as INES level 2 on October 12, 2009.

6.8.3 Lessons learned

A frequent problem during decommissioning is the relaxation of safety culture due to complacency or the wrong perception of decommissioning as a less challenging process than nuclear operation. This case study highlights increased chances of a criticality event. Similar cases are given in Boxes 6.4A and 6.4B.

In transitioning from operations to D&D, the following points should be taken in due account:

- Ensure full understanding of the impacts of criticality accidents by all involved employees.
- Ensure controls are in place and strictly observed for operations with fissile materials. As a prerequisite, this requires the workers clearly understand why the controls are necessary and important.
- Ensure adequate monitoring of operations with fissile material by supervisors, management, and regulators.

BOX 6.4A Violation of criticality rules at decommissioning EURODIF enrichment plant, Tricastin site, France (WISE, 2020)

A few relevant cases are collected in the following.

On July 29, 2014, humid air was incidentally introduced into a group of gaseous diffusion chambers being decommissioned. This could have presented a criticality risk and the risk of formation of hydrofluoric acid.

On September 15, 2015, AREVA revealed that ashes resulting from the burning of oil containing traces of uranium were being collected in bins with a capacity of 200 L instead of the permitted 30 L bins.

On October 16, 2015, ASN announced that during reconditioning of waste containers held at the Areva Tricastin site since 1996, excessive contents of ^{235}U were found in two containers. Note that the origin of this event predates Eurodif decommissioning, although it was discovered during the decommissioning phase.

All of the above events were rated level 1 on the INES scale.

Although these cases posed no direct impacts to either the workers or the public, they could be representative of poor safety culture, which is potentially hazardous especially when fissile material is dealt with.

BOX 6.4B Estimation of fissile material quantity in a dissolver, France (IAEA, 2004)

Decommissioning process; problem encountered; impacts

A continuous process was used for the dissolution of reactor spent fuel in the reprocessing plant. The basic decontamination parameters stemmed from values obtained many years before when a dissolver was replaced and dose rate measurements were taken. A first rinse using nitric acid was conducted without any prior chemical analysis of the deposits at the bottom of the dissolver: The quantity of plutonium removed turned out to be much higher than the estimated amount.

Analysis; solutions

This discrepancy in the estimation of fissile material in the dissolver, when it was discovered, led the operative to quickly redefine a characterization program to determine the physical, chemical, and radiological features of the deposit and the remaining quantity of plutonium. Nondestructive assays such as active and passive neutron measurements, as well as video and sampling analysis, were used.

Lesson learned

The problem encountered during the rinsing of the continuous dissolver came from a poor knowledge of the initial state, mainly from a lack of quantitative and qualitative characterization of the deposits inside the dissolvers. The approach was then taken to monitor the initial state of the dissolver and to carefully characterize the deposits. If done initially, this should have allowed an accurate estimate of the amount of residual fissile material and to adjust the decontamination solutions and setup to the circumstances. The use of incorrect initial parameters affected the level of trust of the regulatory body supervising this project. There was also a considerable impact on the rescheduling of activities.

6.9 Predecommissioning work in a gas-cooled reactor fuel pool, Vandellós Nuclear Power Plant, Spain (IAEA, 2015)

6.9.1 Decommissioning project; problem encountered; impacts

The Vandellós I Nuclear Power Plant (NPP) was first coupled to the electricity grid in March 1972. The NPP was permanently retired from service as a result of the fire that occurred in the turbine alternator in October 1989. After the plant's permanent shutdown, the owner performed the defueling of the reactor, the removal of spent fuel, the management of the operational waste, and the removal of waste stored in graphite silos. They also took apart the CO_2 tanks and the turboalternator. The incident described in this case study belongs to these activities (1991–97). The second stage of decommissioning was conducted by a new licensee (ENRESA) in 1998–2003 and involved the dismantling of all nonradioactive and radioactive structures, components, and systems (SSC) except for the reactor pressure vessel, which was sealed and encased in an ad hoc protective structure. The spent fuel pool building was dismantled during July 1999–February 2002. The plant reached in 2003 a safe enclosure state: Final dismantling operations and site release are postponed to 2028.

At Vandellós I NPP, work was underway on the reactor spent fuel pool in February 1996. The work comprised the emptying of the fuel pool water, the removal of the pool's items, and the decontamination of the surfaces and the building. To avoid airborne contamination, routine monitoring by a particle filter was in use, and the workers were protected with particle filter masks. Notwithstanding the use of masks equipped with filters, internal contamination due to ^{241}Am was discovered in 9 out of 13 workers attending decontamination work: Four workers received internal doses up to 40% of the annual limit, whereas five workers received lower doses. The event was discovered 4 days after, when air sampling filters were checked. After this event, cutting and decontamination work required to be completed inside a confinement, with the workers wearing bubble suits. In addition, surface contamination monitoring was installed at the exit of the building, and the workers were directed to take a nasal smear to ensure the absence of α contamination.

As more material was removed from the pool and the water level went down, airborne contamination increased again. Therefore, all workers entering the building were required to wear bubble suits.

According to point IV-5.1.4 of the INES User Manual part IV (Section 4.1) (IAEA, 2013) when the radiation control and administrative procedures are insufficient and workers are subjected to unplanned exposures (internal or external), such events may be rated level 1: In this case, there were remarkable failures in work procedures that rated this event at level 1.

In August 1996, nasal smears showed ^{241}Am contamination. This incident occurred when a radiation protection worker was helping coworkers remove their protective clothing. This incident prompted the adoption of procedures requiring all radiation protection staff to wear protective masks when helping workers undress after working in controlled areas.

Another event is dated March 26, 1997. Within the framework of Vandellós decommissioning work in the spent fuel pool hall, a specialist worker using autonomous respiration equipment changed the vinyl cap of an underwater radiation detector. During the routine contamination checkout at the exit, a nasal smear indicated 0.048 cps α. The worker was treated with DTPA (diethylene triamine penta acetate) for reducing internal contamination. DTPA is a drug that can bind to plutonium, americium, and curium to reduce the time to expel these radioactive elements from the body. Afterward, he passed through a bioassay program that took several months and concluded that (1) effective dose to the whole body was estimated at 31 mSv and (2) internal integrated dose to bone surface was estimated at 576 mSv. Dose (1) is well within statutory limits; regarding dose (2), bones are organs unspecified in ICRP-60, the standard Spain adhered to.

The following justification was given for the INES rating. The on-site criterion is not applicable as no statutory limit was reached. The event was rated INES level 1 on the basis of "defense-in-depth" criterion, in application of bullet A.6 of the INES User's Manual Addendum "Events requiring medical treatment of workers." The related point of IAEA (2013) states where any surgical or medical treatment (in the current case, DTPA) is necessary to prevent an exposure significantly in excess of an annual dose limit, the event should be rated at least at level 1 under the "defence-in-depth" criterion (LAKA, 1997).

In April 1997, after a positive nasal smear of several workers, personnel were prescribed not to reuse the bubble suit when exiting the building and to retain a protective mask until full skin monitoring was completed at the area boundary. The bioassays associated with the incident showed no internal contamination.

To counteract the increased airborne activity, the ventilation system was also altered. The building was originally designed to operate at overpressure conditions. During decontamination operations, environmental surveys showed no contamination outside the building. However, as a measure of caution, the flow of air was reversed. Prejob dose estimates had been 60 person-mSv, but the actual doses were much higher, namely 240 person-mSv, due to the aforementioned events.

6.9.2 Lessons learned

Important lessons learned from this event include the following:

- Actinides are very radiotoxic; a 200 Bq intake is equivalent to 50 mSv (at the time of the events described, the annual dose limit for professional workers).
- Airborne actinides require the use of enhanced respiratory protection equipment.
- The undressing phase is vulnerable to internal contamination incidents if it is not managed accurately.
- Nasal smears on exiting radiation controlled areas are effective to aid early detection of contamination.
- Monitoring low activity concentrations in air can be difficult due to interference by radon and other airborne naturally occurring radioisotopes.

The possibility of airborne generation of radioactive particles should be considered at all times during decommissioning work, and protective measures should be adequate to the working conditions and environmental contamination.

6.10 Worker alpha contamination halted pond work, Trawsfynydd Nuclear Power Plant, United Kingdom (IAEA, 2015)

6.10.1 Decommissioning process; problem encountered; impacts

Trawsfynydd NPP is a defunct Magnox power station located in Wales, United Kingdom. The plant became operational in 1965 and was shut down in 1991. Work to fully decommission the plant and release the site is planned to take almost 100 years. In 1993, the spent fuel rods started to be removed from both Trawsfynydd reactors and dispatched to Sellafield for reprocessing. Fuel removal was completed in 1997. Intermediate-level waste (ILW) has been remotely removed by robots and is stored in an ad hoc building on-site pending the long-term availability of an ILW disposal facility. The reactor cores will not be removed until 2071. By 2083, the area is expected to be available for new uses.

Decommissioning of the Trawsfynydd NPP cooling pond was temporarily halted in May 1999, after workers in respirators scraping contaminated concrete surfaces were discovered to have inhaled α-contaminated dust. The affected workers were subjected to whole body monitoring, and the use of old and perhaps slightly contaminated respirators was forbidden.

The bioassays from the two most affected workers showed that they had received a dose of around 4 mSv, less than 10% of the annual dose limit. No radioactive substances were released outside the pond building.

6.10.2 Analysis; solutions

The regulator, Nuclear Inspection Inspectorate (NII), issued a preliminary charge notice on the then licensee, British Nuclear Fuels Limited (BNFL), based on procedural mistakes and concern that the incident was otherwise going to be repeated. BNFL signaled the NII that contamination had been found on respirators after they had supposedly been cleaned in the laundry. Provisions were made by the licensee to procure only brand new respirators pending further inquiries. The BNFL related the respirator incident to poor barrier procedures and cross-contamination in the laundry. The NII did not take further action because the licensee had satisfactorily reacted.

6.10.3 Lessons learned

This case reiterates that the environment surrounding the spent fuel pond is prone to produce airborne contamination, especially during decontamination work, and protective measures should be planned, enforced, and periodically audited.

6.11 Unplanned exposure during fuel pool decontamination, West Valley, United States (IAEA, 2015)

But times do change and move continually.

Herbert Spencer (1820–1903)

6.11.1 Decommissioning process; problem encountered; impacts

Workers received unplanned doses from the uptake of airborne contamination while decontaminating fuel receiving and storage (FRS) pool equipment. While components were being lifted from the pool, they were sprayed with utility water to remove attached contamination: However, this method caused the spread of airborne contamination. On May 10, 2002, as a lift rack was being removed from the pool, loose sediment was seen in the bottom channel of the lift rack. Similarly to the work already carried out the month before on the bottom section of the gate, a worker used the utility hose to spray off the sediment. The spray was oriented toward the bridge crane operative, who was raising the rack. A radiation protection technician and other workers were nearby. By checking the strip chart, it was observed that all three continuous air monitors (CAMs) exhibited an increase of 200 counts/min around the time the spraying was recorded. However, this increase was not high enough to exceed the low CAM set points.

On May 24, 2002, during the yearly total body count, ^{137}Cs was found in the radiation protection technician who worked in the FRS plant during the aforementioned incident. Moreover on June 17, 2002, a worker who had worked in the FRS in May was also found to have ^{137}Cs during his yearly total body count. On July 2, 2002, West Valley Nuclear Services Company received data pointing to two decommissioning workers having received a single dose exceeding 1 mSv from airborne contamination, which was below West Valley's annual administrative limit. Preliminary estimates indicated an exposure of around 2 and 1.65 mSv. Although the exposure prompted an abnormal occurrence report, it was well below the West Valley Demonstration Project (WVDP) annual administrative limit of 5 mSv.

6.11.2 Analysis; solutions

Because of the special aspects of the incident and the difficulties in identifying the specific cause of the doses, an independent assessment was conducted to determine direct, contributing, and root causes. The assessment focused on the activities in the FRS plant during the period when a rise in airborne readings was recorded by CAMs and during which workers received doses as measured through yearly total body monitoring. It was considered most probable that the exposures occurred either during the removal of the lift rack or during decontamination of the pool gate, which were similar activities. This conclusion was also confirmed by an analysis of the physical and radiological parameters of the pool water.

6.11.3 Lessons learned

The following actions should be taken in a similar case:

- Strengthen hazard assessment and monitoring when changes in foreseen conditions are encountered (e.g., modifications in pool water clarity or components with structural features accumulating sediments).
- Use low pressure jets to rinse (not decontaminate) components to reduce the chances that contaminated water is transported and goes airborne.
- Make sure that all decontamination be effected underwater.
- Prescribe respiratory protection for operations where pool water is stirred (e.g., spraying, equipment removal), unless a specific assessment proves that respiratory protection is not needed.

Changes in environmental conditions that may alter the hazards involved are often ignored. In this case study, historical data from sample findings, experience from previous work, and foreseen working conditions were assumed sufficient to draft the work procedure. Attention was not paid to the changing water chemistry and the possibility of suspended contamination from components being lifted from the pool. Consequently, the work procedure had no requirement prescribing that decontamination was carried out only under

BOX 6.5 Personnel got contaminated in Waste Handling Plant (Caithness.org 2002)

An investigation was launched after radioactive contamination in a Dounreay waste handling plant was discovered.

A worker's routine self-monitoring showed contamination to his protective footwear. Further checks detected contamination on the protective footwear of some more 20 employees. When the checks were extended to all facility staff, contamination was detected on the hands of two workers; one of them also had contamination on his face.

Both workers were escorted to the Dounreay Occupational Health Department for decontamination. The facial contamination was totally cleansed although some modest contamination could not be removed from their hands. The workers were sent home wearing gloves and were checked again on the following day.

A controlled building evacuation was conducted. The monitoring instruments on the ventilation system detected no radioactive discharges to the environment. Apparently the contamination was limited to the building floor. A controlled reentry into the building allowed a deeper search of the incident causes.

water. Underwater work would have eliminated the spread of airborne contamination.

Cases of worker contamination are pretty common in decommissioning projects. One of them is mentioned in Box 6.5.

6.12 Poor work package description causes radiological uptake, Hanford, United States (IAEA, 2015)

6.12.1 Decommissioning process; problem encountered; impacts

Most of a FSB structure was demolished in compliance with a generic demolition work package. It was known that the water chiller system, a tube heat exchanger, and related piping, among other components, contained internal contamination. The work approach selected was to crush the chiller piping, which caused an airborne spread of contamination. Three individuals at work nearby received minor radiological uptakes.

6.12.2 Analysis; solutions

At the beginning of this operation, the chiller piping was wet when it was cut off the chiller unit, and this wetness minimized the release of the internal contamination during the cutting. However, the contamination dried over a long exposure to the summer conditions. Therefore, by the time the pipe was crushed, the internal contamination had become much more mobile. While the radiological control team was aware of the rising risk, the project management did not take the needed action.

The upper tier radiological plan particularly pointed to the chiller piping and other equipment as ill-suited for aggressive dismantling, e.g., crushing of hollow components. Unfortunately, this criterion was not incorporated into the generic work package in use.

While there was a tacit determination to set up a distinct dismantling work package for the chiller unit and related piping, preparation of such a specific work package had not been finalized by the time the incident occurred.

The generic work package did not clearly define how the chiller piping would be removed, nor did it forbid the crushing of the pipe. The project management understood that a range of removal methods were possible and believed that the piping could be crushed to eliminate voids and dispose of the piping by making use of the facility's containers.

A detail in the work package allowed in-field supervisors to make decisions on dismantling and contamination control with no radiological assessment. This inadequacy allowed to carry out work that was not previously assessed by the ALARA methodology.

When the contamination was detected, the RCT who discovered the contamination radioed a supervisor for assistance, and the supervisor and another RCT reached the incident area in ordinary clothing. They had assumed that radon was the source of the radiation readings. This underestimation of the case may have led to two additional individuals receiving doses, whereas the adequate response should have been to evacuate the suspect area, dress appropriate PPE including respiratory protection, and monitor the suspect area from a known safe location.

6.12.3 Lessons learned

While the very nature of decommissioning work demands some flexibility under expected conditions, management must ensure that scope determination, precautions, mitigations, and restrictions are clearly established and enforced lest decisions are made that could lead to unverified operations.

Those responsible for the drafting of work procedures need to conduct a thorough cross-check of top-level analyses and detailed plans to explicitly confirm the identification and enforcement of radiological and nonradiological controls.

Adequate communications among field managers, work planners, and support teams should never be taken for granted. Management at all levels need to make sure that all premises, prerequisites, and criteria are formalized and well understood by all parties associated with work package planning and implementation.

BOX 6.6 Cutting of pipe results in unexpected worker's contamination, France; June 26, 2013 (IRSN, 2015)

A worker was contaminated during work involving the removal of a pipe module that had contained radioactive liquids. Due to the pipe having contained radioactive liquids, the removal operation was carried out under a vinyl "tent," in a containment airlock. Unfortunately, a pipe connector had not been spotted during the preparatory inspection of the worksite. This connector, which was heavily corroded, caused a leak while being handled and contaminated the worker.

In this case, the pipe was the first containment "barrier." However, the deteriorated connector, which had not been noticed beforehand, weakened the barrier.

This event illustrates the poor preparation of a D&D worksite.

Project management should verify planning assumptions during in-field work to confirm that actual conditions match those assumed and prompt early identification of any issues.

All workers should make sure that all indications of contamination are taken as real and that appropriate actions are taken as needed.

Poor work scope determination in generic work packages can lead to operations being carried out without proper hazard assessment and controls. Adequate response to identified contamination is essential. A similar event is described in Box 6.6.

6.13 Missing communication causes entry into an unposted high contamination area, Hanford, United States (IAEA, 2015)

The world's a scene of changes, and to be constant in Nature were inconstancy.

Herbert Spencer (1820–1903)

6.13.1 Decommissioning process; problem encountered; impacts

On April 11, 2012, Washington Closure Hanford workers at the 100N area, Hanford Site, were taking away the concrete lids of the 1303N silos. The area was classified as contaminated (CA), and due to the lid removal operations, access was also posted in regard to the fall hazard. The perimeter surrounding the adjacent FSB was posted as an HCA/HRA/ARA. A prejob briefing was held to move the HCA/HRA boundary by placing a new fence to delineate the HRA and portions of the HCA for the FSB. The area between the fences was to be decategorized to CA. An RCT carried out a contamination survey in the area to decategorize the spaces between the fences to CA. During the survey, the RCT discovered a spot in a corner of 1303N measuring above HCA levels. On 17 April, relocation of the water cannons and the "old" HCA fence were briefed. The work was carried out based on the RWP for 1303N. Workers went into the CA and took away the water cannons, not knowing they crossed the unposted HCA identified a few days before. The RCT supervisor and field work supervisor (FWS) assumed that the fall hazard posting would rule out entry into the uncharacterized area, unaware that workers had already entered. At the prejob briefing on the following day, the FWS reported work around the silos. At the briefing, the unposted HCA by the silos was mentioned. The RCT supervisor reviewed the survey document and concluded there had been an entry into an unposted HCA.

6.13.2 Analysis; solutions

The HCA information was not explicitly passed on to the workers installing the new HRA fence nor was it forwarded to the radiological status boards for either the FSB or 1303N. The 100N area is getting smaller as work is nearing completion in a few areas. This evolution causes overlapping work boundaries, distinct work teams being close to each other, and employees working overtime in unknown areas. All these changes demand additional monitoring to verify that access and work controls are complied with. Workers should concentrate on the ongoing work rather than look at the work that lies ahead. If the workers discover abnormal conditions, they should alert their coworkers and supervisors as soon as possible. When boundaries need to be changed, all personnel in the affected areas should be informed of the new boundaries prior to accessing these areas. Changes to boundaries and postings should be validated by the organization in charge of the posting.

6.13.3 Lessons learned

Regular exchange of information and coordination between work teams is imperative, especially when teams work side-by-side and when ownership of work areas is transferred.

6.14 Connecticut Yankee Nuclear Power Plant, Connecticut, United States (TLG, 2012)

Every cause produces more than one effect.

Herbert Spencer (1820–1903)

6.14.1 Decommissioning project; problem encountered; impacts

Connecticut Yankee Nuclear Power Plant (CY) was situated at Haddam Neck, Connecticut. It started commercial operation in 1968 and permanently shut

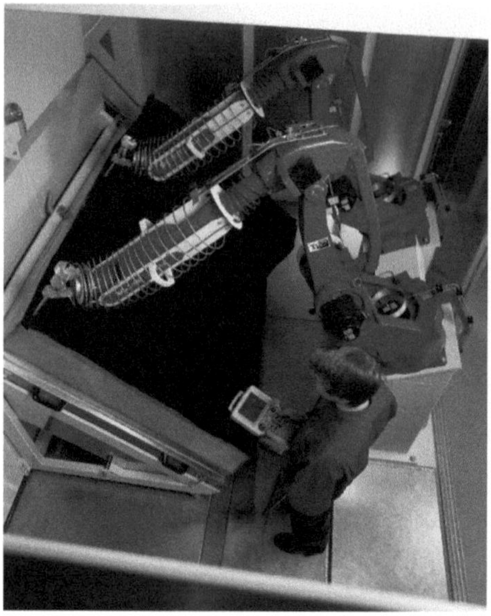

FIGURE 6.3 A water jet cutter: a tool capable of slicing into metal or other materials using a jet of water at high velocity and pressure, or a mixture of water and an abrasive substance. *(Credit to US Department of Defence, Wiki Commons).*

down in 1996. The nuclear decommissioning project ended by 2004. Demolition of the containment dome was completed in July 2006.

CY used the ultrahigh-pressure abrasive grit blasting water jet (Fig. 6.3) for segmentation of reactor internals. These highly activated items contained over 3×10^{16} Bq of greater-than-class-C (GTCC) materials. Thin strips of GTCC were segmented and loaded into 64 fuel assembly sized storage containers. The containers were then transferred to the spent fuel pool for storage until the away-from-reactor dry spent fuel storage facility became available.

The segmentation method posed significant problems including production of secondary waste filtration of the reactor cavity water, elevated collective doses (as much as 0.01 person-Sv/day), and low productivity.

The following extracts only information relevant to the scope of this book.

The ultrahigh-pressure waterjet (UHPWJ) (up to 340,000 kPa) heated the pool water, producing thermal streams which brought high-activity fine chips afloat. This side effect resulted in high doses to the workers operating from the cutting platform above the pool.

Following the cutting activities, the pool required a major cleanup. A robotic arm was designed and deployed to remotely high-pressure wash and vacuum the pool. A new filtration and shielding system was installed to maintain pool clarity and cleanliness during this aggressive decontamination activity.

6.14.2 Lessons learned

UHPWJ is widely used in nuclear D&D. However, in regard to personnel radiological and nonradiological safety, it presents challenges, which should be taken in due consideration in planning and implementing D&D activities, including the following:

- When used underwater, it tends to generate upward currents, which lift radioactive debris to surface.
- A large amount of fine particles and grit generated by the process may quickly saturate the water purification filters.
- Production of particles may impact water visibility to the point that the cutting has to be halted periodically to allow time for the filters to restore the water clarity.
- The main nonradiological hazards related to the use of UHPWJ are injection and fatigue. UHPWJ can cut skin and bones at close distances and inject water into the body or even cause internal damage. Microorganisms can enter through the wound into the body. A strike from a water jet at close distance to the belly or neck can be lethal. Another safety hazard is a hose rupture. Therefore, full PPE should be worn at all times when using UHPWJ equipment.

6.15 Multiple personnel contamination events, Hanford, Washington, United States (Seattle Times, 2019; DOE, 2010b)

6.15.1 Decommissioning process; problem encountered; impacts

For decades, the 300N area was the hub of Hanford's radiation research and nuclear fuel fabrication. Building 324, perhaps the most contaminated area 300 plant, is a three-story structure that has one floor belowground. It operated from the 1960s until the 1990s for materials and chemical process research. It is one of the last buildings remaining to be demolished in the 300N area.

The laboratory operations, enclosing nearly 10,000 m^2, began in 1965 and ended in 1996. The building—and unfortunately the soil underneath—are highly contaminated, and the years-long effort to demolish it proves the technological hindrances and hazards that are faced in this work (Fig. 6.4).

The difficult cleanup of a Hanford research laboratory located in the Building 324 met with more complications in 2019, as some unexpected contamination was discovered, spread farther than foreseeable, and appeared on a worker's personal clothing.

Workers, when they were ready to demolish the building in 2010, discovered that radioactive material had leaked into the soil some 300 m from the Columbia River. Apparently the Building 324 sits on a thick soil layer of radioactive cesium and strontium. This discovery greatly complicated the project, which was put in abeyance as studies were started to devise a new strategy.

It has been proven that the major source of contamination is a hot cell called Radiochemical Engineering Complex B-Cell. During the removal of radioactive grout from the B-Cell's sump, a breach was observed in the steel liner at the floor of the sump, which is likely to have occurred during the facility's operation. Some contaminated substances could have leaked (e.g., following a 1986 spill of concentrated cesium and strontium solutions) through that breach and/or concrete floor due to corrosion or through an expansion joint or crack in the concrete floor. However, other pathways are also possible.

Remote sample tubes ("Geoprobes") were pushed into the soil under the B-Cell's sump to radiologically characterize the soil. Radiation levels of several dozen Sv/h were measured, which may be correlated to a total activity of the order of a 100 TBq. The high radiation rates coming from the soil underneath required setting aside the strategy of dismantling the building as the first decommissioning step. Instead the removal of the contaminated soil has now priority.

Eventually, a plan was developed by the contractor CH2M Hill Plateau Remediation Co. that requires the reinforcement of the building's foundation, so as to stabilize it while the soil is excavated and removed. Since radiation levels are very high, remotely controlled equipment must be used to handle the contaminated soil. The workers started by drilling holes for the placement of cameras assisting the work.

FIGURE 6.4 The Building 324 as seen from the air before some of the demolition took place on ancillary facilities. *(Credit to DOE; Hanford Images).*

A March 15 report by Defense Facilities Board inspectors disclosed that contamination unexpectedly spread beyond drilling tools, making the workers evacuate the area.

On March 29 report, the inspectors reported that efforts to place the cameras halted again when surveys identified contamination levels that exceeded the RWP limit. Through April, a new "shroud" device was designed and tested to contain contamination during the drilling process.

In November CH2M halted again work at the Hanford's Building 324—one of several temporary halts to work there in 2019. The stop followed a letter from the DOE's Richland Operations Office (DOE-RL) requiring CH2M to put an end to a long series of contamination events. Last of these was a November 14 incident in which a worker suffered skin contamination at the Building 324 soil excavation project.

In January–November 2019, there have been 11 personnel contamination events—eight involving clothing and three involving skin, according to DOE data. The DOE stated in its letter to the contractor that only minimal work needed to maintain safety would be allowable on and around Building 324. Some 60 people worked in contaminated areas at the 324 Building when works were suspended.

Regardless of occasional (though relatively frequent) cases of personnel contamination, the main concern is a possible migration of the contaminants into groundwater. However, this mechanism would require a source of water to push the contamination, which has not been observed. As a precautionary measure, intrusive activities resulting in water discharge to the soil are being strictly controlled not to create a new migration route for the soluble contaminants (DOE, 2010b).

6.15.2 Lessons learned

See also the Cintichem case in Section 7.17.

At a number of old nuclear installations, contamination has been spreading undetected for decades to the soil underneath and threatening to migrate to the wider environment (the groundwater and the Columbia River in the above-mentioned case). This soil contamination has been often identified only at the decommissioning stage and has reflected in extra characterization efforts, unplanned hazards to the workers—often to be tackled with ad hoc, uncertain solutions—and environmental damage (not to mention the costs).

6.16 Human factors, Lawrence Livermore National Laboratory, California, United States (DOE, 2005)

Nothing can be done at once hastily and prudently.

Publilius Syrius, First century B.C, Maxim 557

6.16.1 Decommissioning process; problem encountered; impacts

An LLNL researcher attempted to dislodge a (supposedly low-risk) sealed radioactive source from its metal housing. The researcher employed (in time order) hand impacts, a rubber hammer, and tape, but all endeavors failed. The researcher then brought the item to a mechanic for help. Although it was late evening and the mechanic was about to leave, he was willing to help.

The worker first tried to remove the source out of the assembly by using a precision mallet—to no avail. He then heated the back of the housing with a torch, trying to exploit differential thermal expansion to disengage the source, which was assumed to be of aluminum. After a few seconds, the worker noticed powder on the work bench on which the housing had been positioned. Guessing the powder could be radioactively contaminated, the worker at once alerted the Environment, Safety, and Health (ES&H) Unit for assistance. Regardless of the potential impacts, nobody was contaminated by this near-miss.

6.16.2 Analysis; solutions

Investigators concluded that the researcher and the worker did not properly define the goal and extent of the work and analyze the impending hazards before beginning the job. Human factors can take different aspects and are often elusive: They may be invisible as workers concentrate on solving a problem or fulfilling a request. LLNL has gained a well-deserved reputation for recruiting the individuals most skilled at solving problems, while collaborating with a variety of experts and highly qualified staff. In this cultural environment, however, it can be easy for human factors, such as time pressure, overconfidence in prior experience or estimates, or complacency, to affect the decision-making and inadvertently make workers do what looks safe without enough consideration of all safety requirements.

6.16.3 Lessons learned

In the aforementioned event, the following human factors affected the decision-making:

- The work was requested toward the end of a shift, inducing time pressure and an attitude to solve the problem there and then.
- Because the source was assumed to be inherently "safe," the researcher and the mechanic started work neglecting safety requirements, among them the fundamental fact that radioactive work was not permissible in that area.
- Inadequate assumptions about each other's qualifications and skills made up a misplaced trust that the work was safe. The researcher, knowing that the mechanic was highly skilled in his job, did not feel like questioning the mechanic's techniques. Likewise, the mechanic considered the researcher as an expert on radioactive sources and did not question the researcher's judgment.

In summary, the wish to fulfill a request, meet professional expectations, and solve a difficult challenge, time pressure, and mutual overconfidence in each other's qualifications and expertise were human factors that influenced this event. In detail, the worker did not appreciate that his attempts, which looked safe, had extended the work beyond the scope defined in the safety documents. The worker should have halted work and request assistance from supervisors or the ES&H Unit as soon as it became clear that nonstandard measures were needed. All of the aforementioned human factors confused the worker's judgment and his understanding that he was not complying with safety requirements.

6.17 Contamination event at the UP2-400 plant at La Hague, France (IRSN, 2010)

To develop foresight, you need to practice hindsight.

Jane McGonigal (1977)

6.17.1 Decommissioning process; problem encountered; impacts

The UP2-400 plant at La Hague had been transitioning to decommissioning from late 2003. The main activities of the HAO/South facility included the cutting of fuel elements, dissolving uranium and plutonium oxide present in the cladding and clarifying the solution prior to its transfer to the U and Pu separation and purification processes.

During the operation-to-decommissioning transition phase, the operating organization of the HAO/South facility had been draining the process fluids and flushing equipment. On October 6, 2006, the plant organization started cleanup work by flushing the process equipment in the clarification unit consisting of two settling tanks.

The flushing of one of the two settling tanks required the operation of an ejector to agitate the solution inside the tank. To this end, the steam feed line (SFL) had to link the ejector and the facility steam system, the link being ensured by a hose underneath a ventilated hood. The end piece of the ejector SFL, locked out by a padlock, was plugged when the line was inactive. Moreover, a starter in the ejector SFL was used to clean this line with pressurized air discharged into the settling tank via the steam ejector.

Two workers were carrying out the ejector link operation on October 20, 2006. The worker who removed the plug felt a burst of air indicative of a line overpressure, which made him reinsert the plug at once. Startled by this event, he repeated the operation; he reinserted the plug when the overpressure reoccurred. The two workers then halted the operation and left the area to verify whether they had happened to locate the wrong line. After some 20 min, the air monitoring of the room with the ventilated hood activated an alarm.

Later on, when leaving the facility and undergoing radiological checks, the workers found to be contaminated.

As estimated by the plant organization and the Institute for Radiological Protection and Nuclear Safety (IRSN) based on the findings of radio-toxicological analyses, the committed effective dose over 50 years did not exceed the regulatory limit (20 mSv/y) for either worker. This event, which had no environmental impacts, was rated at level 1 on the INES.

6.17.2 Analysis; solutions

The investigations found that the overpressure was caused by a plug remaining inside the ejector for more than 10 years, thus impeding the regular flow of cleaning air. The plant organization believes that this plug had not been removed after the flushing of a settling tank (in other words, the normal configuration had not been reestablished after the flushing). The plant organization also admitted to the absence of a detailed risk analysis before the lockout padlock was removed; such an analysis would probably have raised doubts over the actual conditions of the ejector after a long time of disuse. The plant organization also affirmed that the internal contamination of the workers and the area contamination were mainly due to violations of radioprotection regulations (e.g., not wearing respiratory protection, not following procedures on radiological checks, etc.). The return of experience from this event prompted the plant organization to undertake a number of operational improvements (e.g., introducing an assessment of changing risks during end-of-life activities, with a specific focus on compliance of the plant with the original operating situation, measures to be taken if the normal plant configuration could not be adhered to, radioprotection training for workers, etc.).

Following the event review, the regulator IRSN concluded that the measures taken by the plant organization, including human, organizational, and technical factors, were generally satisfactory and issued a few more recommendations.

6.17.3 Lessons learned

This event highlights the critical role of an accurate preparation of special activities during the operation-to-decommissioning transition (e.g., the need for a detailed risk analysis and a check of the actual state of the plant). Inadequate preparation of special activities often causes undesirable incidents; preparation is particularly necessary when tools/equipment unused for long periods are going to be used again or where the plant organization has limited information on SSC to be acted upon.

6.18 Fire event in the high-level laboratories, Saclay, France (IRSN, 2010)

6.18.1 Decommissioning process; problem encountered; impacts

The case described here happened during the decommissioning of high-level laboratories at the CEA Saclay site. A containment made of vinyl was arranged in a room to contain radioactive substances during the size reduction of disassembled materials. The segmenting activities took place in a cell inside the outer containment; the cell floor was covered with steel plates and the lateral walls and ceiling were made of fire-resistant, spark-mesh panels. The lateral walls ended 20 cm from the ceiling to let natural light into the cell. There was also a system to take in fumes during the cutting (at first carried out with a plasma torch), which discharges air after filtration into the mobile containment, and an air extractor equipped with a prefilter and an HEPA filter, which discharges the air from the mobile containment into the general air ventilation of the plant.

On November 8, 2006, as the plasma torch had stopped functioning on the day before, a grinder was used to cut some metal ventilation ducts inside the cell: The fire permit had not been checked before the work began.

A second worker, who was watching the operation from outside the containment, noticed a glimmer at the prefilter inside the containment. He warned the worker doing the cutting and then halted the air extractor and left the area to call the local safety unit (LSU). A fire detector was set off just before the worker's call.

The worker doing the cutting in the mobile containment tried in vain to extinguish the prefilter fire with a powder extinguisher. The fire diffused across the entire room. The LSU reached the area 7 min after being called: They managed to control the fire with water jets; finally, the fire was fully extinguished 40 min after the beginning of the incident.

The fire caused serious damage in the room: The cell was destroyed, and the outer containment became unusable due to rupture of the windows and the impact of LSU's water jets.

This event had no radiological impacts on the workers or the environment because the segmented ducts had no loose contamination. Radiological surveys of the premises after the event showed no measurable contamination. The water employed to extinguish the fire remained contained in the room until removal. On the INES, the incident was rated 1.

6.18.2 Analysis; solutions

The causes of the fire were the following:

FIGURE 6.5 Sparks from grinding of metallic pieces, SCK-CEN, Mol, Belgium. *(Photo by M. Laraia)*

- Sparks were thrown out of the cutting cell through the empty space between the spark-mesh panels and the ceiling; these burning particles were bigger when cutting with a grinder than with a plasma torch (Fig. 6.5);
- The powder extinguisher was ineffective to extinguish the fire; the fire was fed by the air flowing through the prefilter (shutting down the containment ventilation did not help because this was linked to the general air ventilation of the plant) and through combustible materials (vinyl, plastic gloves, etc.) nearby.
- The investigations showed that the work cell was unfit for the work carried out: The containment was too small for the cutting activities; the ventilation system was too close to the cutting zone to ensure protection from burning particles; and the spark-mesh protection was incomplete.

6.18.3 Lessons learned

This incident indicates that a specific safety assessment is required for each type of cutting and each single environment; generic or old fire permits should be avoided; and the assessment should be redone in case modifications to the operating conditions were relevant to the assessment.

The decommissioning team must ensure that operating conditions and protective measures comply with the assumptions and terms of the safety assessment. It is essential that combustible materials that are not indispensable be removed when working with hot particles.

6.19 Overpressurized drum event at Idaho Accelerated Retrieval Project (DOE, 2019a)

Life is its own journey, presupposes its own change and movement, and one tries to arrest them at one's eternal peril.

Laurens Van der Post, 1906–1996

6.19.1 Decommissioning process; problem encountered; impacts

The Subsurface Disposal Area, a 39-ha area at Idaho Radioactive Waste Management Complex, has been used as a disposal site of many types of waste in pits and trenches since the 1960–1970s. Most of the transuranic (TRU) waste buried there was produced at the Rocky Flats Plant; however, many waste packages originated from other waste generators. Since the early 2005, the cleanup contractor has been removing waste in an environmental remediation project especially intended to protect the near Snake River Plain Aquifer. On a larger scale, this project is part of the overall decommissioning program at Idaho National Laboratory (INL).

The buried waste removal is expected to go on until 2028, when construction of a protective barrier enclosing the entire disposal site is completed.

Waste handling and packaging takes place in a series of retrieval enclosures: Work is currently taking place in the ninth and supposedly final enclosure (ARP IX), which started operations in 2019. Fig. 6.6 shows a detail of ongoing work.

FIGURE 6.6 Workers in full PPE complete packaging of retrieved waste at Idaho ARP. *PPE*, personal protective equipment. *(Credit to DOE).*

On April 11, 2018, an incident occurred in the ARP V facility: A thermal reaction caused the energetic release of radioactive substances from four 200 L drums of suspect TRU waste placed in the ARP V airlock drum storage area. No workers were nearby when the reaction occurred, and the drums were in an HEPA-filtered area. No environmental release was detected.

Initially, INL firefighters went into the plant and noticed one drum with a lid off; no flames were seen but smoke was spreading out of the top of a drum. The firefighters failed to put out hot spots in the drum in question, removed the drum from a group of prepared drums, and left the plant. Round-the-clock surveillance, radiological inspections, and air monitoring were established to keep conditions under control. On the following day, workers went again inside and found four drums had undergone exothermic reactions, violently throwing away the lids and portions of their contents. The ejection caused TRU waste to spread all over the area (normally manned by workers): As a result, the plant surfaces were highly contaminated, and airborne activity was also present.

Although the event related to TRU waste, it is considered to be applicable in principle to many types of radioactive waste.

6.19.2 Analysis; solutions

The investigations identified two mechanisms likely to have caused the incident: (1) oxidation of a pyrophoric form of depleted uranium (DU) powder and fines (the heat source that triggered secondary reactions) over some 8 hours and (2) production of large quantities of methane by hydrolysis of beryllium carbide. Historical documents identified the operations at the Rocky Flats Site that had produced DU and beryllium carbide. While the presence of these components in the waste was not ignored, plans and controls did not account for their potential reactivity, the timespan of primary chemical reactions, and the onset of secondary reactions.

The direct cause of this incident was the rupture of four waste drums in the ARP V facility due to an oxidation beginning during waste repackaging when waste containing reactive DU was mixed with additional material and exposed to the air. The heat source that started the reaction was the oxidation of the uranium metal, which in turn supported secondary reactions (hydrolysis of beryllium carbide, which generated significant amounts of methane gas): It was the generated gas that pressurized and ruptured the waste drums. In this case, the methane gas did not ignite; however, the generation of flammable gases in packaged drums is a recognized hazard throughout the DOE complex.

The investigations identified two root causes for this incident: (1) Site management failed to establish and enforce adequate controls for handling waste of undocumented origin and (2) site management failed to maintain safety culture over many years. Contributing causes include lack of a change management process to identify weaknesses; lack of a comprehensive disposal registry; and inadequate return of the experience acquired during a previous pyrophoric reaction.

6.19.3 Lessons learned

The following lessons learned apply to waste management incidents in the course of decommissioning/environmental remediation programs:

1. Review causal analysis reports from similar events and include lessons learned into relevant site procedures.
2. Review accident scenarios for waste retrieval and repackaging operations.
3. Assess training of staff in charge of authorizing waste processing activities to verify they are properly qualified to take decisions about the safe implementation of these activities. Review training materials to ensure they cover evolving conditions and include realistic cases of pyrophoric substances, metal carbides, and other "difficult" waste types.
4. Evaluate current processes for managing legacy waste of uncertain origin to confirm adequate controls are enforced to safely receive, segregate, treat, package, and store waste prior to its shipment and disposal.
5. Review procedures for existing and foreseen drums containing repackaged waste (1) to verify adequate controls are enforced to avoid or mitigate outbursts; (2) to accept individual containers having suspect reactive materials for waste treatment processes; and (3) to provide sufficient conservatism about the time span of triggering reactions and the onset of secondary reactions.
6. Review response procedures to ensure safe entry conditions for areas with airborne radiological hazards.
7. Maintain and improve a site-wide safety culture that encourages a questioning attitude by all personnel levels toward new or difficult operations regardless of cost and schedule.

6.20 Potential criticality accident at the 233-S Facility, Hanford, Washington, United States (DOE, 2003)

6.20.1 Decommissioning process; problem encountered; impacts

The 233-S Plutonium Concentration Facility (233-S Facility) was constructed in 1955 to increase plutonium production by further concentrating the plutonium nitrate solution produced by solvent extraction by the REDOX station (Fig. 6.7). The facility was in operation until July 1967 when it was listed in the DOE Surplus Facility Management Program as a redundant facility.

The facility remained in a progressive state of deterioration since its closure. For almost 30 years, Surveillance and Maintenance ensured the confinement of the facility's considerable amounts of plutonium contamination. The harsh winter of 1996 hastened the building degradation and augmented the risk of staff exposure to plutonium and of environmental releases. Based on risk increase and related maintenance expenses, the DOE

FIGURE 6.7 Historic view of West Façade of Plutonium Concentration Facility (Building 233-S), Reduction-Oxidation Building (REDOX-202-S) to the right, looking east, ca 1967. *(Credit to National Park Service, Wiki Commons).*

took the decision to demolish the plant. This demolition project was the first open-air demolition of a Hanford severely contaminated plutonium plant. It was also the first DOE-owned plutonium plant to have been demolished prior to first decontaminating surfaces to quasi "free release" criteria. The actual demolition started in October 2003 and ended in April 2004. Most of the building's structures were classified as low-level waste (LLW) and disposed of at Hanford's Environmental Remediation Disposal Facility (ERDF). A minor percentage of the demolition debris was classified as transuranic (TRU) waste; this portion of the waste was packaged for storage at Hanford's Central Waste Complex and was eventually transported for final disposal to the Waste Isolation Pilot Plant (WIPP) in New Mexico.

The initial nondestructive analysis (NDA) of the 233-S Facility process hood assumed the plutonium inventory was located within the piping, vessels, filters, and ducting. This has led to an estimated 2.8 g plutonium inventory on the process hood floor and was based on smear survey data. To deduce the inventory value for the floor, it was assumed that the highest measured smear survey data were evenly distributed across the whole floor.

Prior to beginning D&D activities, a sample of material located on the 233-S process hood floor was collected. Information from the NDA team pointed to a maximum of 20 g of ^{239}Pu in the sample, much more than previously assumed. The new data resulted in an unreviewed safety question (USQ) and a successive unusual occurrence report. The sample was moved to the 222-S Laboratory for further analysis. The sample was a brownish dry mix of sand

and dust with some rustlike stuff. There were also gray slivers that looked like paint chips (possibly old fixatives). Analysis established that the 48.6 g sample included 6.33 g of ^{239}Pu. Therefore, a single sample of material taken from the 233-S process hood included almost three times the amount of plutonium formerly assumed to be contained in the entire process hood (not including vessels and piping).

Engineering assessments and change analysis determined the event occurred when NDA of a process hood sample indicated a maximum sample value of 20 g of plutonium, which exceeded the 2.8 g assumed previously in the safety analysis for the fire accident analysis. Because the newly discovered condition altered the fire accident scenario, the resulting dose impacts were increased, and a USQ was declared. As the increased inventory required reanalysis and it was likely that additional loose material would be found during planned decommissioning operations on the process hood, the project team decided to include more conservatism into the safety analysis. The resulting evaluation was aimed to encompass any future discoveries of loose materials and provide the needed flexibility. This evaluation estimated that 2255 g of plutonium was at risk in the fire accident scenario. This figure is based upon the highest NDA measured value from the process hood (1530 g plutonium).

The increased inventory originated a new safety evaluation that included a revised criticality analysis, safety analysis, and related dose impacts.

Moreover, a criticality alarm system was installed as a "defense-in-depth" provision. The facility's technical safety requirements were revised, and criticality safety postings were installed throughout the facility.

The process hood work activities in the 233-S Facility were delayed for some 5 months at an additional cost of around $250 000. Additional expenses related to the discovered condition included the revision of the criticality analysis, placing and maintenance of a criticality alarm system, procedural changes, installing nuclear criticality safety postings, and training. The overall cost of this event exceeded $ 1 M.

6.20.2 Lessons learned

As conditions continually change from operations to decommissioning and through all phases of decommissioning, and the amounts and locations of fissile materials are difficult to determine, periodic assessments are recommended to establish current conditions and identify impending problems before they occur. Based on the concept of Integrated Safety Management (ISM), decommissioning organizations should periodically assess the validity of their criticality safety programs. Specific inquiries should be pursued frequently.

Assessments should be managed by criticality safety experts and should be conducted as actual "field assessments," including facility surveys, monitoring of actual work, and staff interviews.

As a facility moves from operations to D&D activities, there is a concrete chance for the staff to lose attention to criticality safety. Attention can be lost for a number of reasons including the recruitment of D&D personnel with little or no knowledge of criticality safety; loss of operations knowledge, familiarity with the plant, and experienced personnel taking new jobs as D&D begins; the (false) perception that a criticality accident is impossible in a decommissioning plant; overall reduction in criticality safety staff number; and a cultural change between operations and D&D. Another potential criticality case at Hanford is presented in Box 6.7.

BOX 6.7 Decommissioning process; problem encountered; impacts (DOE, 2003)

Fissile material was found in contaminated equipment located within a canyon storage facility at Hanford's U Plant. (A point of terminology follows. Several chemical separation plants at Hanford look similar. They are over 100-m long, some 25-m t high, and 20-m wide. Going inside one of these huge plants is like going into a long warehouse, a lot of open space with high walls. It is like being on the floor of a deep narrow valley, where you raise your eyes and see the mountains on either side. Because of the similarity to a natural canyon, the Hanford workers called these plants "canyons" and the name stayed.) The facility was built in 1944 as one of three chemical separation plants but was never used for its planned objective. In 1952, U Plant was turned into the tributyl phosphate process plant. In 1958, U Plant was placed in standby, and since then, it has been used to store contaminated equipment. Some initial demolitions in this plant took place in 2010 and 2011. A cementitious grout was pumped into the interior of the U Plant which filled much of the canyon and locked any contamination in situ pending demolition of the whole facility in the future.

During demolition planning, a project engineer was reviewing facility documentation to make sure that characterization activities could be adequate to final demolition. During the review, he noticed that a tank in a canyon cell contained a material that could exceed the limits for fissile material. The document illustrated a tank in a canyon cell that contained a "dark green liquid." The engineer's previous experience at the Hanford Site made him suspect that the liquid contained Pu. Matching his identification of a problematic residue with a measured nuclide concentration, the engineer suggested that the material was most probably TRU and roughly estimated the amount of ^{239}Pu in the tank could range 400–1000 g, which was more than expected.

Lessons learned

The aforementioned cases stress the need for accurate planning, deep documentation assessment, a questioning attitude, and good process knowledge. This will minimize surprises and will allow for the correct execution of controls. The need for clear and prompt communication of potential issues to management should not be underestimated either.

6.21 Legacy issues that can impact D&D activities, Rocky Flats, Golden, Colorado, United States (DOE, 2003)

During D&D activities, unexpected issues will inevitably arise. To minimize the number and impact of such incidents, management must accurately identify legacy issues. Several D&D activities at Rocky Flats have experienced events associated with legacy issues.

The following causal factors were identified at Rocky Flats as having caused potentially criticality conditions:

- Abnormal operating conditions—incidents (potentially) occurring with a material inventory underway, initiation of a shutdown operation, decontamination works, and inauguration of a new process.
- Fissile material transfer—inadvertent transfer of material from safe to unsafe configurations.
- Equipment change or instrument malfunction—new or altered equipment, defective equipment, and instrument failures.
- Unregarded procedures—failure to abide by specified procedures has led to missing scheduled maintenance, samples not analyzed in a timely manner, wrong valve alignments, and missing controls.
- Fissile material in unexpected places—unknown leakages and lack of a well-defined management of activities led to build-ups of material in unexpected places.

Shutdown conditions produce causal factors. Termination of operations within sectors of facilities or whole facilities produces abnormal working conditions. Some processes may be conducted to reach stable conditions or start facility decontamination.

Transfer of fissile material in the form of solids, solutions, and drums takes place during the shutdown phase.

Historical inventory discrepancies may imply that material sits in unforeseen places. Use of degrading equipment or instruments, or equipment renewal may occur at this stage. New procedures are being drafted to manage the abnormal activities. High personnel turnover can happen soon after final closure of operations and new teams are likely to be involved in follow-up stabilization, cleanup, and decommissioning.

In the 4 years following closure of operations, mass and volume violations at Rocky Flats decreased due to material not being processed. Instead, the number of unauthorized activities increased. Prevailing causal factors for procedural violations continue to be inattention to complying with procedures, nonroutine activities, and equipment defects. The relative contribution of abnormal events has gone up. However, none of these events involved a considerable mass approaching the risk of an actual criticality.

To avoid accident under shutdown/decommissioning conditions, it is vital that a well-defined criticality safety program be enforced (Croucher, 1994).

A few events associated with legacy issues at Rocky Flats are illustrated in detail in (DOE, 2003). More cases specific to the application of NDA techniques are given in Section 6.22.

6.21.1 Decommissioning process; problem encountered; impacts

Two wooden TRU waste crates were transported to a room where TRU waste storage was not permissible. The crates were shifted to that room to make D&D activities possible in another area. The D&D crew that shifted the crates believed that all wooden crates were LLW. The two crates contained legacy items planned to be segmented.

These crates had no markings, and the related waste container content logs (known as Travelers) were not observable. One traveler was hidden under a sheet of cardboard, and the other traveler was not on the crate. The workers should have looked for and reviewed the travelers before shifting the crates. After the incident, all legacy crates were reviewed for proper storage and recording. All D&D supervisors were imparted training on the procedures and requirements for handling legacy crates.

In another case, management requested safeguard employees to carry out gamma scans in view of deactivation operations. From the scans, the following problems emerged:

- Glovebox with a safety limit of 400 g had instead 596 g of Pu.
- Glovebox posted as "No Fissile Material Allowed" had 174 g of Pu.
- Glovebox with a safety limit that did not allow Pu had 71 g of Pu.
- Glovebox with no safety limits had 1000 g of Pu.

It is expected that Pu amounts will be unexpectedly found, as deactivation of Pu facilities goes on. Locating Pu remaining from past operations in filters, equipment, and ventilation ducts is fundamental in cleanup and facility stabilization. Pu can be then safely removed, conditioned, and packaged for long-term storage. Locating plutonium stresses the need to accurately assay and assess each single glovebox before carrying out "hands on" decontamination.

In another case, during the deactivation process, a safeguard team was conducting measurements, when a fair quantity of fissile material was located in a glovebox filter. The historical knowledge of the system and the past operations led to suspect that the glovebox could have considerable amounts of plutonium in several places. Because the plutonium was discovered area previously unreviewed by criticality specialists, no criticality safety limits were available for the material in that location. To correct this situation, criticality safety staff conducted an assessment and provided guidance about the admissible amounts.

The instrumentation used to carry out the scan indicated via a spectrum, a qualitative display of the plutonium amount. When the filter was first scanned, there was a "large spectrum visible." This indication of considerable amounts was not reported quickly to management. As an improvement, the employees performing the scans were advised to make sure that new information (even preliminary information) is taken to the attention of the decision-makers. This responsibility concerns also the management. Preliminary information may or may not affect decisions; however, in some circumstances, it can be essential.

6.21.2 Lessons learned

The following lessons learned should be drawn from Rocky Flats experience during the closure phase and in view of decommissioning. Criticality safety awareness training on the causal factors of the criticality events should be imparted to the technical teams so that they can understand the ongoing plant conditions and inherent risks.

Facilities should be inspected for fissile material to confirm its locations are clear before decommissioning. A major reassessment of the criticality safety limits for the whole shutdown facility is advisable. At least, an assessment view of criticality safety limits should be conducted to ensure incorporation of double contingency and human factors for safe management of planned activities. A survey of the process areas by involved workers, safety unit, and management should be conducted to locate and amend weak points in the technical and administrative procedures. Supervision of the ongoing activities is mandatory. And—last but not least—lessons learned from each activity should be considered and made good use of in future activities and projects.

6.22 Personnel should understand the differences in and use of different NDA numbers produced during D&D, Rocky Flats, Colorado, United States (DOE, 2003)

Management must make sure that D&D teams understand the differences and use of various non-destructive analysis (NDA) numbers. NDA values can be produced for different purposes (e.g., safeguards and security, criticality safety, planning, etc.) and are given in different formats. Frequently, this information does not have the quality required for criticality safety purposes (e.g., 2σ, 95% confidence values) and therefore should not be used in criticality safety. If the D&D team does not understand the different purposes of NDA numbers, problems can occur as illustrated in the following.

6.22.1 Decommissioning process; problem encountered; impacts (DOE, 2003)

Personnel were assessing data for a waste drum. From NDA scan sheets about items in the drum, it surfaced that the drum was loaded with more than the

200 g of fissile material allowed by the criticality safety limit. The "traveler" (an inspection and certification document, see Glossary) for the drum was obtained and reviewed together with the NDA scan sheets. It was established that the drum was loaded with 270 g of fissile material. In fact, the "traveler" only showed a figure of 194 g because it was based on nominal scan values, not the 95% confidence numbers mandated by the criticality safety limit.

Prior to the items being placed in the drum, an NDA team had scanned the items and established assay values. Unfortunately, the D&D team did not have the scanned values when they filled the drum. Instead, the D&D team supervisor received values from a manager who happened to have an email that itemized values; however, the email was intended only to respond to nuclear material control requirements for total permissible amounts of attractive specified material and not for the packaging.

Nuclear material control requirements are based on nominal scan values. The D&D manager assumed that the email numbers were given at the 95% confidence and directed the D&D supervisor to use them for the packaging.

During the filling of the drums, the D&D supervisor and the workers mistakenly did not ensure that the mandated 95% confidence values were adopted. The team assumed that the information given by the D&D manager was right and did not check it with the specific documents. Actually the supervisor first objected to the use of the values received; however, the manager reassured him that the numbers were indeed right.

Corrective actions included the following:

- Review the event with all D&D supervisors and briefing all D&D teams to ensure they confirm verify 95% confidence numbers for packaging.
- Modify the NDA scan sheet to better label the column containing the 95% confidence numbers used in packaging.
- Ensure that procedures provided adequate direction for use of right gram estimation values and that values are clearly documented for criticality safety purposes.

In a second case, the site management was informed that NDA measurements of a standard waste box (SWB) were above the 325-g limit for fissile materials. The NDA measured 695 g (2σ, 95% confidence number).

Review of the gram estimation for the SWB disclosed that the waste had not been correctly gram-estimated when loaded. The waste packaging team supervisor received NDA values from an NDA summary holdup letter for the building in question. This information was entitled "Max Grams/sq. ft." The supervisor erroneously understood this to be the total gram value assigned to the glovebox. A process reviewer checked the number put down on the waste "traveler," but checked only that it was below the gram limit, not that it was right.

Corrective actions included the following:

- o Revise an operations order (providing guidance to waste crews on gram estimation) to require that any incomplete or unclear detail concerning 95% confidence be notified to management before the waste was packaged.
- o Establish an independent review process for waste packages.
- o Assign one or two members of each crew to be trained for the gram estimation process and make sure that training was sufficient and well understood.
- o Add TRU waste experts to a technical response unit, and make them available to respond to questions from waste crews.

6.22.2 Lessons learned

NDA procedures must include the rigor and clarity to ensure repeatable execution and provide a record of work carried out. Processes controlling NDA must include quality assurance requirements regarding the application of standardized methods and independent review of calculations. The technical basis for NDA practices must be fully and clearly established. Recording procedures must document the process in such detail to ensure repeatability and ensure that NDA activities may be taken to the end without close oversight by the specialists.

Adequate measurement controls must be enforced to the instruments. Measurement of a standard should be executed and recorded daily, including checks of resolution and energy calibration. In detail:

- NDA Equipment must be fit for the type of measurement.
- Reported measurement uncertainties should include more than the statistical factors. Other factors generating uncertainty can be at play and should be properly taken into account and reported.
- NDA equipment must be properly recalibrated after maintenance, repair, or in a new position.

6.23 Contaminated wound event, Sellafield, United Kingdom; May 28, 2009 (ONR, 2013)

6.23.1 Decommissioning process; problem encountered; impacts

A series of concrete bunkers that once stored plutonium-contaminated materials (PCM) are in line to be demolished. The radiation and contamination levels are so high that the workers must wear air-fed suits or robots are employed.

Located at the low-level waste repository near Drigg, Cumbria, the bunkers were known as magazines and stored PCM produced from Sellafield operations in the 1950–1960s.

The work started in April 2013 and was planned to take 10 years: Actually, the decommissioning program was reduced to 6 years. In 2019, the PCM decommissioning program was nearing the end.

The remaining five magazines will eventually be demolished, making space for future development projects. Managing the PCM legacy is a priority for the site.

During preparatory PCM decommissioning activities in Building B277, a worker in an air-fed suit was using an angle grinder to segment equipment that was heavily contaminated with plutonium, when he was wounded on his thumb by the angle grinder (see similar injury in Fig. 6.8). The circumstances of the events are not clear. However, the worker was wearing the expected PPE for such work.

The operating organization promptly took medical action to deal with the plutonium in the wound. This required the cutting of the wound. The medical action was successful and resulted in a worker's dose below legal limits. This event was rated 1 on the INES.

Just at the time of the incident, Sellafield Ltd. was taking action to respond to an improvement notice that was issued further to a previous (unrelated) wounding incident in B277.

There is a fairly high risk of injuries during the carrying out of segmenting activities. And, when plutonium is involved, the impacts to the affected worker can be serious. Sellafield Ltd uses similar techniques for the decommissioning of plutonium plant to other UK and foreign organizations. Segmenting is aimed to optimize waste filling into the standard 200 L drum, and this is considered the key element for decommissioning plutonium plants. This event,

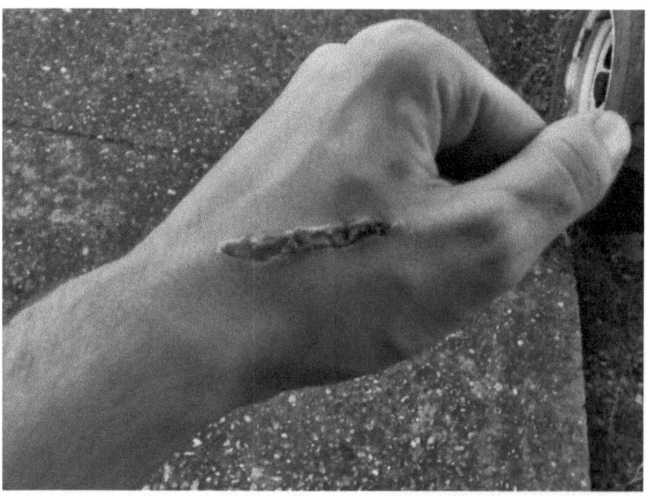

FIGURE 6.8 Angle grinder injury (a week after the event). *(Credit to Lewis Collard, Wiki Commons).*

however, stresses the perception that the need to produce a standard waste form can overrule occupational safety considerations.

Following this incident, Sellafield Ltd. and the regulatory body were cooperating on reaching an optimal trade-off between waste form and occupational safety.

6.23.2 Lessons learned

Working in a radiologically contaminated environment on rough, sharp-edged surfaces exposes the workers to dual challenges. In addition to protecting from contamination, PPE should be also designed to take account of industrial hazards as needed.

6.24 Personnel contamination incidents during the decommissioning of 61 plutonium gloveboxes in Building 212, Argonne National Laboratory, Illinois, United Stated (ANL, 1996)

6.24.1 Decommissioning process

Since programmatic research had come to an end, the decommissioning of 61 plutonium gloveboxes in nine labs of Building 212, Argonne National Laboratory-East (ANL-E) was proposed by the site organization and funded by the DOE. The gloveboxes had been in operation from the 1960s to 1989 for R&D on nuclear fuel (Fig. 6.9). The decommissioning project was aimed to exclude the risk of plutonium release from the contaminated gloveboxes and to reuse the building for new productive applications. The project objectives included the removal of the gloveboxes, the decontamination of the labs to unrestricted release conditions, and the disposal of the radioactive waste generated. The work began in November 1992, and the final radiological survey was completed in January 1996. A few radiological contamination incidents occurred during this project. A selection of cases is given in the following.

6.24.2 Problem encountered 1; impacts

This event of June 12, 1993, was the discovery, by using a hand and shoe monitor, of 16.7 Bq α/100 cm^2 on the shoe of a contractor as he checked out. The contamination was promptly removed with tape, and a second survey showed no contamination. Surveys of the laboratory and surrounding areas showed no contamination; the source of the shoe contamination was not identified.

A second case was the contamination of the shoe soles of two ANL-E employees and of a section above the knee on the work trousers of another worker. The event occurred on July 24, 1993. The workers were retaping a

FIGURE 6.9 ANL-E, Building 212, D wing shutdown plutonium gloveboxes, decontamination and decommissioning project. *(Credit to DOE, Wiki Commons).*

leaking vacuum pump line of a plutonium glovebox. They wore full-face respirators, lab coats, and gloves. The maximum contamination discovered was 26.7 Bq α/100 cm^2 on the work trousers. The contamination was removed with tape to unmeasurable levels. A survey of the work area indicated some contamination on the floor near the vacuum pump, which was decontaminated to unmeasurable levels. Work procedures were revised to include hold points for assessment and signoff by Health Physics. A walk-through review of procedures was to be conducted prior to the start of work. Bioassay sample counts were negative.

6.24.3 Problem encountered 2; impacts

The abnormal occurrence of July 22, 1993, consisted of the contamination of a worker's hand during handling and weighting of a heat-sealed plastic bag of waste. The bag of TRU waste contained metal shelving that had been removed from a glovebox. The hand contamination of 33.3 Bq α/100 cm^2 was discovered by a routine hand and shoe monitor check after the weighing of the bag. It was later noticed that there was a 0.3 cm rip in the bottom of the bag, which had not been noticed earlier. The workers wore a lab coat but no respirator and no gloves. The contamination was promptly washed off with soap and water to an unmeasurable level. The rip in the waste bag was sealed, and no contamination was then found on the bag, on the weighing scale, or in the area. Corrective actions included the following:

- The procedure for taping sharp edges and bundling of materials in a glovebox prior to their removal from the glovebox was reinforced.
- A radiological survey was conducted of all work areas, and mobile contamination was removed.
- Remaining fixed contamination was sealed and posted.
- RWP procedures were stressed for all radiation work as needed.

- Worker training was broadened with a run-through and step-by-step signoff of procedures involved.
- The radiation monitoring facilities were upgraded.

6.24.4 Problem encountered 3; impacts

The abnormal event of March 30, 1994, was caused by contamination of the protective clothing and skin of a contractor working through glovebox gloves to break down glovebox components. The worker was wearing toe caps, a lab coat, two pair of latex gloves, and scrubs under the lab coat. On surveying his gloves with the lab survey instrument, he discovered the left glove was contaminated. He took off his gloves and put them in a radioactive waste bin. He left the lab, and after removal of his lab coat, the hand and foot monitor indicated that his left hand was contaminated. The worker requested assistance and the Health Physics helper measured 46.7 Bq α on the little finger of the right hand and of the left wrist. The contamination was promptly washed off at a decontamination station, and the protective garment was put in a radioactive waste container. The lab coat had a total of 83.3 Bq α on either lower sleeve. Investigation determined the origin of the contamination was two rips, 7.6 and 2.5 cm long, in the glovebox glove the worker had used. The rips could have occurred when the glove had been stretched across the glovebox while using pliers. Follow-up bioassay results indicated that the 50-year committed dose from this incident would be less than 0.5 mSv. Corrective actions included the following:

- The box gloves were replaced prior to further work.
- Procedures were amended to focus on prejob inspection of box gloves.
- A revised procedure was added directing not to stretch box gloves and to request immediate aid to contain any contamination discovered.

6.24.5 Problem encountered 4; impacts

This incident relates to the intake of activity detected in fecal samples. The event lasted over numerous months: It was assigned the date of September 19, 1994, when bioassay analyses were completed on seven contractors who had completed their tasks and left the project. The fecal samples indicated low levels of ^{239}Pu, ^{240}Pu, and ^{241}Am. Urine samples produced negative findings. It is estimated internal doses of less than 1 mSv 50-year committed dose were due to intermittent ingestion or inhalation of radioactive substances. Corrective actions included the following:

- Increase stress on "good housekeeping" including frequent removal of trash and other redundant items and mopping of workplace and adjacent areas.
- Identify trends through workplace contamination surveys and air samples.

6.24.6 Lessons learned

Although the overall achievements of this project were fully satisfactory, a few incidents highlight the need that even minor procedural details be carefully evaluated in planning and subsequent implementation. As they say, "the devil is in the details." See more contamination cases in Boxes 6.8 and 6.9.

6.25 High dose during the removal of cabling from Heavy Water Components Test Reactor Vessel, Savannah River Site, South Carolina, United States (DOE, 2012b)

6.25.1 Decommissioning process; problem encountered; impacts

The Heavy Water Components Test Reactor (HWCTR) was a pressurized heavy water test reactor operated to test fuel designs for heavy water–moderated power reactors (a reactor type that was never commercialized in the United States). It housed a reactor vessel, two steam generators, and other miscellaneous SSC; it was in operation from 1962 until 1964 when operations ceased and the reactor was placed in standby condition. The reactor and spent fuel pool were defueled in the 1970s. The decommissioning strategy selected

BOX 6.8 Multiple cases of contamination at Rocky Flats, Colorado (ehstoday 2004)

During decommissioning, there are frequent cases of contamination spreading within plant areas. A brief overview of occurrences at Rocky Flats follows.

In 2004, the DOE issued a preliminary notice of violation (PNOV) with a proposed fine of $522,500 to the Kaiser-Hill Co. LLC, the decommissioning managing contractor at DOE's Rocky Flats Environmental Technology Site. The PNOV was attributed to multiple violations of DOE nuclear safety regulations and to Kaiser-Hill missing to operate within DOE-approved safety limits and controls. The violations referred to a fire in a glovebox being decommissioned; a ventilation airflow reversal that spread contamination across several rooms; and an activity spread from a poorly secured containment sleeve. Several workers were radiologically exposed although below legal limits. There were no radioactive discharges outside of the plants.

The DOE investigation observed inadequacies in regard to radiological controls, procedural compliance, personnel training, and missing to learn from previous similar cases and take corrective actions.

Reduction of the fine was applied by the DOE for one violation due to the contractor's thorough and timely corrective action. No reductions were proposed for the other fines due to the contractor's inadequate understanding of the problems, inattention to notifying the issues to DOE or ineffective corrections.

BOX 6.9 Dounreay contamination during the transfer of a glovebox (ONR, 2016)

On September 19, 2016, during the removal of an excess glove box to the manipulator store, some radioactive material poured onto the floor of the plant. A Health Physics surveyor's shoe and a worker's shoe and coverall leg were contaminated. No airborne activity monitors were activated, and no radioactive intake was found in any of the people involved. As soon as the incident was revealed, a preliminary scoping survey was undertaken to circumscribe contaminated areas; the plant was made safe and assessed for a controlled return to service. Later on, a recovery operation to locate and clean up the contamination was completed. Monitoring activities showed that the contamination was contained entirely within the plant. The case was rated 1 on the INES.

The licensee conducted an investigation to ascertain the direct and root causes of the incident and circulated this information across the Dounreay site. The regulator (ONR) reviewed, and was satisfied with, both the immediate response of the licensee, the subsequent investigation, and the resulting action plan. The enforcement of the proposed licensee's actions is monitored as part of ONR's inspections.

for HWCTR was in situ decommissioning with the preliminary removal of reactor vessel and steam generators from the reactor building (see a detail of the removal operations in Annex I-1.3 Fig 3).

Removal of the reactor vessel required first the removal of several shielding sleeves that shrouded each of the instrumentation sleeves entering the bottom of the vessel. To do the removal, the "friction fit" shielding sleeves had to be slid over the end of the instrumentation sleeves. This required as a preliminary step that all tubing/cabling linked to the bottom of the reactor vessel be removed.

On November 2, 2010, work was underway to remove the instrumentation. As the instrumentation was being pulled out, one of three small ion chambers was removed from the instrumentation sleeve. The ion chambers had not been noted in the prejob phase. A higher than anticipated dose rate was detected when one of the ion chambers came out of the reactor vessel. The workers' personal dosimeters alarmed right away. The workers at once exited the work area as required by procedure. The dose rate 80 mSv/h was measured at 5 cm from the ion chamber. The three workers received, respectively, a whole body dose of 0.025, 0.027, and 0.056 mSv.

6.25.2 Analysis; solutions

During the work planning, the job-specific radiological work permit (JSRWP) was developed to identify and mitigate the expected hazards. Elevated contamination levels were identified as requiring mitigation. Prior to the removal of the instrumentation, the potential for higher dose rates was recognized and subjected to specific radiological protection considerations. The JSRWP placed controls for monitoring high-dose rates as the tubing was being removed. The personal dosimeters were set to alarm at 1 mSv/h. Although the level of the dose rate was not expected so high as it happened to be, some controls to mitigate the unanticipated event were installed.

The ion chambers were not identified in planning, and no specific measures had been taken to counteract their impacts. In assessing the incident, it was determined that one member of the project team was aware of the presence of the ion chambers but missed to share this information with the other team members before leaving SRS 2 weeks prior to work beginning. At the planning stage, a strategy to remove the shielding sleeves without removing the instrumentation had been considered at a point but discarded later. The worker's failure to share this important information is a blatant violation of the feedback component of planning. Besides, the presence of the three ion chambers was not noted regardless of the team's review of over 1400 drawings. These two major inadequacies resulted in an incomplete hazard assessment of the impending work.

6.25.3 Lessons learned

The decommissioning of nuclear plants that have been unused for many years requires considerable preliminary research in support of planning work, especially when personnel familiar with the plant are no longer available.

This research must include a complete search of all available drawings. However, personnel should be aware that drawings for old, unused installations may not always be available and may not always reflect updated, as-built conditions. Therefore, the work planners should not rely completely upon the outcomes of drawing searches but should also consider interfering conditions and learn from experience acquired in similar projects.

Any issues that are known to individual members of the project team should be shared with the others to ensure during the planning stage that the issues are fully reviewed by competent people, the related hazards assessed, and proper controls implemented.

For example, hazard controls should always assume a risk of inadvertently handling an activated component during the removal of internals from a reactor vessel. In this case, the technical personnel should evaluate the potential dose associated with the handling of those components and install proper controls to avoid or mitigate exposures.

6.26 Worker exposure from a puncture wound

You can never plan the future by the past.

Edmund Burke (1729–1797)

6.26.1 Decommissioning process; problem encountered 1; Sellafield Centre; February 2017 (IOSH, 2019)

Routine glovebox work was being carried out to empty a glovebox sump to clear a level alarm. From the information available to the writer, it is unclear whether the event occurred during actual decommissioning operations; however, the case could be typical of decommissioning.

The worker initially used the normal plant methods; with this approach, no equipment needs to be removed from the sump.

As this approach was unsuccessful, the operative started to remove and clean the probe, a task he had performed before. The probe as supplied has to be adjusted to the right length, and the item should be made "glovebox ready." This includes the removal of any sharp edges.

While the probe was being replaced, a cable got jumbled, which prevented the probe assembly from being positioned properly. Trying to remove this cable, the operative felt a sharp pain and requested at once the assistance of Health Physics.

This event caused a wound and internal radiation exposure.

The worker was wearing two pairs of rubber gloves and a commercial ambidextrous glove for the removal and cleaning of the probe. No additional protection (i.e., cut-resistant gloves) was required for this task because the probe was not supposed to have any sharp edges.

Follow-up monitoring by Health Physics a few weeks after the event identified a small area of contamination, which required further medical help and had to be removed.

An interim dose assessment indicated that the worker had received an internal dose of approximately 80 mSv, which is four times the equivalent annual limit for total body effective dose of 20 mSv. This event was rated at INES level 2.

6.26.2 Decommissioning process; problem encountered 2; Karlsruhe Research Centre (Pfeiffer and Goerinch, 2006)

In the scrapping installation of Karlsruhe Central Decontamination Department, a compacted drum got jammed in the compactor discharge unit. While a worker was removing the material by hand, he slipped and the top side of his ring finger was slightly injured. The wound was found to be radiologically contaminated. The event was rated 0 on the INES and N (normal) in terms of notification priority.

The incident can be attributed to the following causes:

- Although being available, the tools suitable for loosening the jammed drum were not employed.
- The prescription forbidding removal of a jammed drum by hand was ignored.

The following actions were taken:

- Better gloves were procured to protect workers even in case of wrong behavior
- Additional safety instructions were provided the workers taking account of this event.
- As the event had disclosed first aid inadequacies, specific training was provided on how to treat an open wound properly. Besides, a program was launched to regularly check the inventory of first aid boxes and install additional orientation signs to nearest boxes as needed.

6.26.3 Lessons learned

The two events presented in this section highlight a common problem in nuclear decommissioning. Due to having accomplished the same task many times, a worker can feel a sense of overconfidence or complacency resulting in poor identification of workplace hazards. In particular, items presenting sharp edges or other hazardous surfaces may not be immediately recognized.

6.27 Protection of adjacent facilities during dismantling (IAEA, 2005)

6.27.1 Decommissioning process; problem encountered; impacts

Many vulnerable buildings that are manned and that contain nuclear and process equipment are located in the shadow of the Windscale Pile stacks. Debris or tools dropped from the stack during its decommissioning could perforate the roof and cause serious personnel injury or even catastrophic impacts. As it was realized that it would be unfeasible to render the adjacent buildings unpassable for all objects that could fall from the stacks, it was decided to restrict the weight of items handled outside the edge of the stack working platform to 25 kg (the weight of a 7-m-long scaffold tube); moreover, a protective timber decking was installed above the buildings to reduce a possible damage (IAEA, 2005).

A similar case can be taken from the ongoing predecommissioning activities at the severely damaged Fukushima Daiichi NPP, Japan. When the nuclear disaster hit in March 2011, highly contaminated vapor was released through the stack of the Units 1–2. Metal poles used to support the stack were found to have been damaged due to a hydrogen explosion at Unit 1. The exhaust stack, which is 120 m tall and 3.2 m in diameter, was used for the Unit 1 and 2 reactors.

Immediately after the Unit 1 meltdown, radiation levels at the base of the stack exceeded 10 Sv/h. In 2015, radiation levels were still around 2 Sv/h, the highest value among all outdoor areas of the Fukushima site. These radiation fields are too high for human access, so the stack dismantling work must be remotely controlled. There were also concerns that the unstable stack could collapse, possibly impacting on nearby buildings and facilities.

The dismantling equipment sliced sections of the stack top-down. The segmented pieces are stored on-site. The licensee (TEPCO) dismantled the upper half of the stack (60 m) in April 2020.

A remote control room was established in a large readapted bus placed some 200 m from the site. Workers maneuvered the cutting equipment while watching the work area from 140 video cameras. A 750-t crane was used to lift the dismantling equipment over the stack (Asahi, 2019).

6.27.2 Lesson learned

As it is not always feasible to fully eliminate a hazard, a trade-off can often be identified through a combination of engineering and administrative measures. Several methods are described in IAEA (2005) to prevent damage to surrounding buildings or personnel during the dismantling of nuclear stacks, especially when entire stacks are toppled by explosives in one go.

6.28 Internal contamination of workers through injury due to personal protective equipment deficiencies, French installations, 2009/2010 (IRSN, 2012)

Two events involving internal contamination of decommissioning workers occurred in France in 2009–10 and are reported below. The most significant event occurred at the MAU plant, AREVA NC La Hague site, where a worker had his hand punctured by a contaminated metal wire through a protective glove, resulting in serious internal contamination. This event and the one that occurred at the CEA-Grenoble LAMA facility are illustrated in the following. Both events highlight the importance of the prejob stage to identify all hazards and take preventive and mitigation measures.

6.28.1 Decommissioning process; problem encountered; MAU unit of the UP2-400 plant, AREVA NC La Hague Site, France; November 19, 2009

The MAU (this French acronym means intermediate-level uranium) unit is part of the first UP2-400-irradiated fuel processing plant at the AREVA NC La Hague. The UP2-400 plant has been in a shutdown phase since late 2003.

MAU separated the uranium and plutonium contained in a nitric acid solution, and then it purified uranium. Plutonium removed in the aqueous phase was sent to the MAPu unit at the same plant.

On the day of the accident, a contractor was contaminated during a cell dusting operation. He was at work in a leak-tight ventilated suit when his right hand struck a metal wire with a pipe identification tag to a pipe. The metal wire perforated the contractor's protective gloves and pricked the worker's hand, resulting in internal contamination.

The contaminated employee was treated at once by the site's radiation protection and medical departments. Initial evaluations indicated that the worker had received internal α-contamination. Tests estimated that the effective dose received was over the statutory annual dose limit (20 mSv). Based on the impacts of this event, ASN rated it as INES level 2.

The risk of piercing PPE had been considered by the responsible organization in planning for the cell dusting operation. In particular, workers had protections adapted to projecting corners of equipment, piping, and cable supports, as well as to identification tags for piping and securing wires. The organization determined that the event was due to the metal wire not being noticed during the workplace preparatory phase. Being the wire placed behind the piping, it was not readily visible. AREVA NC identified the following causal factors: (1) the protective gloves were not fitted to resist pricks; (2) many obstacles within the cell which hindered access to piping; and (3) the inadequate workplace lighting which explains why the wire was not clearly visible. AREVA NC discontinued this decontamination work at the MAU unit to establish better working conditions. In particular, AREVA NC looked for an "antiprick" glove, which would be better fitted to the activities planned. Tests found that the gloves' resistance-to-puncture index, which is a standard indicator, poorly correlated with the actual resistance to pricking. AREVA NC also determined that kevlar gloves, while being more resistant to tearing and cutting, ensured much less prick protection than leather gloves. In conclusion, AREVA NC mandated the use of leather gloves for dusting operations in cells classified as restricted "red" areas at the MAU unit and broadened this order to all decontamination activities at La Hague. Furthermore, AREVA NC took action to improve working conditions, e.g., the lighting in work areas. IRSN stresses that the effectiveness of leather gloves against pricking may be limited, as shown by the incident of June 14, 2010, at a solid waste store, SRS, South Carolina, United States. In this incident, a worker wearing several pairs of gloves, including leather gloves, was pricked with a contaminated metal wire during an inventory activity. The effective dose received was estimated at over 20 mSv (the same order of magnitude as the MAU incident). This event was related to the worker missing to follow working procedures but proved also that leather gloves may be unfit to avoid pricking with pointed items.

6.28.2 Decommissioning process; problem encountered; impacts: CEA-Grenoble LAMA Facility, France, April 28, 2010

The CEA-Grenoble Active Materials Analysis Laboratory (French acronym is LAMA) is an ex-laboratory where uranium- or plutonium-based fuels and structural reactor materials used to be examined after irradiation. In September 2008, the licensee CEA was authorized to perform final shutdown and dismantling of this plant. In accordance with the provisions of this authorization, CEA carried out the dismantling activities of LAMA from 2009 to 2015. The CEA sent the French regulator (ASN), a delicensing application for LAMA in March 2015. On the basis of its inspections and its assessment of further elements provided by CEA in 2016, ASN determined that the LAMA had been properly decommissioned, and the project objectives had been achieved. In view of the very low remaining radiological impact of the site, it was not considered necessary to establish any active institutional control. The draft site delicensing resolution was made available for public consultation in March 2017. ASN delicensed LAMA in August 2017.

On April 28, 2010, a worker was injured during operations to decontaminate very-high-level chamber no. 2, which presents contamination and exposure hazards (it was classified as a limited stay area). The worker was wearing full PPE including respiratory protection and injured his left buttock on a protruding metal piece while he was entering the cell. Initial evaluations indicated that the worker had been internally contaminated. It was estimated that the effective dose received was below the annual statutory dose limit (20 mSv). ASN classified this event as INES level 1.

The plant organization determined that this incident was due to poor visibility in the work area: The visibility issue in turn was caused by low lighting, many obstacles, and a violation of operating procedures involving the use of the tool that injured the worker. These causes were similar to those that led to the aforementioned MAU event.

To avoid recurrence of this event, CEA established specific lighting conditions in the work area. Measures were also taken to limit the number of obstacles and eliminate sharp tools. Finally, the plant organization stressed to workers that the safety instructions must be observed.

6.28.3 Lessons learned

The two aforementioned events highlight the high risk of serious internal contamination during D&D activities in α-contaminated environments. These events indicate that even mild injuries may result in significant internal contamination. Risk management of these operations relies mostly on the planning phase. The feedback from the two events given in this section shows that, while plant organizations should be attentive to the selection of the best fit PPE, it is also crucial that they pay attention to work conditions (lighting, obstacles, protruding tools, etc.) and to training workers and maintaining a valid safety culture during the entire D&D process.

6.29 Internal contamination to dismantling workers due to improper practices for preventing the risk of occupational exposure (IRSN, 2015)

6.29.1 Decommissioning process, Fontenay-aux-Roses Centre, France

The PETRUS complex, currently under decommissioning, was used for R&D activities on transuranics. The main components of PETRUS were a shielded cell line, a room with old tanks for liquids, an off-loading area, and connecting passageways. The room on the second basement floor had two vulcathene tanks, which once stored HLW, and three tanks which once stored very high-level waste (VHLW). (Vulcathene is a readily (de)mountable piping system for safely transferring chemical waste from laboratories in research and industrial facilities, etc., in worldwide use over a 50+-year period. As a material, vulcathene is manufactured from copolymer polypropylene with 3% carbon black ultraviolet stabilizer).

The appearance of holes in the vulcathene tanks in 1974 rendered them unusable and caused severe contamination of the room, making it off-limits to workers.

Prior to PETRUS dismantling, some equipment had to be removed, including the sampling piping for the vulcathene tanks located above the tank room, and the chemical injection system on the shielded cell line roof.

6.29.2 Problem encountered 1; impacts: February 2013

The incident scene was the personnel airlock intended for the removal of the vulcathene tank effluent sampling system. A worker wearing a protective suit and an air-purifying respirator (APR) went down on one knee to the floor to take contamination smears. The worker had to take this posture because the personnel airlock was very small (mainly due to the lack of in-line removal of earlier waste). Later on, α-contamination was measured on the protective suit, the worker's knee, and on the APR filter. Following decontamination of the suit still, some contamination remained on the worker's knee. The medical department took over, and the radiotoxicological findings indicated a committed dose (internal contamination transferred via the skin) at about half the regulatory annual limit. Tests carried out on the personnel airlock showed smearable ^{238}Pu and ^{137}Cs contamination.

6.29.3 Problem encountered 2; impacts: March 2013

During preparation for dismantling of the chemical injection system on the roof of the shielded cell line, a radiological mapping of the parts targeted for removal indicated some smearable α-contamination. The loose contamination was cleaned up, and the fixed contamination painted over. During the dismantling of the piping supports, the gloves and APR of a worker were discovered to be contaminated. The APR still being worn by the worker was decontaminated, and a new filter was substituted; this operation was carried out on the shielded cell line roof where the use of an APR (with filter in place) is compulsory. These activities were conducted without informing the radiation protection unit.

When the work was over, a worker's nasal sample revealed persisting contamination, and radiotoxicological findings estimated a committed dose less than 25% of the regulatory annual limit.

The regulatory body rated both events as INES level 1.

6.29.4 Lessons learned

The assessment of the first event showed that the responsible organization had insufficient operational records of the vulcathene tank effluent sampling system: The contamination, which had come from behind the sampling system structure, was supposed to originate from an unrecorded leak or a waste spill. Moreover, as there were no remaining liquids in the personnel airlock and no smearable contamination was detected, the protective suit that was initially considered (a ventilated leak-tight suit) was replaced by the combination of a protective suit and the APR. This change, which was less fitted to the working environment, proved inadequate in the second aforementioned case; actually, several previous contamination incidents on the roof of the PETRUS shielded cell line had been reported.

The radiological mapping carried out before the work started was not complete due to the inaccessibility of some work areas. To compensate for this lack of information, radiation controls of the work areas had been performed as the project moved on and new areas became accessible to detect in a timely manner any contamination missed at the start of the project. However, despite the facility's history, which indicated a concrete chance of old contamination, no continuous air monitoring was used or measures taken to limit the spread of contamination (e.g., the installation of a ventilated personnel airlock).

Management's corrective action to avoid the recurrence of these events was to increase the level of awareness of those in charge of these operations including the timely alerting of the radiation protection unit when a radiological event occurred or a deviation from routine activities was being considered. Management indicated also that it planned to

- reconsider the practice of replacing an APR filter in the case of contamination and, if necessary, write a specific procedure;
- establish radiation protection procedures addressing each task planned to be carried out on the PETRUS roof.

The following generic lessons can be learned from these two events:

- When the operational records for the equipment being addressed are insufficient, or when access to the work area is problematic, radiation readings should be taken as the work moves on; this is especially important for confirming that the radiation control measures taken are consistent with the risk as assessed.
- If it is proposed to change the PPE initially selected for the radiation work, the change must be supported by a specific risk assessment and confirmed by radiation protection experts.
- Changing the APR filter in an area where an APR must be worn should not be allowed as this operation could contaminate the worker concerned. One exception to this policy can be granted if the filter is clogged and the worker's respiration is made difficult; for this special case, an ad hoc procedure should be made available, and clear instructions and training should be provided to the worker.

6.30 Contamination through inhalation, CEA Cadarache Site, France, problems encountered 1 and 2; Saint-Laurent-des-Eaux Site, problem encountered 3 (IRSN, 2017)

6.30.1 Decommissioning process

The plutonium technology plant (ATPu) at Cadarache fabricated uranium and plutonium mixed (MOX) fuels for sodium and light water reactors in 1962–2003. The ATPu was closed in June 2008. Over the 40 years the MOX plant operated, it recycled more than 50 tonnes of plutonium. Operations to clean up and dismantle the MOX plant began in 2009 and primarily concerned the ATPu and the Chemical Purification Laboratory. The decommissioning decree was issued in March 2009. The licensee undertook the cleanup and dismantling of equipment, the reconditioning of operational waste, etc. Since then, more than 460 glove boxes, 30 tanks, and 4 km of piping have been segmented, removed, and packaged for disposal. Areva announced in March 2017 that around 60 emptied process areas have now been handed back to the CEA, the Cadarache licensee.

6.30.2 Problem encountered 1; impacts: May 2015

Two workers were carrying out cutting activities in a work airlock classified as a high-risk area for contamination. To avoid the contamination risk through inhalation, the workers were equipped with suitable PPE (gloves, sealed ventilated vinyl suit with air supply or respirator mask, etc.). Upon completion of the work, the workers left the work airlock and were undressing, when the radiation monitors in the undressing airlock and the first monitoring airlock signaled contamination.

The radiological controls conducted by the radiation protection service disclosed contamination spots on the clothing of both workers. On the other hand, no contamination was detected on their bodies.

Radiotoxicological tests were prescribed by the occupational health physicist; the findings showed for one of the two workers a committed effective dose—due to internal contamination—above 25% of the annual regulatory limit for workers (20 mSv/year).

6.30.3 Problem encountered 2; impacts: June 2016

During a regular check (total body counting), internal contamination was measured on an operative working in controlled areas of the ATPu plant. Further tests estimated a committed effective dose between 25% and 100% of the annual regulatory limit.

Problems 1 and 2 were both rated level 1 on the INES.

6.30.4 Decommissioning process; problem encountered 3; impacts: June 2016

The Saint-Laurent-des-Eaux site has two graphite-moderated gas-cooled (GCR) reactors, which were commissioned in 1969 and 1971 and were generating until 1990 and 1992, respectively. Their permanent shutdown of both reactors was announced in April 1994, and the decree authorizing their decommissioning was issued in May 2010.

The event in question was reported in June 2016 following the discovery by a medical lab of α-contamination on 25 workers: These workers had worked from March 2014 to April 2016 on preparatory dismantling and removal of legacy α-contaminated waste. For 4 of the 25 workers, the estimated dose ranged 0.5−1.34 mSv, which is less than 1/10 of the annual limit of 20 mSv; for the other 21, the results were not significant.

This event was rated level 0 on the INES.

6.30.5 Analysis; solutions

6.30.5.1 Problem 1:

The source of the active particles, which produced the internal contamination of the two workers, could not be firmly identified: Two possible sources were the presence of mobile contamination in the undressing airlock or the workers' PPE. Management believed that active particles could have been suspended in air when the airlock doors were opened and closed, disturbing the ventilation flow and creating a blow of air contamination.

The likeliest contamination path assumed by the licensee is inhalation. This could have occurred during the removal of the worker's PPE if a seal flaw in the respirator mask led to air inhalation. This seal flaw could have been originated by an improper or sudden movement during the undressing.

6.30.5.2 Problem 2:

Full-range investigations were conducted to determine the cause of the internal contamination. The investigations searched workplace logs, record data for the areas at contamination risk, interviews with workers and licensee staff, and all radiological reports issued while the worker was at the plant. From this documentation, it was precisely identified where and how long the operative had worked between the date of the last negative radiotoxicology test and the date when the event was disclosed.

The historical analysis indicated that the inhalation originated in all likelihood during the repair of the protective vinyl cover on a work airlock when the worker did not have a respirator mask. The work was carried out without a mask because a prior check had shown no contamination on the exposed side of the vinyl cover being replaced.

6.30.5.3 Problem 3:

The analyses conducted by the licensee highlighted PPE inadequacies as follows:

- The air supply of the ventilated PPE used in some projects has certain design flaws: Consequently, some workers did not wear the suit correctly.
- The cartridge masks were assigned to individual workers without complying with best practice: For example the adherence of the mask to the wearer's face features was disregarded.
- It was not possible with some mask models to fit the adjusted straps in place using the buckles.

Additionally to the PPE design and use problems, the workers found it difficult to understand the undressing procedures. The available procedures were not specific enough, and the workers had not been given any regular training in these procedures. Furthermore, there were no monitoring procedures when leaving the airlocks, which implied that the bad practices were not singled out.

6.30.6 Lessons learned

In the nuclear industry, when working in areas subject to contamination risk, workers wear PPE to be protected from contamination.

The aforementioned events point to several problems, including the management of radiological cleanliness in the undressing airlock (radiological surveys, periodic decontamination, waste removal, etc.); the management of dressing and undressing activities; and the fitting of respirator masks.

Concerning radiological cleanliness, the cleanup and radiological checks of undressing airlocks should be performed at regular intervals. Besides, the radiological cleanliness of the airlocks also depends on control of the dispersed contamination that may deposit on workers' suits: For this control to be effective and prevent active substances from becoming airborne, it is good practice to apply a fixative to the PPE before the workers undress. However, attention to details is imperative here (e.g., when too short drying time is allowed to the fixative).

Regarding the undressing itself, there can be several mechanisms causing the internal contamination of the workers when their PPE is removed: haste due to complex PPE design (e.g., loss of air supply to air-fed suits), improper movements (tight airlocks, ambiguous or incomplete procedures, lack of specific or refreshing training, complacency about the undressing operations etc.), and poor monitoring (e.g., the monitoring person is not familiar with good and bad practices).

The work airlock and the undressing areas should be designed to facilitate operations (e.g., transparent panels to ease monitoring).

Regarding respiratory equipment, any changes in the wearer's face features shape (e.g., if they have not shaved for several days) can affect the fitting of the mask. A systematic seal test specific to each individual mask before work and the fitting of the mask to the worker's face features should be more widely understood, enforced, and shared by all workers involved.

6.31 Fire Event, Monts D'Arrée Site, France, September 23, 2015 (IRSN, 2017)

6.31.1 Decommissioning process

The EL4 nuclear reactor at Monts D'Arrée Site in Brittany was a prototype power reactor, fueled with low enriched uranium, moderated with heavy water, and cooled with carbon dioxide gas (CO_2). The reactor was generating in the period 1967–1985, when it was finally shutdown. Decommissioning Stage 2 began in 1997, consisting of

- decontamination and demolition of the buildings except for the reactor vessel (RV);
- removal of the nuclear waste; and
- isolation of the RV.

On December 13, 2000, a rise in the groundwater level caused a flood in the chemical processing building (an incident similar to that described in Section 7.2). In January 2001, a fire broke out in the joint inner building. In 2005, stage 2 was officially complete.

As part of the partial dismantling of the plant, authorized on July 27, 2011, two heat exchangers employed for cooling the reactor with CO_2 during its operational lifetime were dismantled in the period October 2013–March 2015. The cleanup phase of this worksite, including the treatment and conditioning of resulting waste, was performed in 2015.

6.31.2 Problem encountered; impacts

On September 23, 2015, toward the completion of cleanup activities, some workers began to cut up a tool that was considered waste, inside a reactor containment work airlock. No special radiological risk was foreseen in this activity because of the very small radiological inventory.

After the waste had been cut up with a grinder, one of the workers noticed that a fire had started where decontamination wipes wet with decontaminating and degreasing solvent had been placed. The workers sounded the alarm and tried to extinguish the fire unsuccessfully. They then evacuated the airlock.

With the fire raging, the on-site emergency plan was activated, which mobilized the public services and the regulator's emergency response.

The local fire brigade quickly put out the fire.

Radioactive contamination checks conducted on all workers discovered no contamination.

However, four workers required medical attention due to smoke inhalation.

The plant's ventilation system immediately shut down. No radioactive contamination was detected around the plant; the lack of environmental impact was confirmed by in situ measurements made by the regulator's mobile unit.

Because the fire was quickly put out and no significant contamination was found in the work airlock, the on-site emergency was recalled on the following day.

The regulatory body rated this event at INES level 1. The event is similar to those described in Sections 5.28, 5.33, etc.

6.31.3 Analysis; solutions

Following this event, the regulatory body conducted an inspection of the site to investigate the causes of the fire and the corrective measures taken by the licensee (Électricité de France, EDF). EDF provided further details during the assessment of the event.

Because of the modest quantity remaining of the decontaminant (3 L), which was insufficient to make up a liquid waste package (200 L), the workers soaked wipes in the decontaminant and left them to evaporate on the floor inside the airlock, near an air intake. As per the "no-free-liquids" policy, the wipes would be then qualified for solidification. This practice had already been used in other EDF's sites.

The solvent was not marked "flammable" (there was no pictogram); therefore, the workers believed that there was no risk from placing the wipes a few meters from the cutting operation. However, the decontaminant had a "flash point" close to the threshold at which a product is defined "flammable." The "flash point" of a product (defined under standard conditions) is the lowest temperature at which a given organic compound gives off sufficient vapor to ignite in air when in contact with a flame or hot spot. A product is classified "flammable" when its "flash point" is below 60°C. The "flash point" temperature is stated in the product's safety data sheet but not on its tag.

Besides, due to the difficulties with cutting the said waste with a sabre saw as initially planned (cold cutting), it was decided to use a grinding machine (hot cutting see Fig. 6.5) but without formalizing an ad hoc hot work permit. The worker assumed that the generic risk assessment for the cold dismantling activities was enough. Instead, sparks ejected from the grinding machine to the combustible wipes started the fire (possibly strengthened by the air intake).

The identified causes include

- the available operational procedures covered the main activities of heat exchanger dismantling except its decontamination;
- the cutting machine produced hot spots: the grinding technique had not been planned in advance and the required hot work permit had not been obtained;
- the practice of using up remaining solvent by wetting wipes in it and letting the solvent evaporate increased the combustible inventory in the airlock (the low flash point of the solvent contributed to this condition).

Further to the event, EDF revised its procedures to give more focus to the specific conditions of the cleanup phase of a dismantling worksite including related hazards.

In consideration of the fire hazards associated with the combined use of wipes and solvent (especially during hot work), EDF forbade the evaporation practice on all its sites and reminded the workers of the need to keep the use of flammable products to a minimum.

EDF also launched a national consultation on what to do with unused amounts of liquid chemical waste, which are treated as radioactive waste and carried out tests to validate the use of a polymer encapsulating small amounts of liquids. In the longer term, other options are being evaluated, e.g., nonnuclear disposal.

6.31.4 Lessons learned

The aforementioned event reinforces the need to evaluate risks in all activities of a D&D project, even when the activities are of short duration and modest radiological risk. Decommissioning often includes many parallel activities, each with its own risks (spreading of radioactive materials, fire, etc.) and constraints (confined spaces, noise, etc.). Vigilance must be exerted at all times.

This event also raises a potential issue for low-risk sites, where few operatives attend the activities.

In compliance with the on-site emergency plan requirements for the site in question, only four EDF staff ran the site's contacts with all the parties involved in emergency response, namely, EDF's corporate management, local authorities, and regulators.

The site organization set up to deal with emergencies did not attract special concerns or questions by the regulators, in application of the graded approach to risk. However, this case discloses the limitations of the graded approach. Eventually, EDF arranged for on-site emergency drills in 2017, which the regulatory body attended.

6.32 Radiological contamination event during Separations Process Research Unit demolition, September 29, 2010, Niskayuna, New York, United States (DOE, 2010c)

6.32.1 Decommissioning process; problem encountered; impacts

SPRU, situated at the Knolls Atomic Power Laboratory (KAPL), was in operation 1950–53 as a pilot plant to investigate on the laboratory scale the REDOX and PUREX processes to extract uranium and plutonium from spent fuel. Its research work supported operations at the Hanford Site and SRS.

Construction of Buildings H2 and G2 was completed in 1949. The research work contaminated the SPRU installations and land and led to the decision to remediate the site. KAPL went on using Building H2 into the late 1990s for waste treatment. Practically, all areas of H2 were under radiological control.

In late 2007, the DOE awarded a 4-year contract to Washington Group International (WGI) to accomplish deactivation, demolition, and removal of the SPRU installations (Buildings G2, H2, the Tank Enclosures, and the connecting tunnel); cleanup and environmental remediation of the underground and adjacent contaminated soil; and the decontamination of the piping linking the SPRU installations to other installations in operation. Demolition of Buildings H2 and G2, removal of concrete debris, verification that the remaining soil complies with cleanup criteria, backfilling the excavations, and restoring the work area had been completed by late 2018. A small amount of transuranic waste containing hazardous chemicals (mixed waste) remains on-site. This waste is to be disposed of at the Waste Isolation Pilot Plant. Fig. 6.10 shows the H2 basement excavation at the completion of the work.

Open-air demolition of H2 started on September 23, 2010, using heavy equipment. Within a week, demolition of Building H2 had accomplished the demolition of the roof structure, the stack, and nearly all the external and internal walls. Some of the demolition debris had been put into intermodal containers for disposal (an intermodal container aka ISO container is a large standard transport container, suitable for intermodal cargo transport: These containers can be transferred across different modes of shipment—ship, rail,

FIGURE 6.10 H2 basement excavation. *(Credit to DOE).*

truck—without unloading and reloading their cargo each time). Other debris was lying on the H2 platform or where it had been generated. Demolition crews had started to remove six evaporator system columns. A worker was segmenting the condenser/columns and was halted when an electrician and a radiological control technician (RCT) observed a white whiff coming from one of the condenser/columns. The RCT checked the area for explosive gases and then allowed the team to resume demolition. No radiological surveys were carried out at this time. Several components (flash and separator columns) were laid down on the H2 footprint platform. The water spray operative noted the wind was so strong that he had to considerably redirect the spray nozzle to compensate for the pressure of the wind.

While removing a separator column from its evaporator cell, a bolt was thrown into the air and hit an electrical transformer outside the construction area. The bolt was located, surveyed, discovered to have fixed contamination, and removed. Demolition operations were stopped, and the safety area was enlarged beyond the demolition area. The RCT who performed the survey went on vacation: His supervisor subsequently recorded a fixed contamination less than 1 Bq/cm^2 β-γ. The use of a wrong instrument resulted in underestimation of the surface contamination (the actual value would have been around 4 Bq/cm^2). A bit later, due to high background radiation readings, the RCT attending the operations at that time ordered the demolition team out of the affected area and conducted scanning with portable instrumentation, finding contamination on both boots of each of the four workers. The highest levels recorded were over 10 Bq/cm^2 β-γ (below reportable levels per DOE procedure). In response to the event, the area surrounding H2 and between G2 and H2 was classified as contaminated. and an airborne radioactivity area. The demolition area of the H2 platform was surveyed, and contamination levels were found up to 100 Bq/cm^2 β-γ and 2 Bq/cm^2 α. Further contamination surveys in the following days identified more areas of contamination, which were classified accordingly. In preparation for an incoming tropical storm, a pile of debris off the H2 platform was pushed back onto the platform with heavy equipment. The workers sprayed encapsulation material (fixative) on the debris piles and the separator columns on H2 platform. Other preparatory work to control storm water included an interim berm and the covering of debris with tarpaulins. Those preparations and the extremely heavy rainfall that followed worsened both the contamination event and the possibility to locate contamination and its original source.

A few days later on-site contamination levels on the H2 platform were found close to 280 Bq/cm^2 (40 μGy/h) removable β-γ. Contamination (albeit at low levels) was detected over an area of around 10,000 m_2 on the KAPL site as a consequence of the incident.

6.32.2 Lessons learned

Though preliminary indications showed that low levels of contamination had been found on workers shoes and on KAPL property next to the SPRU work activities, the extent and implications of the contamination were not well determined and understood by the SPRU project management for many days. Based on the foreseen cost to remediate the accident and specific circumstances, a type B investigation was launched by DOE.

The Investigation Board determined that this incident was avoidable.

The direct cause of the event was the open air demolition of the evaporator system components.

The root causes of the accident were as follows:

- The management failure to completely understand, characterize, and control the radiological hazard.
- The management failure to implement a work control process that secured facility conditions as the work moved on.

The Board identified the following contributing causes:

- There was no plan for application of encapsulation materials. The workers applying fixative did not know the inner construction of the vessel(s).
- There was no attempt to verify the surface coverage of the fixative.
- There was overreliance on the effectiveness of the fixative to stabilize contamination.
- It was not clear how fixative was to be used when handling vessels, tanks, etc. with internal geometries.
- Decontamination was not used during removal of the evaporator system vessels.
- The work package did not incorporate the hazard controls identified in the job hazard assessment.

- The work package implementation did not guarantee all process components in H2 were identified and characterized.
- RWPs were written generically and were not tailored to the specific task.
- Tasks regarding the identification of hazards were not completed.
- The working level documents were approved without full assurance that the hazard controls were identified.
- The work plans were not adequate to implement proper radiological controls.
- No criteria was defined and approved to assign a color code to components for special handling.
- The Project did not understand the relevance of operational history and system knowledge.
- Full characterization of SPRU for D&D purposes was not completed.
- DOE's supervision of SPRU D&D did not assure programmatic inadequacies were identified and corrected.

6.33 Personnel contamination due to congested environment and other factors, Building 324, Hanford Site, June 24, 2019 (DOE, 2020)

6.33.1 Decommissioning process; problem encountered; impacts

Hanford's Building 324 supported research on highly radioactive materials during its operational lifetime 1966–1996. Demolition operations were further deferred in 2010 after significant contamination was discovered under a section of the building, which originated from leakage of HLW inside the building. Removing that contamination prior to demolishing the building is a high priority for the DOE due to the plant vicinity (300 m) to the Columbia River.

The design, testing, and procurement process of remotely operated equipment and building modifications are in progress to remove the highly contaminated soil as a precursory activity to the demolition of the plant (see also Section 6.15).

On the day of the incident, a worker carrying out decontamination work in room 18 activated the alarm of the installed personnel contamination monitor while exiting from the work area. RCTs conducted surveys and measured worker's arm contamination of around 6 Bq/cm^2 β-γ: The skin contamination was higher than the maximum contamination acceptable for exit from a CA, 0.16 Bq/cm_2 β-γ. Other contamination incidents at Building 324 are described in Section 6.15.

6.33.2 Analysis; solutions

The analysis identified one direct cause and seven contributing causes.

The direct cause was due to PPE being inadequate for the work carried out.

Contamination penetrated through the worker's semiimpermeable PPE when it came directly in contact with contaminated surfaces. Due to tight spaces in and around the excavation assembly, work being performed required physical contact between PPE and contaminated surfaces while carrying out decontamination.

Contributing cause 01 was a leak path different from what was expected.

Although engineered changes have been made to the excavation operations to monitor known leak paths and reduce the contamination that is spread from the excavation path, a new leak path, previously unknown, was crossed. On June 18, 2019 (a few days before the incident occurred), drilling operations were being carried out. At 1.2 m depth, the drill worker noticed material spreading out of the drill breakout joint. After release of contamination into room 18 above levels specified by the radiation work permit (RWP), work was interrupted to allow for decontamination. The skin contamination occurred during the decontamination operations of room 18.

Contributing cause 02 is attributed to the radiological contamination being very mobile.

The unexpected mobility of the contamination revealed while drilling increases the migration throughout room 18. This contamination was of a more dispersible form than encountered before. Normally, contamination originating from the 324 Radiochemical Engineering Cell is associated with oily substances, which makes it more readily detected and less mobile (i.e., tends to stay near the floor). On the contrary, the new contamination percolates easily through the dry sand under room 18.

Contributing cause 03 may be poor work practice when undressing.

While undressing, workers remove the outer contaminated clothing in a way to avoid that the inner layers of clothing be contaminated. If the undressing is conducted improperly, there is a chance of cross-contamination between contaminated and noncontaminated surfaces and clothing.

Contributing cause 04 is the insufficient space for undressing at the base of stairwell.

Due to the layout of the plant, the next airspace outside of room 18 was the stairwell landing (approximately 1.4 m × 1.4 m). The small size of this area made it improper for removal of the highly contaminated outer pair of coveralls without touching door, walls, and possibly coworkers. To access a larger area to undress, the workers would have to go up a flight of stairs (Corridor 13) into a Gallery, with a concrete risk of further spreading contamination into the plant and having to classify these areas as high CAs. Due to this dilemma, the outer set of coveralls was undressed inside room 18.

Contributing cause 05 may be the use of nonaerosol hairspray instead of aerosol hairspray.

The nonaerosol (pump bottle) hairspray was applied as a fixative during the undressing process when contamination was detected on individuals outer

PPE. If applied improperly (too close/too heavy), the nonaerosol hairspray can wet the outer set of PPE and allow the contamination to migrate across the PPE.

Contributing cause 06 may be a conflict of driving factors (contamination/dose/industrial hygiene). PPE comparisons based on heat stress show that semiimpermeable Protech PPE has a lower clothing adjustment factor than the impermeable Frham Tex PPE. The latter also requires a more robust heat stress monitoring procedure (monitoring of physiological parameters). Due to the greater heat stress of carrying out work in impermeable PPE than in semi-impermeable PPE, Protech was selected. Although Protech is preferable to reduce the heat stress, it may have contributed to this incident because it is less protective against contamination via capillary absorption in presence of perspiration. Box 6.10 highlights some impacts of unusually (high, rather than low) temperature on decommissioning activities.

Contributing cause 07 is the inadequate control to mitigate environmental conditions.

Work in room 18 was labor intensive, causing the worker to sweat. Temperatures inside room 18 are generally higher than the rest of the plant by

BOX 6.10 Temperature as a physical hazard

The impacts of high (or low) temperatures depend on decommissioning location and the extent of indoor versus outdoor work. Heat and cold stress should be taken into account in the decommissioning plan. A recent, striking case study is provided by the ongoing decommissioning (or rather cleanup) at the accident-damaged Fukushima Daiichi NPP. Ever since the accident, every day about 3000 workers have been busy fixing the situation. The frequent occurrence of heat disorders has been a worry for the workers wearing PPE: Poor ventilation can be an additional problem. Disorders of heat exposure include cramps, exhaustion, and heat stroke. Heat stress can also impact concentration and cause weakness, dizziness, and fainting. Consequently, PPE wearer may make errors leading to physical accidents and even radioactive contamination. Reduced concentration may impair workers' ability to follow procedures including their readiness to respond to emergencies. Coworkers may also be impacted, for example, support workers assisting the undressing, if the PPE is wrongly removed. From March 22 to September 16, 2011, the Fukushima Labour Bureau reviewed 43 cases of cleanup workers who were impacted by heat disorders. Age of worker, date and time of occurrence, and local temperature and humidity were reviewed for each case, as well as the problem severity. Since the accident occurred Japanese institutions have been assessing the heat disorders to find out a solution preventing future heat disorders (Tsuji, 2013). As the result of these efforts, the number of heat strokes at Fukushima went down from 24 in 2011 to just 4 in 2016 (METI, 2017). See also a full discussion on heat stress and other hazards associated with the use of PPE in HSE (2008).

several degrees, and an evaporative cooler in the room may have heightened the room humidity. Excessive perspiration can produce the capillary absorption of liquids, so pushing liquid contamination across the PPE.

6.33.3 Lessons learned

A review of eight similar contamination events occurring at the 324 Facility over a 9-month period was conducted afterwards. The review identified two common elements: inadequate radiological work practices and inadequate PPE. The similarity review could have decreased the likelihood of this event, but the review was conducted too late, and therefore, any corrective actions produced from the review were not implemented at the time of the events.

This point highlights the necessity that in planning for and execution of D&D activities, an advance review of events occurring during similar work is mandatory to make more likely that all hazards are identified and effective provisions are taken in advance.

6.34 Changing conditions during demolition at Plutonium Finishing Plant, Hanford Site, Washington, United States (DOE, 2019b)

Tempora mutantur, nos et mutamur in illis (Times change, and we change with them too).

From Owen's Epigrammata (1615)

6.34.1 Decommissioning process

The Plutonium Finishing Plant (PFP), which started operation in 1949, produced most of plutonium used for US nuclear weapons. Since the plant shutdown in 1989, its remaining nuclear inventory material has been packaged and sent offsite. Demolition of its buildings, gloveboxes, and tanks is underway. PFP demolition, generally considered the most hazardous work in the DOE complex, was almost completed when several airborne contamination events occurred. Demolition work stopped but was resumed after many months (see Section 7.1 for the complete decommissioning history of the PFP).

6.34.2 Problem encountered 1; impacts

Workers were size-reducing a glove box in a room classified as an HCA. Being the glove box three stories high, there was no practical method to enclose the work area within a containment tent. Negative air pressure was employed to reduce the chance of contamination spreading outside the HCA. A control room with a shared door supervised the glove box activities and was also used as an emergency exit from the HCA. This door was not subjected to radiological contamination control. Tape had been put around the door, but the door was not sealed in case it had to be used for emergency exit. The requirements for this door did not allow it to be monitored, and the organization in charge did not realize this area was not under regular surveillance.

When workers went into the contiguous control room to carry out contamination surveys, no contamination was detected up to the door between the external (uncontrolled) area and the HCA. However, direct contamination of nearly 50 Bq/cm^2 α was measured on the tape close to the door handle, and 30 Bq/cm^2 of removable contamination was measured on the floor right in front of the door. The workers at once left the room and notified the situation. α-Contamination higher than 33 Bq/cm^2 requires a room to be classified as an HCA; in addition, enhanced controls should be installed for personnel safety such as augmented PPE including respiratory protection. No personnel contamination was reported.

6.34.3 Analysis; solutions

Apparently at some time, negative air pressure was lost, the tape on the door did not work, and contamination migrated from the HCA to the control room. Several scenarios considering air flow and room geometry were considered during the planning for the HCA work, but clearly they were not adequate to this special situation. No regular surveys of the control room and of the tape on the door were carried out. This caused the dispersion of contamination outside the work area when air flow into the work area failed.

6.34.4 Problem encountered 2; impacts

The affected worker was a member of the crew doing decontamination and demolition of the Plutonium Reclamation Facility (PRF) canyon. The crew was cleaning up concrete chips, metal shavings, and dried chemicals left over from canyon operations. Due to the high concentration of airborne contaminants in the canyon, surfaces were fogged with fixatives before human entry, and portable foggers were employed in certain work areas. During this work, a worker exceeded their annual radiological dose limit to the extremities. The worker's extremity exposure was 160 mSv, which was higher than the Hanford administrative control limit of 150 mSv.

6.34.5 Analysis; solutions

In one area, debris was very hard to remove, and more and more fixative was increasingly employed. The increased amount of fixative caused a foaming reaction in the plates employed to remove debris. This event led to the debris being handled twice, while analysis of the cause of the foaming was conducted. The waste became even harder to work with because it was treated, dried, and retreated. The procedures were reviewed, and successive changes were made. However, no attention was paid to the higher doses to the workers caused by the longer time close to the debris. When the difficulties that led to the minor changes were considered all together, the original assumptions no longer matched the actual conditions of the work.

6.34.6 Lessons learned

The two aforementioned events highlight (1) the need to look at the broader situation during work planning and procedure upgrading and (2) the need to reconsider techniques and procedures to confirm that original assumptions remain valid even under changed conditions. In either incident, management focused on smaller details of immediate interest. By failing to apply a broader view, the work package missed controls that could have avoided the incidents.

6.35 Exothermic metal reaction during converter disassembly, East Tennessee Technology Park, Oak Ridge, Tennessee, United States; June 27, 2002 (DOE, 2002)

6.35.1 Decommissioning process; problem encountered; impacts

East Tennessee Technology Park (ETTP; formerly K-25 Site) was a uranium enrichment gaseous diffusion plant for defense purposes. K-33 Building was built in 1954 and enriched uranium until 1985. At the time of the incident, the site installations were under decommissioning.

In 1997, BNFL was granted by Oak Ridge Operations Office (ORO) a contract for the ETTP Three-Building (K-29, K-31, and K-33) Decontamination and Decommissioning and Recycle Project.

This project's mission was to dismantle, remove, and dispose of the process equipment and to decontaminate the three buildings. The work involved heavy construction-type dismantling/disassembly equipment. The demolition work and identification of hazards were made more difficult by the lack of precise as-built drawings for 50-year-old buildings. Demolition of all ETTP gaseous diffusion plants was completed in 2017.

On the day of the accident, an exothermic metal reaction (hereafter reaction) occurred during disassembly of a converter using a plasma arc torch (Fig. 6.11). K-33 had 632 converters weighing 33 t each. The converters had sintered nickel tubes that, under some conditions, can produce a reaction when exposed to the kerf from the plasma arc cutting. By "kerf," it is meant here the white-hot sparks and molten metal splatter that diffuse beyond the cut made by the torch.

The reaction occurred in the tube bundle during the cutting of a support. The reaction noted by glowing hot metal generating white smoke. "Exothermic metal reaction" is the term that best specifies the event because there are no flames, although, in practice, it is a fire.

FIGURE 6.11 Maintenance mechanics train in plasma arc-cutting prior to removing km of piping at the Paducah Site. *(Credit to USDOE, Wiki Commons).*

The reaction is self-sustaining in that it does not require oxygen from the air to propagate. This, and the complex geometry of the tube bundles, makes firefighting difficult.

When the tube bundle reaction was noticed, the cutter and the fire watch tried to extinguish it with dry chemical powder and CO_2 fire extinguishers. Notwithstanding the work instructions that required the workers to leave the area, they removed the tube bundle and attempted to dismantle the reacting metal. The removed tube bundle came in contact with a flame-retardant security curtain. Once the supervisor arrived, he directed the evacuation of the area and activated the fire alarm. After the evacuation, two overhead sprinklers discharged because the security curtain was on fire. Two previous reactions had occurred during BNFL's disassembly activities in 2000 and 2001.

The event resulted in no injuries to the workers or environmental impacts, but it is quoted in this chapter because of its potentially involving radiologically contaminated areas. In fact, it highlights that radiological and non-radiological impacts are both possible in the course of one and the same event.

6.35.2 Analysis, solutions

The observed cause of the reaction was the kerf from the plasma arc torch intersecting the tube bundle.

The Investigation Board determined that this event was avoidable. It highlighted weaknesses in the core functions of ISM. Although the workers and supervisors understood that the vicinity of the plasma arc torch presented an increased risk, work controls were not properly implemented to minimize the likelihood of the reaction. Different work shifts had taken various precautions to prevent the reaction, but these measures were not incorporated into the work control process or shared with other shifts.

The Board found two root causes for this incident:

1. The corrective actions from the previous two reactions in 2000 and 2001 did not frame these reactions as an avoidable result of decommissioning operations.
2. BNFL's management systems and processes were not effective in avoiding the tube bundle reaction and providing emergency response.

The following contributing causes were identified by the Board:

1. The enhanced work planning (EWP) for the converter disassembly was not observed.
2. Communications and transfer of lessons learned between shifts were inadequate.
3. The hazards related to converter dismantling were not properly identified and assessed.
4. The workers took improper emergency actions, with inadequate results.
5. The response to emergency conditions at Building K-33 was inadequate.
6. EWP training was insufficient.
7. The Field Change Notice (FCN) process and the EWP change process did not identify the higher risk of a tube bundle reaction.
8. ORO's and BNFL's leniency contributed to inadequate implementation of the core functions of the ISM program.
9. The BNFL Corrective Action Program was ineffective.

6.35.3 Lessons learned

The Board focused on strengthening the work controls and the supervision of the controls to properly implement the requirements of the work package.

Additional to work control improvements, the BNFL approach in applying FCNs to work packages needs greater managerial control to ensure that hazards produced by the changes are appropriately evaluated and actions are taken to minimize their adverse impacts. A better emergency response to the tube bundle reaction event needs to be developed and fully transferred to the workers.

Corrective actions planned after the 2001 event were not implemented. BNFL had found problems with implementation of its Corrective Action Program that demanded program amendments. BNFL had admitted to issues regarding timeliness and prioritization in its nonconformance report.

In summary, the Board considered the BNFL Corrective Action Program to be ineffective and recommended the establishment and enforcement of a revised, more rigorous corrective action program.

References

Argonne National Laboratory, September 1996. Decontamination and Decommissioning of 61 Plutonium Gloveboxes in D-Wing. Building 212 Argonne National Laboratory – East: Final Project Report, ANL/D&D/TM-96/3. https://inis.iaea.org/collection/NCLCollectionStore/_Public/28/043/28043367.pdf.

Pfeiffer, W., Goerinch, H., 2006. Minor incidents during the decommissioning of prototype operations and research facilities of the Karlsruhe research centre. In: Lambert, F., Volkov, Y. (Eds.), Safety Improvements through Lessons Learned from Operational Experience in Nuclear Research Facilities. Springer, pp. 67–78. ISBN-10 1-4020-3887-9 (PB). https://studylib.net/doc/14394125/safety-improvements-through-lessons-learned-from-operatio.

The Asahi Shimbun, August 2, 2019. Work Begins to Topple Dangerous Exhaust Stack at Fukushima Plant. http://www.asahi.com/ajw/articles/AJ201908020037.html.

Nuclear Safety Authority (in French, Autorité de Sûreté Nucléaire), Incident on the ATPu Nuclear Facility (Cadarache CEA site): The ASN Rates Incident as Level 2, Draws up a Report and Suspends All Dismantling Operations, October 20, 2009. http://www.french-nuclear-safety.fr/Information/News-releases/Incident-on-the-ATPu-nuclear-facility-Cadarache-CEA-site.

Investigation of Contamination in Waste Handling Plant, November 12, 2002. http://www.caithness.org/fpb/dounreay/12nov2002.htm.

Croucher, D.W., June 1994. Criticality Safety Aspects of Decontamination and Decommissioning at Defense Nuclear Facilities, CONF-940602-3, 35th Annual Meeting of the American Nuclear Society, pp. 11–16. New Orleans, LA, US. https://inis.iaea.org/collection/NCLCollectionStore/_Public/25/058/25058500.pdf?r=1&r=1.

US Department of Energy, 2005. Lessons Learned Database. Lesson ID: LL 2005-LLNL-14 (available upon DOE authorization).

US Department of Energy, Office of Health, Safety and Security, Lessons Learned Database, Radiological Work Permit (RWP) Suspension Guide Exceeded – 105-R Disassembly Decon Room (ARRA), Lesson ID: 2010-SR-SRNS-0016, 10/26/2010a (available upon DOE authorization).

US Department of Energy, July 2002. Type B Accident Investigation Exothermic Metal Reaction Event during Converter Disassembly in Building K-33 at the East Tennessee Technology Park on June 27, 2002. DOE/ORO-2132. https://www.energy.gov/sites/prod/files/2014/04/f15/BNFLTypeBReport-final.pdf.

US Department of Energy, November 2010b. River Corridor Closure Project, 324 Building B-cell Contamination. https://www.hanford.gov/news.cfm/DOE/E1011025_2.pdf.

US Department of Energy, September 29, 2010c. Type B Accident Investigation Report – Radiological Contamination Event During Separations Process Research Unit Building H2 Demolition. https://www.energy.gov/sites/prod/files/2014/04/f14/Type_B_AI_Report_SPRU_0.pdf.

US Department of Energy, August 2003. Criticality Safety Lessons Learned in a Deactivation and Decommissioning Environment. DOE/RL-2003-3. https://it.b-ok2.org/book/2061635/81398b.

US Department of Energy, January 2012. Accident Investigation Report- Plutonium Contamination in the Zero Power Physics Reactor Facility at the Idaho National Laboratory. https://www.energy.gov/sites/prod/files/2014/04/f14/INL_AI_Report_11-08-2011.pdf.

US Department of Energy, Lessons Learned Data Base, ID No.: RCCC-2012-0014 (Available upon DOE authorization).

US Department of Energy, Unanticipated, High Dose During the Removal of Wire Flux Monitor Cabling from the HWCTR Reactor Vessel, SDD-2012-00029 https://www.dndkm.org/DOEKMDocuments/LessonsLearned/62-SDD-2012-00029_EFCOG%20Lessons%20Learned%20006_Unanticipated%20High%20Dose%20HWCTR_FINAL2.pdf.

US Department of Energy, September 2019a. Over-Pressurized Drum Event Involving Reactive Materials and Flammable Gases in Radioactive Waste, OE-2: 2019-01. https://www.energy.gov/sites/prod/files/2019/09/f66/OE-2%202019-01.pdf.

US Department of Energy, April 4, 2019b. Lessons Learned on Changing Conditions at Plutonium Finishing Plant. https://www.dndkm.org/DOEKMDocuments/LessonsLearned/119-LLB_April_2019_PFP_Demolition_03.22.19.pdf.

US Department of Energy, Occurrence Reporting and Processing System (ORPS). https://data.doe.gov/MS/asp/Main.aspx. (Updated 30 April 2020).

ehstoday.com, February 11, 2004. DOE cites Kaiser-Hill Co. for Nuclear Safety Rule Violations. http://ehstoday.com/news/ehs_imp_36839.

Fellhauer, C., March 1998. Experimental Boiling Water Reactor Decommissioning, IAEA/ANL Interregional Training Course on Decontamination and Decommissioning of Research Reactors and Other Small Nuclear Facilities. held at Argonne National Laboratory, IL, US, pp. 9–27.

Health and Safety Executive, 2008. Physiological Safety of Air-Fed Suit Use During Nuclear Decommissioning, RR 658. https://www.hse.gov.uk/Research/rrpdf/rr658.pdf.

International Atomic Energy Agency, 2004. Planning, Managing and Organizing the Decommissioning of Nuclear Facilities: Lessons Learned, IAEA-TECDOC-1394. IAEA, Vienna.

International Atomic Energy Agency, 2005. Dismantling of Contaminated Stacks at Nuclear Facilities, Technical Reports Series No. 440. IAEA, Vienna.

International Atomic Energy Agency, 2006. Decommissioning of Underground Structures, Systems and Components, Technical Reports Series No. 439. IAEA, Vienna.

International Atomic Energy Agency, 2013. INES- the International Nuclear and Radiological Event Scale- User's Manual, 2008 Edition. IAEA, Vienna.

International Atomic Energy Agency, 2015. Decommissioning of Pools in Nuclear Facilities, Nuclear Energy Series No NW-T-2.6. IAEA, Vienna.

International Atomic Energy Agency, 2016. Managing the Unexpected in Decommissioning, Nuclear Energy Series No NW-T-2.8. IAEA, Vienna.

Institution of Occupational Safety and Health, IOSH Magazine, Sellafield Worker Exposed to Plutonium Had Hand Punctured by Corroded Probe. https://www.ioshmagazine.com/2019/04/03/sellafield-worker-exposed-plutonium-had-hand-punctured-corroded-probe.

Institute for radiological Protection and Nuclear Safety (in French, Institut de Radioprotection et Sûreté Nucléaire, ISRN) Nuclear Fuel Cycle Facilities, Laboratories, Irradiators, Particle Accelerators, Under-Decommissioning Reactors and Radioactive Waste Management Facilities Safety, May 12, 2010. Lessons Learned from Events Notified between 2005 and 2008, DSU Report 215. https://www.irsn.fr/EN/publications/technical-publications/Documents/IRSN_bilan_incidents_2005-2008_va.pdf.

Institute for Radiological Protection and Nuclear Safety (in French, Institut de Radioprotection et Sûreté Nucléaire, ISRN), Safety of Laboratories, Plants, Facilities Being Dismantled, Waste Processing, Interim Storage and Disposal Facilities – Lessons Learned from Events Reported in 2009 and 2010, 2012. DSU Report 248. https://www.irsn.fr/EN/publications/technical-publications/Documents/IRSN_Report_LUDD-events_2009-2012_DSU248.pdf.

Institute for Radiological Protection and Nuclear Safety (in French, Institut de Radioprotection et Sûreté Nucléaire, ISRN), Safety at Civil Basic Nuclear Installations Other Than Nuclear Power Plants in France-Lessons Learned by IRSN from Significant Events Reported in 2013 and 2014, November 2015. https://www.irsn.fr/EN/publications/technical-publications/Documents/IRSN-Report_Safety-Nuclear-Installations-France-2013-2014.pdf.

Institute for Radiological Protection and Nuclear Safety (in French, Institut de Radioprotection et Sûreté Nucléaire, ISRN), Safety at Civil Basic Nuclear Installations Other Than Nuclear Power Plants in Operation in France – Lessons Learned By IRSN from Significant Events Reported in 2015 and 2016, November 2017. https://www.irsn.fr/EN/publications/technical-publications/Documents/IRSN_Safety-at-Nuclear-installations-other-than-NPPs-in-operation-2015-2016.pdf.

Kerr, D., 30 June–4 July, 2008. A Decommissioning Project Manager's Perspective, IBC's 24th Annual Residential Summer School on Decommissioning and Radioactive Waste Management. Cambridge, UK.

Knoxblogs- Atomic City Underground, Multiple Radioactive Waste Spills at SRS Forces Change in Procedure, September 1, 2010. In: http://knoxblogs.com/atomiccity/wp-content/uploads/sites/11/2010/09/multiple-radioactive-waste-spills-at-srs-forces-change-procedure.

LAKA Foundation LAKA (LAndelijk Kernenergie Archief – National Nuclear Energy Archive, Amsterdam, The Netherlands), Worker Contamination for Am-241 Inhalation when Manipulating a Radiation Detector, 1997. https://www.laka.org/docu/ines/event/603.

Los Alamos National Laboratory, 2010. Pre-Job Brief Guide. https://www.lanl.gov/safety/videos/ism/docs/pre_job_guide.pdf.

Japan's Ministry of Economy, March 2017. Trade and Industry (METI), Important Stories on Decommissioning. https://www.meti.go.jp/english/earthquake/nuclear/decommissioning/pdf/20170927_roadmap.pdf.

Office for Nuclear Regulation, 2013. Third Quarter 2009 Statement of Nuclear Incidents at Nuclear Installations. http://www.onr.org.uk/quarterly-stat/2009-3.htm.

Office for Nuclear Regulation, 2016. Events Reported to the NUCLEAR safety Regulator in the Period of 1 April 2001 to 31 March 2015. http://www.onr.org.uk/documents/2016/events-reported.pdf.

The Seattle Times, Unexpected Contamination Reported in HANFORD cleanup, May 20, 2019. https://www.seattletimes.com/seattle-news/setbacks-this-spring-at-a-difficult-hanford-clean-up-site-close-by-richland-and-the-columbia-river/.

Laguardia, T.S., 2012. Decommissioning of western-type light-water nuclear reactors (LWRs). In: Laraia, M. (Ed.), Nuclear Decommissioning: Planning, Execution and International Experience. Woodhead Publishing, ISBN 978-0-85709-115-4.

Tsuji, M., et al., 2013. Worker heat disorders at the Fukushima Daiichi nuclear power plant. Sangyo Eiseigaku Zasshi 55 (2), 53–58. https://doi.org/10.1539/sangyoeisei.b12008. Epub 2013 Mar 5 (article in Japanese, English abstract).

World Information Service on Energy, Uranium Project, Uranium Enrichment and Fuel Fabrication – Decommissioning Issues (Europe). https://www.wise-uranium.org/edeur.html. (Last updated 25 March 2020).

World Nuclear News, March 08, 2017. Areva Completes Cadarache MOX Decommissioning. https://www.world-nuclear-news.org/WR-Areva-completes-Cadarache-MOX-decommissioning-0803174.html.

Chapter 7

Radiological events impacting the public and/or the environment

7.1 The dismantling of the Plutonium Finishing Plant, Hanford, Washington, United States

7.1.1 Decommissioning process; problem encountered; impacts (Tricities, 2019)

The Plutonium Finishing Plant (PFP), Hanford nuclear site, operated from 1949 to 1989; it was also called Z Plant because it was the last phase of production of plutonium before the material was transferred to weapons production facilities elsewhere in the country. The PFP was a complex of more than 90 buildings. In 1991, the site mission changed to stabilization of plutonium-containing material, decontamination and decommissioning, and environmental remediation. Material conditioning, packaging, and storage were completed in December 2009 when the last containers of stabilized material were transported off-site.

The end of production left behind glove boxes and other equipment (ventilation piping, process lines) highly contaminated, asbestos, and other hazards.

The planning process on how to safely remove all of the PFP plutonium took several years before demolition of the buildings began in November 2016. For many years before that, demolition of PFP had been a very high priority on the Department of Energy's (DOE) list of facilities in danger, because of the security concerns it posed.

But the radioactive inventory of PFP posed a serious challenge to DOE staff and demolition contractors not to expose themselves, the environment, or the public to the radioactive residues.

Minor spreads of contamination had been noted as early as January 2017. During summer 2017, bioassay findings disclosed 31 workers had been radiologically exposed after dismantling work spread radioactive contamination around PFP. Demolition went on until December 2017, when air samplers

Nuclear Decommissioning Case Studies. https://doi.org/10.1016/B978-0-12-823700-7.00007-2
Copyright © 2021 Elsevier Inc. All rights reserved.

detected high radiation levels. These events caused growing tensions between workers and management and even resulted in strikes.

CH2M Hill Plateau Remediation Co., the contractor in charge of PFP demolition, called a stop work at that time to perform surveys and samples. These surveys highlighted that contamination had spread from PFP demolition site to not only the demolition workers but also areas external to the demolition site. Areas outside of administrative offices nearby as well as parked cars exhibited some low-level contamination.

Further to the December contamination event, some 300 workers were subject to bioassays, and it was proved that 11 of them had inhaled or somehow absorbed radioactive contamination. In total, in 2017, 42 Hanford workers had internal contamination in consequence of working on or near the PFP demolition site. CH2M Hill maintained the stop work for 8 months, while requesting an independent assessment team to demonstrate the ability to safely manage demolition activities including, among others, a review of the monitoring and surveillance program (CH2M Hill, 2018a). In summer 2018, the US DOE completed their own review (DOE, 2018a). From December 2017 to September 2018, no major demolition activities were allowed by DOE until the site completed stabilization activities (e.g., causal analysis, extent of condition) and successfully completed the aforementioned self-assessment.

The DOE had several remarks about CH2M Hill's practices. Remarks included failure in expected performance in regard to decision-making, less than rigorous use of processes, and poor work force communications. Other issues included demolition work rates exceeding the work control specifications derived from computer models of release plume migration and diluting fixatives when applying the contamination control substance (a practice discouraged by the manufacturer). Moreover, hazards and risks related to emerging and changing circumstances were not properly reviewed, while the tolerance of deviations became a continual, general practice.

The lack of alerting performance indicators (e.g., no airborne contamination detected) gave the management a fictitious sense of security, allowing it to accelerate the rate of demolition activities whose work control hazards were insufficiently evaluated, recorded, or controlled.

However, the DOE's review appreciated that—following the disclosure of aforementioned incidents—CH2M Hill had improved its control (e.g., use of water and fixatives to keep down airborne contamination), communications, and survey practices.

Eventually, CH2M Hill was authorized to carry out lower-risk demolition activities on PFP; work resumed in September 2018 and was completed in October 2019. Before CH2M Hill could resume higher-risk work, an independent assessment had to be satisfactorily completed.

Higher-risk work to complete PFP demolition included extracting pipelines underneath the processing facility and removing debris with the residual parts of the PFP: It started in November 2019. In February 2020, it was announced that the PFP decommissioning project was practically complete. While the demolition phase is now complete, more site cleanup work remains. In early

2021, cleanup is scheduled to resume the removal of debris from the final PFP building – the 236-Z Plutonium Reclamation Facility (PRF). Some of the high-risk activities remaining include debris size reduction, debris packing, and loadout for disposal. The PRF will restart cleanup in January 2021, and upon completion, will be followed by characterization of the remaining slab, application of the clean soil cap, and establishment of surveillance (DOE, 2020a). Figs. 7.1 and 7.2 are aerial views of PFP before and after demolition.

A dose assessment conducted on the most exposed workers concluded that they had received doses around 0.1 mSv over 50 years. For comparison, the average resident receives about 3 mSv a year from natural background.

During PFP demolition, water suppression and fixative controls were effective until January 2017, when a spread of airborne contamination was detected to the west of the PFP site. A noted failure in water suppression during debris removal was viewed as the likely cause of the event. Process controls were better determined, and radiological boundaries expanded.

Demolition went on until June 8, 2017, when another spread of contamination and airborne radioactivity was noted. The impact of the June 2017 event exceeded the January 2017 situation, and major operational changes were made to the controls, radiological boundaries, and demolition program. New controls involved the setup of exhausters, new procedures in fixative

FIGURE 7.1 The Plutonium Finishing Plant's main processing facility at the Hanford site, November 2016, before demolition. *(Credit to USDOE).*

FIGURE 7.2 The Plutonium Finishing Plant's main processing facility at the Hanford site, after the completion of demolition, February 2020. *(Credit to USDOE).*

application, and revised use of foggers. Additional radiological monitoring locations were included.

A summary of the radiological impacts following the contamination discovered outside posted radiological boundaries in December 2017 is given in the following.

➢ 281 workers requested bioassays
  - 270: nondetectable dose
  - 2: doses less than 10 μSv
  - 8: doses 10–100 μSv
  - 1: dose 100–200 μSv

7.1.2 Analysis; solutions

Root cause evaluation (CH2M Hill, 2018b):

- **Root cause 1:** Overreliance on arbitrarily selected empirical data acquired during, and following, demolition was adopted in taking decisions on the rate and techniques of demolition. For example, the controls and detection methods for the spread of radiological contamination during demolition did not consider the potential for larger, nonrespirable particles.
- **Root cause 2:** Risks and consequences associated with emerging and changing conditions were not adequately reviewed and evaluated.
- **Contributing cause 1:** Previous success of the application of fixative was assumed to provide equivalent protection to containerized debris or covering the debris with soil.
- **Contributing cause 2:** Radiological indicators near the demolition site did not indicate the need to expand fixative applications or perform surveys in addition to the established monitoring plan.
- **Extraneous conditions adverse to quality 1:** Prompt notification to management or the Department of Energy Richland Operations Office (DOE-RL) Facility Representative did not occur because personnel at the facility did not implement a proper procedure.
- **Extraneous conditions adverse to quality 2:** PFP management did not adequately react to all employee concerns and suggestions.

  In response to the root cause evaluation report (CH2M Hill, 2018c), 131 corrective actions were undertaken by the contractor.

7.1.3 Lessons learned

Lessons learned of general relevance can be drawn from this case study. Based on the technical details given in CH2M Hill (2018c), lessons learned include the following:

- Conduct a comprehensive root cause investigation following incidents that have resulted in workers or environmental contamination.
- Do not compromise safety due to unrealistic time pressure.

- Maintain transparent communications channels with decommissioning workers at all times.
- Do not take assumptions concerning the planned decommissioning procedures that are later found to be inappropriate for technical reasons, requiring a strategy change with consequent delay to the decommissioning project.
- Process changes must be assessed against baseline assumptions to ensure that controls still work satisfactorily. Changes based on empirical data may result in unexpected risks.

7.2 Flooding of underground areas at Garigliano Nuclear Power Plant, Italy

7.2.1 Decommissioning process; problem encountered; impacts

Garigliano Nuclear Power Plant (NPP) was in operation during 1964–78. In November 1980, when the flood occurred, the plant was not declared in permanent shutdown yet. This old, prototype reactor was officially shut down 15 months later, The case study reported in the following, although strictly speaking it did not occur during the active decommissioning phase, is, however, typical of events that may occur so long as site decontamination activities are not complete.

The underground tanks referred to in the description of the accident contained 360 m^3 of three types of liquid waste: spent ion exchange resins, filtering sludge, and evaporator concentrates.

During the flood of the Garigliano River in November 1980, water entered and left the tank room, carrying contamination to the environment.

Initial estimates indicated that the amount of activity released to the environment was of the order of 1 Ci (3.7×10^{10} Bq): Later, more realistic estimates reduced this amount to few tenths of the initial figures. Water samples taken from monitoring wells in the area exhibited radioactive concentrations of the order of 1/10,000 of drinking water maximum permissible concentration. No contamination was found in the groundwater (Parliamentary Acts, 1981).

7.2.2 Analysis; solutions

The waste tanks, located in underground areas, were lidless. This situation, taking into account that the pipes through which the concentrate was sent were eccentric and that the condensate was pushed by steam, favored the formation of splashes that could deposit on the floor and on the walls of the premises; additionally, the tanks were not placed in a "tray," intended to contain leaks. The fact that water went in and out of the contaminated areas showed that the room did not possess adequate sealing.

The main root cause of the incident was eventually eliminated in the 1990s with the removal of the source term (the stored waste). To this end, an extraction and conditioning campaign of this operational waste was performed beginning in 1994 in a purpose-built facility on-site (Garigliano Extraction and Conditioning of the Operational waste, GECO). In total, 1671 400-L drums were generated; the occupational exposure resulting from this work was 160 person mSv (JC, 2008). The inventory of the conditioned waste was 32 TBq (JC, 2008).

7.2.3 Lessons learned

Old nuclear plants often present design and construction deficiencies that are conducive to incidents. The decommissioning program should identify such weak points and give priority to their elimination.

7.3 Poor information on components being dismantled resulted in release of noxious substances (IAEA, 2016)

7.3.1 Decommissioning process; problem encountered; impacts

The Paducah Gaseous Diffusion Plant (PGDP) is a nuclear installation located near Paducah, Kentucky, United States, which produced enriched uranium in the period 1952–2013. The US DOE is the owner. PGDP produced low-enriched uranium, originally for military purposes and later for nuclear fuel for commercial power plants. Since 2013, DOE has been managing Paducah's D&D. There were initially over 500 facilities to be demolished; of them, a few dozen have been demolished. The Paducah cleanup program is likely to extend beyond 2040 (Fig. 7.3).

FIGURE 7.3 Heavy equipment demolishes an ancillary building (foreground) at the Paducah Metals Plant in summer 2010. The conveyor system immediately behind the equipment also was demolished. *(Credit to DOE, Wiki Commons).*

A case study is illustrated in the following. While releases of noxious substances at Paducah complexes were being investigated, some workers voiced concern that they had poor information on components being dismantled. For example, in one case, the workers did not realize that the insulated rectangular ducting enclosing a UF_6 pipe contained instrument lines. As a consequence, the instrument lines were inadvertently sliced, and UF_6 was discharged, which resulted in the release of UO_2F_2 and hydrofluoric acid (HF) to the containment.

7.3.2 Analysis; solutions

In decommissioning of old nuclear plants, recorded information about construction and operation of a system is often incomplete, and it fails to provide an accurate technical basis to identify hazards. Site surveys are not sufficient to provide complete information on hazards lying ahead. Discussions with former staff or designer may allow the D&D team to offset for the lack of written records and lead to a more clear knowledge of the decommissioning hazards. Experienced staff may know of alternative uses for the equipment that are not recorded. For example, nitrogen lines used to cleanse UF_6 systems may be contaminated with UF_6 because the nitrogen line was used to bypass plugs in the UF_6 line. Having this knowledge, the D&D team can be ready for any off-gassing of HF when the nitrogen line is breached.

7.3.3 Lessons learned

When performing decommissioning activities in buildings that have been unused for several years, it is critical to be well aware of how the building systems were formerly used. It is especially recommended to

- associate former designers and personnel with the decommissioning planning;
- question the workers (both current and former staff) to determine who has best experience of the system(s) planned for dismantling, and consolidate their knowledge in standardized documents for use in planning.

7.4 Contamination identified in a nonradiological area, East Tennessee Technology Park, Oak Ridge, Tennessee, United States (IAEA, 2016)

7.4.1 Decommissioning process; problem encountered; impacts

The US DOE manages the environmental cleanup at the Oak Ridge Reservation including decommissioning, environmental remediation, waste treatment and disposal, and other cleanup-related operations. The site includes the

Y-12 National Security Complex, Oak Ridge National Laboratory, and East Tennessee Technology Park (ETTP). This case study refers to the latter.

The DOE's objective is to complete all demolitions and major cleanup at ETTP by 2020. When this target is achieved, for the first time in history, a uranium enrichment complex will be fully cleaned and removed. Within ETTP, K-25 Building was one of the site's five huge gaseous diffusion buildings that were once the world's largest construction. The DOE intends to turn ETTP into a multiuse industrial park, national park, and conservation area.

The area in question was next to a radiological area north of Building K-1015 laundry installation: It was formerly classified as contaminated but had been decategorized after a survey was conducted in November 1995. During August to September 1995, all four laundry lint collectors located north of building K-1015 were within a contaminated area. In November 1995, the four lint collectors and the area around were surveyed for decategorization. Following this survey, lint collector numbers 3 and 4 and the area around were decategorized to a nonradiological area. Lint collector numbers 1 and 2 and the area around were decategorized from contamination area to fixed contamination area. In June 1997, during a DOE's safety management review of the ETTP program, it was requested that a survey be conducted on the decategorized areas. During this survey, many contaminated spots were detected along the building's foundation and in cracks by a concrete pad.

7.4.2 Analysis; solutions

Three possible scenarios or a combination of them may have caused the unexpected appearance of contamination. First, after a careful review of the decategorization surveys, the contaminated spots may have been missed. The survey does not specifically show readings taken in the specific locations where the contamination was later detected. The radiological control technician (RCT) who performed the decategorization survey had left the job in December 1996, so it could not be ascertained when the cracks in the tarmac appeared; the cracks may not have been there at the time of the decategorization survey. Thus, the tarmac may have shielded the contamination from the survey instruments. Second, based on the specific location where the contamination was detected, leaching of contamination in the soil to a level detectable by the survey instruments was possible. Third, there had been some 20 rainy days in a row before the contamination was found. A fixed contamination area was next to the area where the contamination was discovered. A survey of the filaments inside the lint collectors within the fixed contamination area indicated that the lint was contaminated above the limit for a fixed contamination area.

Right away, as an interim measure, the open ends of the collectors were wrapped in plastic and fixed with adhesive tape to seal the filaments. Besides, the fixed contamination area was recategorized as a contamination area. Of the three possible causes highlighted earlier, the third one has been verified in the field. Surveys of the soil, filaments, and vegetation in the area all showed relatively high activity levels.

A full survey of this facility had been carried out on the outside of K-1025. Findings showed that the building was rightly categorized, except for a small section north of the sludge pit which showed dose rates over twice the background. The boundaries of the contamination area were extended by some 30 cm on the north side of the pit to include the area with higher dose rates. The area categorization around the four lint collectors north of K-1015 was changed to reflect the actual radiological situation. The collectors that were included previously within the fixed contamination area (numbers 1 and 2) were posted with internal contamination labels. Lint collectors that were in the nonradiological area (numbers 3 and 4) were posted with possible internal contamination labels. The building operatives were involved in this work, and actions were taken to impede the spread of contamination from the lint collection system.

Regular surveillance was established after rain to make sure that contamination had not moved from the contamination area. Besides, a barrier was installed along borderlines to block any lint that moving water may have carried out of the lint collection system.

7.4.3 Lessons learned

Areas that have been decategorized to a safer status (e.g. nonradiological) should be periodically resurveyed for contamination if cracks or other changes are visible.

7.5 A buried wastewater pipe leaked activity to the environment (IAEA, 2016)

7.5.1 Decommissioning process; problem encountered; impacts

The Advanced Test Reactor (ATR) Complex has been a focal point in the Idaho National Laboratory (INL) energy research mission for many decades. Three major test reactors have been in operation at the ATR Complex: the Materials Test Reactor (MTR) (1952–70), the Engineering Test Reactor (ETR) (1957–82), and the Advanced Test Reactor (ATR) (1967–present). The MTR is currently being demolished. Demolition of the ETR has been completed. The operation of the ATR is the primary activity at the ATR Complex. This reactor is used to assess the effects of neutrons and gamma fields on fuels and reactor materials used in nuclear reactors. The DOE has completed demolition of over 20 facilities at the ATR Complex (Fig. 7.4).

FIGURE 7.4 Advanced Test Reactor Complex—aerial view. *(Credit to DOE, Wiki Commons).*

On October 9, 2001 while digging soil the crew came across wet soil next to an underground 10 cm diameter radioactive wastewater transfer pipe. The soil turned out to be radiologically contaminated. Further digging showed that the pipe had cracked following an inspection digging carried out in 1997. The margins of the broken pipe were corroded, which proved that the crack had been there for quite some time. There was no worker contamination or contamination spread outside the posted borders. According to the Occurrence Reporting and Processing System (ORPS) guidelines, the soil could be contaminated to a depth greater than 3 m below the pipe break and could remain so for undefined time.

The crew dug some 1.8 m below grade and unearthed the pipe. The workers saw water trickling from around the pipe, and, eventually a significant pool of contaminated water appeared in the pit around the pipe. It appeared that four pipes had ruptured. A survey of the dug soil with a portable detector confirmed the soil contamination. The leak was blocked by closing system pumps, and a fiberglass patch was placed over the pipe fracture.

The pipe wastewater was normally filtered to take away most radioactive contents. It was not clear how long the leak had persisted or how much water had leaked. Worst-case estimates based on the ORPS guidelines showed that no radioactive isotopes in the water would have passed the daily discharge limits.

7.5.2 Analysis; solutions

The cause of this event was an equipment end-of-life failure. The ruptured pipe was some 50 years old. Experts had long viewed the likelihood of pipe failure and the need for a replacement. To this end, this pipe was unearthed and surveyed in 1997. No pipe ruptures were noticed during this inspection. Nevertheless, the condition of the pipe still raised concern since the inspection had disclosed significant inner and outer corrosion and pitting areas deep almost half-wall thickness. It is probable that a combination of the degraded condition of the pipe material and soil settling following this inspection digging caused the pipe failure.

A contributory cause could have been an improper legacy management issue. Based upon technical assessment and funding status, replacement was recommended within 5 years. The pipe was scheduled for replacement within that time span. The leak would not have occurred if the pipe had been removed and replaced soon after the 1997 inspection.

A few corrective actions were taken in consequence of this event. The pipe was leak checked, fixed, and put back in service. The contaminated soil was extracted to a depth of around 3 m and replaced with clean soil. The soil was sampled to ensure that all contamination had been taken away.

7.5.3 Lessons learned

This event highlights the need for timely replacement of underground piping that could fail and leak out radioactive liquids. Deferring replacement of such piping may incur significant risks. Additionally, management should keep in mind that when old underground pipes are unearthed for inspection, the risk of a pipe break grows due to soil settlement.

Events of this type are quite possible over prolonged safe enclosure phases and reinforce the need that active maintenance and surveillance be kept for shutdown facilities.

7.6 Wrong historical records led to the unexpected discovery of contaminated areas (IAEA, 2016)

7.6.1 Decommissioning process; problem encountered; impacts

The New Brunswick Laboratory (NBL) was a US government–owned center specializing in the measurement of nuclear materials. It was initially located in New Brunswick, New Jersey. It was decommissioned and relocated between 1975 and 1977 on the site of Argonne National Laboratory (ANL)—East, Illinois, United States.

Unexpected events occurred during the decommissioning of the NBL. The health physics team of ANL was contracted to conduct a random check of the installations to verify that the categorization/segregation of areas was adequate. It was soon observed that several supposedly clean areas were, in fact, contaminated above the unrestricted use criteria. Moreover, office items that had been set aside as clean were also contaminated.

7.6.2 Analysis; solutions

Former employees of the installation could not recall the use of the radioisotopes that were discovered. Therefore, the origin of the contamination was uncertain. Possible causes for this unexpected case include the following:

- In establishing the contamination of the facility's areas, too much confidence was given to the a *posteriori* reconstruction of the facility's operation.
- Determination of contaminated areas was based on former decontamination criteria and not on the industrial standard in force at the time.
- Improper instruments were used in the contamination survey.
- The previous employees were not much interested in waste management, since they were relocating to a new installation.

Following completion of the final survey and the drafting of the survey report, the ANL health physics team was tasked to verify that the remaining areas were clean and all structures and items could be removed and disposed of in a nonnuclear landfill site. A comprehensive radiological evaluation was performed, and again, it was determined that the remaining parts did not comply with the unrestricted use/release criteria. It was later found that the decontamination contractor previously tasked with the monitoring of mixed α- and β-contamination was not adequately experienced in this job. Specific observations made during the radiological evaluation included the following:

- ^{241}Am was found in the soil where the drain pipes from the former Pu facility had been taken away.
- Several wall and floor joints showed that the contamination had penetrated into the floor base; as concrete was removed, the radiation levels went up.
- Contamination around a storm sewage entry was due to ^{226}Ra and its daughters.
- Apparently, there were leaking drains under workbenches and in pipelines.
- Contaminated soil was found that could not be related to the disposal of pitchblende from the Middlesex Sampling Plant, New Jersey, United States.

As it often happens in research installations, decontamination of one area disclosed another contaminated area that was previously unknown.

7.6.3 Lessons learned

The following lessons learned can be drawn from this case study:

- Specialist skills in measuring mixed α- and β-contaminants are essential for defining the boundaries of the decommissioning project and for performing the radiological evaluation.

- Historical information of past activities in long-shutdown installations is often confusing and inaccurate.
- Full understanding on release and disposal criteria is essential including for complex situations and requires concurrence of all involved parties.

7.7 Metallic paint hid contamination during survey, Idaho National Engineering and Environmental Laboratory (IAEA, 2016)

Procrastination is the thief of time.

Edward Young (1683–1765).

7.7.1 Decommissioning process; problem encountered; impacts

A common, though unexpected event, during the final postdecommissioning survey for site release, is the discovery of contamination that has been shielded during previous surveys by paint or other coatings. An event of this type occurred during the decommissioning of the auxiliary reactor area (ARA) facility at INEEL. ARA-II housed the SL-1, a 200 kW reactor that was being tested for use at remote locations. An accidental nuclear reaction and explosion occurred in 1961 and destroyed the reactor. The initial cleanup and remediation activities ended $1\frac{1}{2}$ year later. A radiological characterization of remaining contaminated areas was conducted in 1983 to provide the technical basis for hazard assessment and waste management. Inside the buildings, radiation surveys were carried out, and smear samples were taken. Surface- and below-grade soil samples were taken and measured. Structural materials such as insulation and metal cladding were also sampled to determine the extent of the hazards and the expected waste streams. The 1983 characterization undertaking found no smearable contamination. However, as the dismantling was being carried out later, it was found that most of the metal constructions had been coated with a thick metallic paint after the reactor accident to fix contamination in situ. It was also found that concrete covers had been spread over the floors in two buildings to fix contamination in situ. These activities had masked underlying contamination.

7.7.2 Lessons learned

The following lessons can be drawn from this case study:

- Records important for decommissioning, especially the radiological characterization, demand timely preparation and sufficient time for accurate review.

- Physical and radiological characterizations done prior to planning for decommissioning are not necessarily indicative of the actual situations that will be found on-site.
- The identification of waste streams for evaluation of management options should be initiated soon after preliminary characterization is available. All parties responsible for waste treatment and disposal should be engaged as soon as possible to determine the need for further waste characterization. These interactions can take time and cause prolonged delays to complete decommissioning.

7.8 Obstructed roof drain spreads radiological contamination during rainfall, West Valley Demonstration Project, New York, United States (DOE, 2020b)

It never occurred to her that if the drain-pipes of a house are clogged, the rain may collect in pools on the roof; and she suspected no danger until suddenly she discovered a crack in the wall, Gustave Flaubert (1821–80) Madame Bovary, II, 5.

7.8.1 Decommissioning process; problem encountered; impacts

The West Valley Demonstration Project (WVDP) site, 48 km south of Buffalo, New York, United States, is part of the Western New York Nuclear Service Centre (WNYNSC) and includes 1335 ha of land allocated for a former nuclear fuel reprocessing plant. The WNYNSC houses the shutdown reprocessing plant that operated during 1966–72 and generated 2271 m^3 of liquid HLW. The WNYNSC also houses contaminated structures and two radioactive waste disposal sites.

As determined in 2010, the DOE's selected alternative to decommissioning the WVDP site consists of a two-phased approach. Phase 1 includes the decommissioning of most site facilities, including demolition of the main plant process building (MPPB) (Fig. 7.5) and vitrification facility and studies to reduce uncertainties associated with decommissioning the remaining facilities. Phase 2 includes the completion of the decommissioning process and long-term management decision-making for the whole site. Phase 1 of the decommissioning approach is under way and approaching completion (NRC, 2019a).

In November 2018, a thorough radiological survey of the Waste Reduction and Packaging Area (WRPA) dock and nearby paved work area revealed contamination up to 17.5 Bq/cm^2 β-γ (based on direct reading). This WVDP area had been previously controlled due to noted contamination from water leaking from the Solvent Storage Terrace (SST) roof and through various

FIGURE 7.5 Work is under way to remove contamination in the off-gas blower room inside former nuclear fuel reprocessing facility at West Valley Demonstration Project. *(Credit to DOE, Wiki Commons).*

locations in the MPPB, eventually coming out of the MPPB through the WRPA dock and flowing outside to nonradiological paved work areas. A worker noticed that the water had a blue tinge, pointing to a contact with a fixative applied to an MPPB installation a short time before. It was determined that the water had flowed outside the WRPA dock, past the work area, and down to a drain pipe that went under a road crossing. Notifications were issued; the area was cordoned off to check the water and impede unrestricted access. Personnel were directed to the MPPB roofs to look for clogged drains. It was found that the SST roof drain was clogged. Once it was freed, water drained regularly. For the previous 5 days, there had been heavy rainfall.

7.8.2 Lesson learned

The lesson to be learned from this event is that the maintenance of idle facilities being prepared for decommissioning must be constant: The maintenance program should be reviewed to take into account the increased risks due to less frequent inspections.

7.9 Transfer of property without proper evaluation of radiological situations (DOE, 2020b)

Be a good neighbor (Marcus Cato, 234-149 BC, On Farming).

7.9.1 Decommissioning process; problem encountered; impacts

On March 31, 2016, during a routine survey of an outdoor concrete pad on the boundary of C-333 (a contamination area, or CA) at the PGDP, several small contaminated spots were found on gravel and asphalt outside the boundary. Surface contamination up to about 150Bq/cm^2 β-γ was found. Total contamination was less than the administrative limits (around 1 Bq/cm^2 total β-γ and around 0.15 Bq/cm^2 totalα). The origin of contamination was established to have derived from the maintenance and storage of contaminated pieces in the period 1952–90, which diffused from the pad to the adjacent area. After surveys discovered an area of outdoor contamination, the concrete pad was classified contaminated (1995–2004). In 2004, the adjacent area was covered with gravel and asphalt. When the title of the PGDP was transferred from the USNRC licensee to DOE, the pad area was classified as a CA. Nevertheless, the radiological situations of the pad and adjacent area were not completely defined, recorded, or notified by the former responsible party to the DOE. It was evident that the material found outside the CA boundary resulted from deterioration of the cover installed over the contaminated pad and had diffused beyond the boundary. This made it essential to monitor the deterioration of outdoor radiological areas before the property could be transferred to the DOE.

7.9.2 Lesson learned

Ownership of a given facility may change over its lifetime. The licensee in charge at any time must ensure the accuracy/reliability of the records and the effectiveness of the record transfer in that they could be claimed responsible for the safety and radiation protection of the subsequent activities undertaken by the new owner and other aspects.

7.10 Communication errors and control process breakdown lead to radioactive contaminated material being taken off-site; Hanford, Washington, United States (DOE, 2020b)

7.10.1 Decommissioning process; problem encountered; impacts

In July 2018, a cross-beam was taken by a site contractor without management approval from a PFP storage area (classified radiological buffer area, or RBA). The contractor had agreed directly with a RCT to carry out a radiological survey and agreed with Mission Support Alliance, LLC (MSA) truck drivers to remove the cross-beam for inspection and placement of inspection labels. However, the RCT surveyed a different cross-beam in the same area, not realizing which cross-beam was being removed. The cross-beam that had not been surveyed was loaded onto a truck and went out of the PFP RBA via the normal service road.

Upon leaving PFP, the truck went from PFP to Building 6290, and the cross-beam stayed in the truck. A few days later, the truck with the cross-beam went to Building 2355 (an off-site facility) for an unrelated duty. The cross-beam remained in the truck. The truck came back to Building 6290 on the same day. On the following day, the cross-beam was off-loaded into the Building 6290 yard. A week later, the cross-beam was moved to the Building 6290 nonradiological area, where it was inspected and labeled.

Fortunately, due to an individual concern that the cross-beam had come from the PFP, a radiological survey of the piece was carried out. This survey identified one spot of loose contamination. The affected section of the beam was bound in plastic and transferred to a radiological material area.

7.10.2 Lesson learned

This case highlights the need to (1) adhere to a specified procedure (including surveys) for removing tools or substances and (2) ensure proper notification and item identification; the tool or substance in question should remain under radiological control until release.

7.11 Contaminated materials found off-site, Hanford site, Washington, United States (DOE, 2014)

7.11.1 Decommissioning process; problem encountered; impacts

Twice in 2 months in 2014, workers discovered small amounts of contaminated plastic material outside of areas posted for contamination at the 618-10 burial ground. It is suspected that high winds blew the material out of the burial ground, which is a 3-ha site that operated between 1953 and 1963 as the main repository for much of the highly radioactive hazardous waste generated on the Hanford site. Workers were excavating the site in preparation for remediation. In the first incident, workers found six pieces of plastic and one piece of metal foil that were contaminated outside the perimeter of the burial ground. In the second incident, workers found four pieces of contaminated plastic outside the burial ground after high winds blew through the area the afternoon before. After both events, workers collected all materials found outside of the burial ground and covered all visible plastic identified with clean soil and a fixative.

7.11.2 Analysis; solutions

The excavation process is fairly destructive to the packaging of severely deteriorated buried waste and often results in ripped and crumpled pieces of plastic. During the sorting process, operatives are looking for drums and other anomalies (e.g., bottles of liquids, sealed containers, or new waste forms). As plastic and other debris is exposed, the operatives attempt to keep it covered with soil and keep it wet in the event the wind picks up. At the completion of the day's activities, the operative visually inspects the excavation area from the cab of the excavator to ensure there is no exposed combustible material or other loose debris that could become airborne. Any combustible material or loose debris found is covered with a minimum 3.75 cm layer of soil. A fixative is applied to the excavation dig face and contaminated solids when periods of inactivity are expected to be longer than 24 h or winds over 30 km/h are forecast. While these actions are generally successful in containing material to the trenches at the burial ground, that is not always the case.

7.11.3 Lessons learned

Be aware of day-to-day changes in conditions by remaining vigilant in observing all aspects of a task. Repetitive work may illude one regarding what to expect from a worksite. While the history of a site may tell us much about what will be encountered, changes can and do occur.

7.12 Remediation of polychlorinated biphenyl-contaminated land

7.12.1 Decommissioning process; problem encountered; impacts

Land contamination may be encountered in decommissioning sites due to known or unknown historical discharges of hazardous materials or due to the presence of idle structures used in the past to store or handle these materials, e.g., underground storage tanks, which have contaminated the surrounding soils or water courses during operations or during the abandonment period.

Having to decommission nuclear installations located in contaminated lands raises a number of occupational, public, or environmental impacts, or all of these jointly. Land contamination may be due to radiological substances or chemicals such as industrial oil, pesticides, heavy metals, polychlorinated biphenyl (PCBs), organic compounds, and asbestos. A number of decommissioning projects complicated by the presence of environmental contamination are given in Laraia (2019): Indeed, the presence of environmental contamination expands a decommissioning project into an environmental remediation project (a subject that is not directly covered by this book). The impacts on a decommissioning project generated by a contaminated land are case specific and include many factors such as nature of contaminants, topographical features, weather conditions, intended land use, and the type of decommissioning activities (e.g., indoor or outdoor). Although environmental remediation of contaminated land (as distinct from decommissioning) is not

FIGURE 7.6 River Pump House site after demolition. As the final phase of the operation, site cleanup crews backfilled the riverbank with soil and rock, shaping it to blend with the natural contour of the shoreline. *(Credit to DOE).*

specifically dealt with in this book, there are several sections (7.16. 7.17, 7.25, etc.) that offer examples of decommissioning vs. remediation interactions (IAEA, 2009). addresses these interactions in full. One example of a combined demolition-remediation process is illustrated in Fig. 7.6; the River Pump House was a pump station located on the bank of the Columbia River, Hanford site, and feeding water to Hanford facilities for multiple uses.

More information is given here about PCBs impacting the Yankee Rowe (YR) decommissioning project. Physical decommissioning of the former YR NPP was completed in 2007. The US NRC notified the licensee in August 2007 that the NPP site had been fully decommissioned in compliance with NRC prescriptions and approved YR's Final Status Survey Report as specified by the License Termination Plan. The nuclear license remains limited to 0.8 ha embracing the Independent Spent Fuel Storage Installation. The final Massachusetts decommissioning-related approval needed to finalize site closure in compliance with Massachusetts laws was granted in 2008 when the State Department of Environmental Protection issued the final Phase II Comprehensive Site Assessment Report.

One environmental remediation project that was completed at YR included the investigation and removal of soil and sediments on the site contaminated by PCBs (Mcgrath, 2018). Building structures on the YR site were painted with a PCB-containing coating. Over time, paint had peeled off due to weathering and had contaminated the soil and, further to groundwater flow and storm runoff, had contaminated on-site surface water. Moreover, spot removal of the PCB-containing coating was required prior to dismantling of the structure for worker protection. YR began in 2000 the investigation of PCBs in the site's soils and sediments. The soil and sediment remediation was completed in 2005.

The occurrence of PCBs is a common problem during decommissioning. Due to their outstanding properties (water insolubility, fire resistance, long service life, chemical inertness, high electrical resistance, etc.) and low cost, PCBs were widely used in industrial plants, including nuclear installations.

The following gives a summary of the typical components/material where PCBs have been employed and are expected to come up in the course of decommissioning:

(a) Epoxy coatings
(b) Oil impregnated electrical insulation
(c) Electrical transformer oils
(d) Hydraulic oils
(e) Oils in cutting machines
(f) Lubricants
(g) Plasticizers
(h) Capacitor dielectrics

PCBs are very stable; therefore, they remain a long time in the environment and transfer to the food chain. PCBs are soluble in fats and can be easily incorporated in fat tissue. PCBs are believed to be carcinogenic. Incorporation can occur by inhalation, ingestion, or through the skin. Chronic toxicity is the main concern (IAEA, 2006c).

7.12.2 Lessons learned

Management of nonradiological hazards during nuclear decommissioning can be as complex as, or even more complex than, management of radiological contamination. Therefore, a structured strategy should be established to minimize the impact of nonradiological contaminants on decommissioning cost, schedule, and site end state. Management of nonradiological issues in nuclear decommissioning is complicated by: site-specific nonradiological contaminants in systems, structures, and environment; the variety and occasional disuniformity of prescriptions issued by different regulatory bodies; how these issues impact planning and implementation of decommissioning activities; interactions with the management of radiological issues; and unexpected impacts on workers, public, and environment.

A basic management strategy to deal with radiologically and nonradiologically contaminated lands should include as a minimum:

- Managing contaminated media with the purpose of protecting the safety and health of site occupants (e.g. workers and office staff), the adjacent community, and the environment during, and after completion of, decommissioning

- Being aware of or reconstructing the historical use of the land concerning the potential presence of hazardous substances prior to the commencement of decommissioning activities
- Preparing plans to respond to the discovery of contaminated media to minimize the risk to health, safety, and the environment
- Preparing a management plan to manage redundant, abandoned, and hazardous substances
- Identifying and cooperating with whoever is responsible for the whole site and/or is liable for the contamination, potential successors to site ownership and use, and stakeholders at large

7.13 Loss of liquids from the Magnox Swarf Storage Silo, Sellafield, United Sates (IAEA News, 2019)

7.13.1 Decommissioning process; problem encountered; impacts

The Sellafield's Magnox Swarf Storage Silo was constructed in the 1960s without a decommissioning plan. It stores 60,000 items of ILW in 22 chambers (mostly cladding material that was removed from spent Magnox fuel and contaminated engineering debris). A precise waste inventory is missing because record keeping during silo's operation was fragmentary.

Engineers have been busy for many years to design, construct, and install three machines to empty the silo. Work has now started to install the second of the three emptying machines. The first piece of the 350-t machine—a transfer tunnel—was raised into the silo in late 2019. By 2024, all three machines should be in operation. The silo is planned to be empty by 2045.

The machines will move on rails over the top of the building's 16-m-deep chambers. Once a machine is installed over a chamber, it will lock on, and tooling will reach in to pull up the waste stored. The material will be contained in boxes, which will be transported out of the building via the transfer tunnel. Later, the boxes will be removed to a new storage building on-site (GOV.UK, 2020).

In November 2019, the Sellafield licensee informed the regulator that liquids lost from the old Magnox Swarf Storage Silo had exceeded expected levels since July 2019. Most of the activity inside the silo (99.5%) is solid waste, while the remaining 0.5% is contained in the water (named as liquor). There have historically been underground leak from the building, and the potential leak paths unfortunately remain open. Water levels in the silos are routinely monitored, and clean water is added to compensate for natural losses (e.g., evaporation).

7.13.2 Analysis; solutions

Analysis has shown that around 1 m^3 of liquid is being lost each day. In activity terms, this is equivalent to the loss of around 8 TBq of ^{137}Cs (the main radionuclide) from July to November 2019. The activity is being discharged into the ground underneath the silo. Monitoring of groundwater wells and in-ground gamma activity in the surroundings of the facility has not shown any release to the environment. It is foreseeable that the soil migration of contaminants will be very slow, therefore posing no risk to the public or the environment. On the other hand, the structural design of the facility (e.g., the lower level of the building below grade) is assumed to preclude any practicable intervention. The licensee is investigating the cause of the event and is trying to better characterize the losses. Facility's structural data continue to be monitored and assessed. Radiological surveys are being carried out to identify changes in the vicinity. There has been no detectable change to the radiological situation in and around the building: Therefore, there are no near-term safety concerns to the public or the environment. The barriers and systems that are intended to protect against accident evolution within the facility operate regularly. The presence of leak paths through the silo structure is not considered to affect the integrity of the reinforced concrete structure. As described in the beginning of this case study, a long-term strategy is under way to remove the waste inventory to a new store. This will eventually allow the decommissioning of the facility. The incident was given International Nuclear and Radiological Event Scale (INES) rating 2 based on the following criteria:

Impact on people and the environment	
Release beyond authorized limits?	Yes
Overexposure of a member of the public?	No
Overexposure of a worker?	No
Impact on the radiological barriers and controls at facilities	
Contamination spread within the facility?	No
Damage to radiological barriers (including fuel damage) within the facility?	Yes
Degradation of defense in-depth	No
Other information	
Person injured physically or casualty?	No
Is there a continuing problem?	Yes

7.13.3 Lessons learned

Uncontrolled liquid releases from old buildings are almost inevitable in most old nuclear building. Prevention is difficult and expensive since construction criteria of many years ago were not so strict as today's.

7.14 Inadequate supervision of contractors: Tower Shielding Facility, Oak Ridge National Laboratory, Tennessee, United States (IAEA, 2006a)

Delay is preferable to error, Thomas Jefferson (1743–1826).

7.14.1 Decommissioning process; problem encountered; impacts

The Aircraft Nuclear Propulsion Program has begun soon after World War II and soon became a major national priority: Since the beginning, it has been clear that the radiation shield weight was critical. In 1952, Oak Ridge National Laboratory (ORNL) proposed to study shielding problems by suspending high in the air the reactor and shield models being tested. Following approval, the Tower Shielding Facility (TSF) was built in a remote part of the ORNL. The TSF (Fig. 7.7) consisted of four 100 m towers, 60 m away from each other in one dimension, and 30 m in the other. Cables connected to the towers could lift the reactors and shields to 60 m above ground. Over time, four reactor types were tested at the TSF. A wide range of shielding materials were tested during reactor operations; moreover, a lot of information was gained about air scattering. During the last years of operation prior to the facility's permanent shutdown in 1992, the reactor was mostly operated within a concrete shield on the ground as a more practical radiation source. The towers were also used to perform drop tests to validate the soundness of spent fuel transport casks. Legacy contamination still remains in some parts of the facility. Defueling the TSF was completed in 2003, but final cleanup is in abeyance (Muckenlhaler, 1992).

FIGURE 7.7 Tower shielding facility. *(Credit to USDOE, Wiki Commons).*

FIGURE 7.8 Tower shielding facility project, metal and concrete sold for recycling. *(Credit to USDOE, Wiki Commons).*

During late 1997 to early 1998, a subcontractor was surveying surplus materials at the TSF. Because many of the metal parts had been exposed to reactor neutrons, the main radionuclide was ^{60}Co. α-Emitters were less significant. Due to time pressure, limited number of qualified staff, too few α-monitors, and a history of modest α-contamination, the subcontractor discontinued the practice of checking α-emissions on all items. The hired RCTs went on filling in "no α detected" on the tags (Fig. 7.8).

In June 1998, ORNL RCTs found ^{241}Am, an α-emitter, on a piece of a box that had been characterized only for β-γ-contamination. The contents of the original box had formerly been segregated to remove the contaminated fraction and the remaining part had been checked only for β-γ-contamination and tagged for unrestricted release. This part of the box had been sold some time before as inactive scrap to a local smelter.

Immediately after, the ORNL management sent an RCT and the TSF manager to the smelter site to resurvey that section of the box. The RCT's measurements led to the conclusion that the section of the box had no α-contamination but discovered α-contamination on another box originating from the TSF. This contaminated box was decontaminated at the smelter site and later shipped back to ORNL.

7.14.2 Lessons learned

Because of insufficient supervision, the prime contractor remained unaware of the loose surveying practices of the subcontractor until after contaminated material had been dispatched off-site as clean material. Since time pressure and other constraints can often encourage subcontractors to shortcut established procedures, adequate supervision should be provided to confirm that regulatory and other contractual requirements are fulfilled by the subcontractors.

7.15 Pneumatic Transfer Tube Removal Project, ANL-East, Illinois, United States (IAEA, 2006b)

7.15.1 Decommissioning process; problem encountered; impacts

The Building 200/205 Pneumatic Transfer Tube D&D Project had the following objectives:

- Remove any radioactive and other hazards related to the transfer tube.
- Survey the tube for external contamination.
- Remove the tube and prepare it for disposal.
- Survey the soil around the tube for contamination.
- Backfill the tube trench.
- Restore the area to its pristine condition.

The structural conditions of the transfer tube and any nuclear fuel samples lost in the tube were the main concerns, while the precise location of the tube was less important. In January 1998, ANL-E Health Physics collected radiological data from the ends of the transfer tube in Building 200/205. Contamination was measured to a level of 1120 Bq β-γ and 333 Bq α (smearable) (ANL, 1998).

While removing the transfer tube system linking two buildings, a problem was water in the trench, which had to be solved on the spot. During the digging, water would spout out of the surrounding sand areas and partly flooded the trench. At one time, water penetrated the cut end of the tube before it could be lifted above the water. The water having entered the tube was found when it spilled out of the tube and onto the ground. Fig. 7.9 shows a detail of the tube removal project.

7.15.2 Analysis; solutions

A barrier was installed surrounding the area and the soil excavated. The area was sampled to confirm that all the contamination had been taken away. The affected area was less than 1 m^2. The remaining water was drained from the tube and gathered.

7.15.3 Lessons learned

A plan should have been prepared in case the water became caught up inside the tube. For instance, a container could have been ready to catch the contaminated water, thus preventing spillage and the need to remove the contaminated soil. Better preparation and job planning would have prevented the incident.

FIGURE 7.9 Dismantling of a buried pipe using hydraulic cutters, ANL pneumatic transfer tube project, Argonne, Illinois, United States. *ANL*, Argonne National Laboratory. *(Credit to DOE).*

7.16 Building 34 decommissioning, ANL-East, Illinois, United States (IAEA, 2006b)

7.16.1 Decommissioning process; problem encountered; impacts

During its operation, Building 34 at ANL /East (ANL-E) was a treatment facility for industrial wastewater from several site instillations. Undetected leakage from lidless concrete tanks, while they were operating, called for soil remediation additional to the building decommissioning project. This extension generated a larger than previously estimated volume of contaminated waste that required packaging and shipment.

7.16.2 Analysis; solutions

Monitoring during tank operations had not disclosed any leakage, and therefore, no allowance had been made for environmental remediation within the scope of the building decommissioning project. This caused the unplanned expenditure of extra funds to cover the costs of the excavation, packaging, and disposal of the contaminated soil.

7.16.3 Lessons learned

After many years of continuous operation and possibly inadequate maintenance, old tanks are often leaking. Besides, there may be construction issues, since the operational requirements for older installations facilities tended to be less strict than for those in current use. Some attention should also be given to sampling in vicinity of all structures containing piping and tanks. Contingency should always be included for unexpected waste, which will require proper management.

7.17 Cintichem research reactor, Tuxedo, New York, United States (IAEA, 2006b)

7.17.1 Decommissioning process; problem encountered; impacts

The Cintichem pool-type, light water moderated and cooled research reactor was licensed to operate up to 5 MWt. It started commercial operation in 1961. Based on its worrisome radiological characterization (see the following), Cintichem shut down permanently in 1990 and began decommissioning at once. The decommissioning project ended in 1997.

A belowground ventilation system for the hot cell section at Cintichem had leaked for years during the conversion of reactor materials into radiopharmaceuticals. In this regard, there was a concern that the soil underlying the area could be contaminated by this leakage. However, soil characterization and sampling were considered impractical. Instead, it was planned to leave the area untouched until the decommissioning project reached the stage of hot cell demolition, which would facilitate access to and remediation of the contaminated soil underneath. The down gradient water flow from the reactor and hot cell intercepted a drinking water reservoir.

7.17.2 Analysis; solutions

Early characterization (including physical intrusion) would have allowed identify the large leakage, which had occurred over the years and had produced a massive amount of contaminated soil requiring removal to prevent the contaminants from reaching the drinking water supply. Instead, inadequate identification of contaminated areas led to the later discovery of unexpectedly larger volumes of contaminated soils. Innovative methods could have been employed to explore the contaminated soil under the hot cells.

7.17.3 Lessons learned

Quite a few decommissioning projects have been more time-consuming and costly than anticipated. One such a case was the decommissioning of the Cintichem reactor. The costs for that project hiked from $25 to 100 million due to incomplete records and poor predecommissioning characterization, which had failed to identify the massive spread of contamination in the soil underlying the reactor (MC Intyre, 2012). The causes of cost overruns in decommissioning projects are variable, but many are due to poor planning. An uncertain end point, modifications to project scope, or unexpected issues such as at Cintichem, all are possible causes of costly changes of course while the project is under way. To reduce the related cost increases, a complete characterization should be conducted to accurately define the scope of a decommissioning/remediation project well before beginning the work.

Like other case studies discussed in this book, the Cintichem case was not originated by decommissioning *per se*, but it was discovered during decommissioning and became a decommissioning issue.

7.18 Hanford F-reactor, Washington, United States (IAEA, 2006b)

7.18.1 Decommissioning process; problem encountered; impacts

The F-reactor was the third of Hanford's three original reactors. It became operational in 1945 and was finally shut down in 1965. The reactor used up to 265 m^3 of water per minute to cool the reactor during operations (Fig. 7.10). The F-reactor was placed in a state of safe enclosure (cocooned in US DOE terminology) in 2003. In 2008, radiation technicians entered the reactor for the first time to inspect the building and verify that no radiation was leaking from the reactor and that nothing was intruding the building. Regardless of exhibiting traces of mice and insects, the building looked in perfect condition.

The F-reactor had a large basin (600 m^2) at the back of the reactor block for discharging irradiated fuel. Fuel pellets were pushed from the front to the rear face of the reactor. These fuel pellets were cooled before being transferred to other Hanford facilities for processing. When the reactor was closed, these basins were not accurately deactivated to a safe, stable, and verifiable condition. It was unclear if any entire fuel elements or fuel pieces remained in the basins after they were backfilled with sand to the top—i.e., a sand layer of about 6 m.

FIGURE 7.10 Hanford F-reactor cooling basins. *(Credit to DOE, Wiki Commons).*

After removal of the filling materials from the basins, the areas had to be thoroughly reexcavated to gain full characterization. As these areas were radiologically monitored, several items (including fuel fragments) were found at the bottom. Innovative technologies were used to help retrieve these items, including radiation mapping (a technology known as laser-assisted ranging and data system, LARADS), remote monitoring, and a remotely controlled excavator.

7.18.2 Lessons learned

Accurate and timely stabilization (deactivation in US DOE terminology) of underground facilities is required before they are placed into safe enclosure. To this end, prior characterization of all affected areas is essential. Innovative technologies can greatly help to remediate poorly stabilized facilities, but they also significantly add to the project schedule and costs. These problems could be avoided if the facility were accurately stabilized.

7.19 Tunnel collapse, Hanford, Washington, United States

7.19.1 Decommissioning process; problem encountered; impacts

As stated earlier, Hanford was the site of US plutonium production operations in the period 1943–87. The site—housing nine reactors and associated processing facilities—is now managed by the DOE's Richland Operations Office (RL), which is tasked with the decommissioning of all remaining structures and the remediation of contaminated areas.

Two tunnels were constructed in the 1950s and 1960s in the 200 East Area close to the former Plutonium Uranium Extraction Plant (PUREX). The highly contaminated PUREX building has remained unused for 20 years. Like the other Hanford structures, PUREX is planned to be decontaminated, dismantled, and partly removed.

The tunnels—made of wood and concrete and covered with some 2.5 m of soil—house railway carriages that were full of contaminated equipment. Tunnel 1, about 110 m long, contains eight carriages. This tunnel is linked to the 520-m long tunnel 2 that contains 28 carriages. The tunnels were sealed in the mid-1990s and were occasionally surveyed.

On May 9, 2017, a section of tunnel collapsed in which contaminated materials are stored: This led to an emergency state declared at the US DOE's Hanford site. No radiological contamination was detected following the incident; however, the potential impacts to the environment and possibly to the workers deserve some attention.

The operating organization declared an alert after workers found that a 6-m section of the shorter tunnel had caved in. An alert is the lowest tier of an emergency state and is announced when an event is not foreseen to impact people beyond of the facility border. Further to the alert, the employees around the PUREX plant moved to a shelter. The sheltering order was canceled soon as no contamination had been detected.

The alert was later broadened to a site area emergency. This condition is declared when the event is impacting or could impact people outside the facility border but not outside the border of the Hanford site. Only employees securing minimal safe operations were allowed to remain on the affected area of the site.

Immediately, workers started preparations to fill the hole created by tunnel collapse: They used 420 m^3 of clean soil to temporarily stabilize the tunnel. The DOE stated, "Before allowing uncontrolled access to the area where the partial tunnel collapse occurred, officials plan to take additional near-term actions to ensure the safety of the workforce and the public" (WNN, 2017).

Following immediate intervention, DOE-RL and its contractor CH2M HILL Plateau Remediation Company (CH2M) used engineered grout to further stabilize the tunnel. Grouting made the tunnel safer in a move toward its end state. At Hanford, grout has been used to stabilize other legacy structures.

The workers accessed the tunnel through existing ports or newly excavated entrances. Hoses were employed to inject the low-density grout in several layers. Grouting had and was indeed finished by November 2017, before winter could further damage the tunnel structure (DOE, 2017a).

To maintain a constant delivery of grout, the work took place at night when trucks transporting the grout could dodge Hanford site's daytime vehicle traffic.

7.19.2 Analysis; solutions

During the grouting of tunnel 1, the adjacent tunnel 2 was under continual monitoring. Tunnel 2—built of steel and concrete—is much larger than tunnel 1. It was used in 1964–96. Although better constructed than tunnel 1, tunnel 2 was also at risk of subsidence and collapse (Fig. 7.11).

DOE assessed a range of options, including covering the structure with a foil of heavy plastic, a tent or a prefabricated building. It was also considered filling the tunnel with expanding foam or with sand or clay. Another option was the controlled collapse of the tunnel, although it could have been hard to contain the contamination spread following the collapse.

The option of removing the waste immediately was also assessed. Although that option would be a permanent solution, it would require the immediate allocation of significant funds and the project would take longer, leaving the tunnel exposed to the risk of collapse.

An expert panel recommended grouting as the preferred option based on overall safety, practicability, and costs, provided this option would allow the future disposal of the tunnel waste. For the final disposal, it is envisaged that

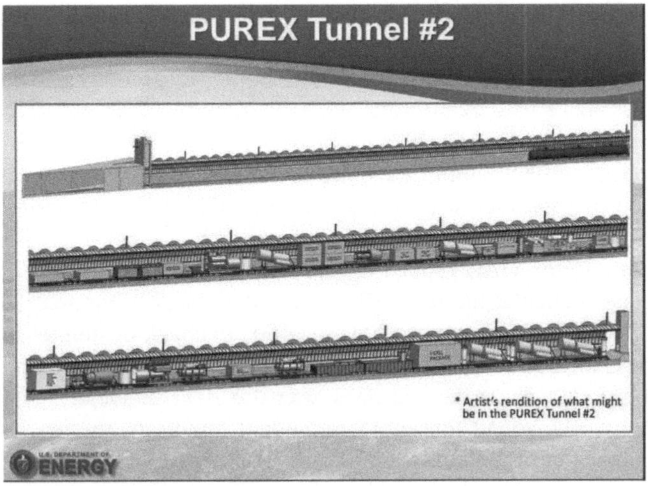

FIGURE 7.11 PUREX tunnel 2. PUREX, Plutonium Uranium Extraction Plant. *(Credit to DOE).*

the grout would be cut into blocks, some of them possibly enclosing a whole rail carriage, and hoisted out of the tunnel for disposal at a Hanford site.

Grouting of tunnel 2 began in October 2018: This timing allowed to consolidate lessons learned from the tunnel 1 project, firmly establish design criteria and work controls, and better appreciate the impact of seasonal variations on the grouting process (Fig. 7.12).

The tunnel 2 grouting mission was declared accomplished on April 26, 2019. The project required the injection of some 40,000 m^3 of grout (Ecology, 2019).

FIGURE 7.12 Grouting underway at Hanford waste storage tunnel 2. *(Credit to US DOE).*

Notwithstanding the success of this project, a report from the independent Government Accountability Office (GAO) criticized the US DOE for not investigating the root causes of the tunnel collapse; according to the report, inadequacies in DOE's monitoring and maintenance of its old contaminated plants are worrisome (GAO, 2020).

Apparently, some buildings have not been inspected for 50 years. The GAO report stated that without inspections, incidents like the tunnel collapse could occur anytime.

According to the GAO report, DOE had known about structural problems including subsidence at the tunnel 1 for decades. Inspections were conducted until 1991, when recommendations were issued to check the tunnel again in 10 years. In fact, the 2001 inspection was not conducted, and no assessment of the tunnel conditions was performed until after the 2017 incident.

The report also found that at 18 redundant buildings, routine inspections are only conducted at "selected facilities." 10 of these facilities have not been inspected due to concern for worker safety.

Finally, GAO had three recommendations for DOE: identify the root cause of the 2017 incident; conduct regular, comprehensive surveys of idle facilities and take timely action as needed; and evaluate the supervision of these facilities. The DOE agreed to these recommendations.

7.19.3 Lessons learned

Redundant facilities that have been unused for decades present inherent radiological and nonradiological risks. While in principle immediate dismantling and site remediation is the preferred strategy for shutdown facilities, there are often important reasons that may render this strategy impractical. If long-term deferral is inevitable, all technical and administrative measures should be in place to ensure that physical deterioration of structures, systems, or components does not pose unacceptable hazards. To this end, regular surveys and assessments are imperative. Another case of subsidence is given in Section 7.44.

7.20 Sodium tank fire, Dounreay site, Scotland, United Kingdom (ONR, 2016)

7.20.1 Decommissioning process; problem encountered; impacts

Dounreay site, near Thurso, Scotland, was the UK's center of fast reactor research and development (R&D) in the period 1955–94 and is now Scotland's largest nuclear cleanup and decommissioning.

As for all other decommissioning sites in the United Kingdom, Dounreay site owner is the Nuclear Decommissioning Authority (NDA), a nondepartmental public entity of the UK Government. Dounreay Site Restoration Limited (DSRL) is the site license company that operates under contract to the NDA.

FIGURE 7.13 Prototype fast reactor, historic photo around 1979. *(Credit to DOE, Wiki Commons).*

The third and last reactor to be built at Dounreay was the prototype fast reactor (PFR). PFR was a pool-type fast breeder reactor, cooled by 1500 t of liquid sodium and fueled with mixed oxide (MOX) fuel. It went critical in 1974, and in 1985, reached its electrical design output of 250 MWe. The reactor was permanently shut down in 1994, so ending the nuclear power production at Dounreay. Currently, the major decommissioning task at PFR is the removal of the sodium inventory to Sellafield (Fig. 7.13).

The PFR sodium tank farm includes four tanks containing sodium residues (contaminated with tritium) from the PFR operation.

On October 6, 2014, following activities in the Dounreay PFR sodium tank farm aimed at removing the residual sodium from an excess tank, the tank was left under surveillance for the night.

Early in the morning of the day after, the on-site fire brigade was alerted by an alarm from the tank farm. Around the same time, a surveillance worker heard loud noises and noticed smoke, and also called the fire brigade.

The fire brigade arrived soon and quickly extinguished the fire. The area was monitored until reassurance was given that reignition was unlikely. The area was kept under observation, until the fire residues were removed. All decommissioning work in the sodium tank farm was halted at once. DSRL undertook an investigation and found procedural noncompliances and inadequate behaviors that contributed to the event and lacked the expected commitments and standards.

While it was confirmed that some tritium had been released via an unauthorized route, this presented no remarkable risk to workers, the public, or the environment. The NDA and the regulators (Office for Nuclear Regulation, ONR; Scottish Environment Protection Agency, SEPA) were informed and launched independent investigations (ONR, 2014). Following its investigation, on November 11, 2014, ONR forwarded an improvement notice to the licensee and ordered that tank farm decommissioning work would not restart until the needed improvements had been implemented.

A dedicated team led by a senior manager was appointed by DSLR: They were made responsible for a safety improvement plan centered on learning lessons from this and other incidents (there had been more sodium fires at Dounreay) and implementing improvements.

In April 2015, ONR performed a planned compliance inspection as part of the Dounreay integrated intervention strategy. The inspection was aimed at notifying ONR's position in regard to the commencement of work to seal tanks 3 and 4 (which were affected by the 2014 fire) at the PFR tank farm. The inspection was performed against license condition (LC) 22 (modification or experiment on existing plant) regarding the modification proposal to seal tanks 3 and 4.

ONR evaluated that the implementation of this modification was compliant with LC 22. In particular, it was evaluated there has been good worker involvement in the planning of the modification, which has produced a good understanding of the work being undertaken among senior management, overseers, and the workforce. ONR also evaluated that the work has been accurately assessed for hazards and risks, was being prepared well, and will establish a safer, sealed state of the tanks.

The needed improvements to comply with the requirements of the improvement notice were completed in June 2015. Improvements regarded: arrangements for control of work; training; procedures; safety performance metrics; and the appointment of a technical steering group to monitor similar work in the future.

Following the fire event and the corrective actions taken by the operating organization, regulatory permission was granted in July 2015 to continue and complete activities in the sodium tank farm (BBC, 2015). The planned work ahead includes sealing the two tanks, on which work had previously begun prior to the fire event, followed by sodium processing with humidified air to destroy any residual sodium. Tank dismantling will start after the sodium processing is completed.

The fire event was rated level 1 on the INES.

7.20.2 Lessons learned

Sodium fire is a well-known hazard at sodium-cooled reactors. However, preventive and mitigation measures are available to reduce the risk. Technical measures should be coupled with administrative provisions; human factors should not be disregarded (in terms of human factors, a decommissioning site is critical due to the required organizational and cultural changes, challenges to personnel motivation, interaction of newly formed teams, etc.)

7.21 Leakage of radioactive effluents from decommissioning site, Sizewell A Nuclear Power Plant, United Kingdom (ONR, 2016)

Observe due measure, for right timing is in all things the most important factor (Hesiod, c.700 B.C.)

7.21.1 Decommissioning process; problem encountered; impacts

The Sizewell A nuclear power station includes two nuclear power reactors of the Magnox type, with electrical output of 210 MWe each. Both reactors started operation in 1966 and were permanently shut down on December 31, 2006.

Magnox Ltd, owned by Cavendish Fluor Partnership, is the management and operations contractor responsible for managing the UK's former Magnox nuclear power plants on behalf of UK's NDA. NDA is responsible for assigning contracts for the decommissioning of Sizewell A.

Defueling began in 2009, with fuel removed from the reactors placed in the site's spent fuel storage ponds before being encased in containers for transport to Sellafield for reprocessing. The Sizewell site was fuel-free in February 2015. In February 2018, Scuba divers started working in the former fuel storage ponds of the plant.

Preparations for a safe enclosure phase are expected to take until around 2034, followed by long-term safe enclosure (called care and maintenance in Magnox terminology) until about 2092. Demolition of reactor buildings and site release will follow.

On January 7, 2007 (just few days after final shutdown), a contractor working on the decommissioning of the station noticed water leaking onto the floor of the laundry where he happened to be washing clothing. The water was found to originate from the cooling pond that houses the reactor's spent nuclear fuel: The pond water had dropped more than 30 cm (against a total depth of several meters). It is estimated that some150 m^3 of radioactive water had leaked from a 4.6-m-long break in a PVC pipe, with some spilling into the North Sea. However, the radioactive amount released into the environment was less than 1% of the amount the power station is authorized by the license to discharge each year.

Alarm systems did not function, and even if they had, operators would have not noticed (see the following). However, operatives on duty reacted promptly to the warning, and the situation was stabilized within 45 min.

This incident was not fully disclosed for more than 2 years after the facts and attracted a lot of public attention due to the belated timing of the information release as well as the potential consequences of the incident.

An out-of-date but working control and instrumentation installed for the system that maintained the water level in the pond had been replaced. However, it was "replaced with one that is a poor design, poorly installed and did not work on the day." The system was set to send an alarm to the Sizewell A control room if the water level was deviating more than 15 cm from the normal level. A second alarm was to be sent if the water level decreased more than 40 cm below normal. Neither of these systems worked on the day of the incident.

Even if the pond alarms had been sent, this would only have activated an alarm that had already been on for 2 days. Operatives had been neglecting this "standing alarm" for a while as a false alarm, and nobody had recorded it or signaled it as a defect. "The tolerance of standing alarms in the main control room after the reactors were shut down raises issues associated with (License onditions on) operator training and their suitability for the job" (WNN, 2009).

The PVC pipe that broke did not meet design specifications, and it is uncertain that it ever did during reactor operation. Furthermore, "No evidence could be found of this pipework ever having been on the Sizewell A maintenance schedule." A question to be answered is why checks failed to disclose the condition of the pipe before it got shattered. The incident raised serious worries over maintenance aspects of the shutdown plant.

As reported by the Nuclear Installations Inspectorate (NII), the regulatory body in charge, without the occasional intervention of the contractor, the pond could have lost much more water before the next scheduled plant inspection (10 h), possibly until the irradiated fuel would have been exposed and caught fire, which could have produced an airborne off-site release. This would have been a worst-case scenario. However, the NII stated (WNN, 2009) that such a fire was incredible. Due to the location of the pipe break, around 60 cm of water would remain in the pond even after the pond had drained. Although this would not be enough to fully cover the fuel or safely shield radiation, it would greatly reduce the heat load of the fuel.

According to the NII, a "conservative" calculation showed the maximum heat load of the pond skips containing spent fuel to be around 5.2 kW, while heat loads below 7 kW would not trigger a fire. Moreover, even at 7 kW, the skips would not reach 600°C for 36 h, well in time for control to be reestablished (WNN, 2009). The event was classified INES level 1 (an anomaly).

The 2007 event prompted a thorough review and rebuild of the plant's design and safety systems. The introduction of new early warning alarms and protection systems is one of the improvements.

In September 2011, a shift worker found slightly contaminated water flowing from the Sizewell A pond water treatment system (PWTS) into a containment area. The responsible team took immediate and effective action to block the leakage.

Magnox's investigation indicated that the pond level decreased by 1.4 cm, which was too little to activate the early warning alarms.

A total of 13.3 m^3 of water leaked over a period of four and a half hours but was retained by the secondary containment, and presented no risk to the remaining nuclear fuel on-site, to the public or environment, and caused insignificant occupational exposure.

The operating organization launched an immediate investigation and established that the leak had been caused by damage to a flow meter following a poorly conducted maintenance operation (S&P Global, 2011).

Recognizing that this event was relevant to other reactor sites, the licensee performed a review of operation and maintenance practices of flow measurement across its Magnox fleet and launched a related program of short- and long-term improvements.

As per its enforcement policy, the regulator (ONR) addressed a formal letter to the licensee defining regulatory expectations (including implementation of needed improvements, transfer of lessons learnt to similar NPP sites, and actions to be taken before the facility could resume routine activities). ONR inspectors monitored the completion of the improvements. The event was rated INES level 0—no safety significance (ONR, 2016).

7.21.2 Lessons learned

Shutdown facilities may suffer from a perceived lack of priority and poor motivation, which may cause complacency, lack of alertness, and impact the quality of maintenance and surveys. This potential situation should be readdressed in a timely manner by the top management. A similar case of workers' inattention to repeated alarms is given in Box 7.1.

BOX 7.1 Inattention to message from the hazard alarm system to the control room, AVR HTR Nuclear Power Plant, Germany; August 26, 2018 (BASE, 2000)

Due to a frequent, repetitive appearance of a fault message, which was not considered relevant for the plant's decommissioning status, but which significantly disturbed the whole monitoring of the plant, it was decided to temporarily suppress the message. To this end, the related cable was disconnected. In the course of an operational process carried out the following day, it was found that an expected message was not displayed on the hazard alarm system. The connection was reestablished, and the alarm system was working properly again. The cause of the incorrect action (disconnection of the cable) was the incorrect interpretation of the frequently occurring fault message. In this case, there was no impact on the dismantling of the plant.

7.22 Leak from cooling pond, Dresden 1 Nuclear Power Plant, Illinois, United States

In order that people may be happy in their work, these three things are needed: They must be fit for it. They must not do too much of it. And they must have a sense of success in it. John Ruskin (1819–1900), Pre-Raphaelitism (1851).

7.22.1 Decommissioning process; problem encountered; impacts

Dresden NPP Unit 1, a boiling water reactor, generated electrical power from 1960 to October 31, 1978, when it was permanently shut down. The plant is now in safe enclosure (SAFSTOR in NRC terminology). The decommissioning plan was authorized by NRC in 1993. No significant dismantling operations are in progress. Following the spent fuel pond leak described in this case study, all spent fuel from Dresden-1 was removed to the on-site independent spent fuel storage installation (ISFSI). During safe enclosure (planned to last until 2027), Unit 1 is subjected to inspection and monitoring. The licensee expects that the active decommissioning of Dresden-1 will last from the end of the safe enclosure period to 2031. A 4-year delay in the remediation of the whole site will follow the major decommissioning activities of Dresden-1 to allow for the decommissioning of Dresden Units 2 and 3, located in the same site as Unit 1, with completion of these activities around 2035. Site remediation is planned for completion in 2036, including the demolition of the residual buildings and removal of any contaminated soil. The final site survey is tentatively planned for late 2036 (NRC, 2019b).

In January 1994, workers at Dresden 1 NPP found over 200 m^3 of water in the lowest elevation of the spherical reactor building. Two days later, at a routine decommissioning meeting, the incident was communicated to the decommissioning team, who soon realized the risk for freezing of a 1-m-diameter fuel transfer tube that connects the sphere to the spent fuel pool (SFP) (Fig. 7.14). Incidentally, the delay between the discovery of the incident and its reporting to a wider audience denotes the poor understanding of such situations and, in general, the lower priority given to shutdown facilities.

After investigating the incident, the NRC issued a "'generic letter"' to licensees of eight shutdown NPP units. The letter directed the utilities to start a program of inspections to confirm the adequacy of SFP containment systems.

The water in Dresden 1's basement originated from a prolonged below zero cold wave that had caused the freezing and bursting of several service water lines; for many years, the containment building had been unheated. Although the water feed system to the fuel pool had not been jeopardized, the lack of containment building heating could under more severe weather conditions have caused the freezing and burst of a tube that would have drained the SFP to about 1 m below the top of the stored fuel assemblies. The decommissioning

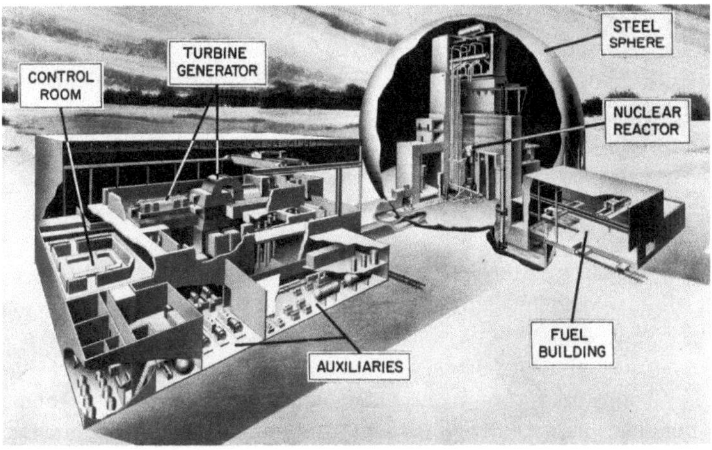

FIGURE 7.14 Dresden 1 inside: see the reactor sphere and the adjacent fuel building with the SFP. *SFP*, spent fuel pool. *(Courtesy of NRC, Wiki Commons).*

team prescribed that the pool gates between the adjacent SFP and the fuel transfer pool (in the Sphere) be installed, and a portable source of heat was installed on the fuel transfer tube.

Due to the long time elapsed from plant shutdown to the incident (15 years), it is highly unlikely that the heated fuel would have released airborne contamination. However, workers could have been exposed to high radiation fields. The NRC letter noted the lack of a leak detection system and cracks in the unlined SFP, which highlighted the potential for pool leakage.

The letter also stressed that the flaws identified were based on the wrong perception that Dresden 1 could not cause any serious safety problem because it was permanently shut down (Monitor, 1995). The Dresden event also discloses the significant reduction in regulatory supervision that occurred after a reactor permanent shutdown. At the time of this event, the Dresden site had two reactor units in operation and one permanently shut down. The NRC had inspectors assigned full-time to the NPP site supported by inspectors from its regional offices and the headquarters. Those inspectors devoted almost all of their time and attention to the operating reactors; otherwise, they could have observed that the licensee turned off the Unit 1 SFP cooling and cleanup 12 years earlier (resulting in the proliferation of microorganisms) or had discontinued heating the Unit 1 containment building 5 years earlier. Dresden 1 did not even appear on the NRC's inspection charts (Lochbaum, 2013).

7.22.2 Lessons learned

Over time, it has become clear that not everything in decommissioning can be confined to technical or organizational processes, although these remain predominant. There is a somehow "hidden" side, which includes motivation, behaviors, mind-sets, and, in other words, "culture" (and "safety culture" as a subset of it). Needless to say, "cultural" aspects are more difficult to frame, measure, assess, and correct than other aspects of decommissioning. Decommissioning as a whole is particularly vulnerable to the impacts to a cultural "deficit." In cultural terms, decommissioning is quite different from such formalized and regulated phases of a plant's life cycle as construction or operation.

Broadly speaking, safety culture can be defined as a pervasive (i.e., running through all management and operating levels of an operating organization) expectation that workers will remain at all times committed to safe and conservative activity, that overseers will encompass personal, industrial, and nuclear safety considerations in their decision-making, that management will include personal, industrial, and nuclear safety margins in their objectives, and that the entire organization will applaud and reward safe behaviors and blame attitudes that compromise safety. A good safety culture should be alert and imaginative in spotting hazards, proactive in planning and enforcing hazard mitigation measures, disciplined in carrying out activities, and attentive to what can be learned from both successes and failures (and near misses).

Uncertainty about the timing, schedule, and end state of the decommissioning project can cause motivational issues. It can be assumed that a long period of same enclosure (such as the one enforced for Dresden 1) is conducive to a false perception of low priority and a lack of interest, particularly when there are competing duties to comply with (in our case study, the concurrent operations of Dresden 2 and 3).

It is up to a plant's top management and to the regulators to counteract such devious tendencies and reinforce safety culture at all levels and ensure that due diligence is applied to dormant phases of decommissioning (LARAIA, 2015).

In terms of decommissioning processes and strategy, the following areas deserve attention:

- SFP water mass balance and inventory as well as the trending in water addition to fuel pool should be kept under constant supervision at shutdown plants.
- Depending on the vulnerability of SFP to leakage, demineralization may be required to limit the total contamination inventory in the pool water.
- A generic lesson learned is that more weight should be given to the immediate dismantling strategy rather than to the deferred dismantling strategy. Or at least the decommissioning plan should include an early transfer of spent fuel elements from the SFP to away-from-reactor dry storage (Anagnostopoulos, 1995).

A lesser event that resulted in temporary loss of cooling to the spent fuel pool is described in Box 7.2.

BOX 7.2 Short-term interruption of the spent fuel pool cooling, Neckarwestheim 1 Nuclear Power Plant, June 30, 2012

At the time of this event, Neckarwestheim 1 (GKN 1) was finally shut down, but the fuel elements were still in the storage pool. Due to a defective electronic card, a pump of the cooling line in operation for the spent fuel pool cooling failed. The electronic card was replaced, and the cooling pump started up again. The report submitted by the licensee states that a clear cause for the shutdown of the pump was not found. The replaced electronic card was discarded. The brief interruption of the pool cooling did not result in a measurable increase in the temperature of the pool water: By the time of the event, the thermal output of the fuel elements in the pool had already declined sharply.

7.23 Transport of radioactive waste wrongly addressed to a landfill disposal site (ONR, 2016)

7.23.1 Decommissioning project; problem encountered; impacts

In April 2010, Sellafield Ltd (SL) sent a routine batch of supposedly exempt waste to the Lillyhall, an ordinary landfill site in Workington, Cumbria, United Kingdom. Exempt waste is defined as material with an activity concentration lower than 0.4 Bq/g. This batch comprised four bags that SL later identified as being above the exemption limit. This event came to light when SL discovered—before the bag was shipped—another bag of waste having activity levels above the exemption limit, but wrongly declared exempt. The bag was quarantined, and operations in the SL waste categorization facility were interrupted. The Department for Transport (DfT) was informed shortly after the event was discovered.

The four bags contained miscellaneous low-level radioactive waste (LLW) (e.g., waste plastic, paper, tissues, clothing, wood, and metal). The bags should have been sent to a specialist facility that deals with and stores LLW. In October 2011, the regulatory responsibility for the transport of radioactive materials was transferred from the DfT to the Office for Nuclear Regulation (ORR), who took over the investigation and enforcement actions. There were several legal violations, including among others: SL's environmental permit and the Carriage of Dangerous Goods and Use of Transportable Pressure Equipment Regulations.

The four bags of waste were safely recovered by SL from the Lillyhall site and shipped to their correct disposal. One bag had been punctured causing minor land contamination. This was remediated, and the subsequent monitoring could detect only background radiation levels. SL stated that no more bags of waste with activity concentrations above the exemption limits had been shipped to Lillyhall. "*A detailed analysis of the waste has shown that* *there were no environmental implications of the event and that the maximum potential dose to members of staff at the Lillyhall site was significantly less than a member of the public would receive from normal background radiation (*WNN, 2013*).*"

Sellafield's investigation found that the error was caused by an inappropriate configuration of the bag monitoring equipment together with shortfalls in the commissioning procedure for this system, which produced the wrong classification of the bags as "general" rather than "radioactive" waste. ONR and Environment Agency (EA) conducted a joint investigation and highlighted several management and operational inadequacies. The regulators also noted Sellafield's failure to learn the lessons from previous incidents of this kind. ONR and EA ordered Sellafield Ltd. to establish improvements to prevent recurrences and also launched a prosecution alleging significant breaches of the law. In February 2013, Sellafield Ltd. pleaded guilty to the alleged offences and was fined £700,000 ($ 864,000) for legal offences. The Carlisle Crown Court also ordered the nuclear company to pay £72,000 ($ 86,000) costs. The organization appealed, but the fine was later confirmed by the Court of Appeal.

Following the incident, before waste operations were reintroduced at Sellafield, additional monitoring measures were enforced to avoid a recurrence of this type. The event was classified at level 0 on the INES.

The reader should note that Lillyhall has had since 1995 an environmental permit to accept small volumes of very low-level radioactive waste, VLLW (not including Sellafield waste). In later years, Lillyhall has been authorized to accept high volumes of VLLW. There are also other landfills sites in the United Kingdom that have been authorized to accept VLLW, so alleviating considerably the burden on the Drigg LLW repository, near Sellafield. Lillyhall is the landfill site closest to Sellafield.

7.23.2 Lessons learned

The inadvertent transfer of unacceptable waste to ordinary landfills or scrap dealers has been recurrent in nuclear history (Sections 5.23 and 7.14). Events of this type are usually due to a combination of monitoring flaws and poor managerial practices.

7.24 Loss of containment of two radioactive packages at Orano's decommissioning uranium conversion plant, Pierrelatte, France (WISE, 2020)

7.24.1 Decommissioning process; problem encountered; impacts

Orano Cycle reported to the French Nuclear Safety Authority (ASN) on August 8, 2018, a significant incident of loss of containment of two packages of radioactive material stored in the uranium conversion plant (INB 105).

Located at the Tricastin site, ORANO cycle uranium conversion plant was primarily tasked with the chemical conversion of uranium into uranium hexafluoride (UF_6) for enrichment. INB 105, shut down since 2008, was used especially to recycle uranium from reprocessing or from fresh nuclear fuel. The packages involved in the event originate from this installation and are stored in a dedicated building. They consist of a metal drum (the first containment), enclosed in an overpack (the second containment). On July 23, and again on August 6, 2018, surveyors observed the leakage of the two radioactive packages. Monitoring devices measured abnormally high levels of activity levels in the uranium store.

7.24.2 Analysis; solutions

Due to the high storage temperatures in summer and the presence of gaseous hydrofluoric acid in the drums, the two packages were subject to overpressure, which caused a loss of tightness and a leak around the cover of the overpack. Pending repackaging, the two packages were moved to a sturdy vacuum chamber to reestablish static and dynamic containment conditions. Investigations conducted by ORANO showed that one of the two overpacks was improperly closed, which caused the spread of contamination outside the storage, as confirmed by environmental monitoring of the site. Moreover, the inaccurate setting of monitoring alarms caused the late disclosure of the incident. Fortunately, due to the lack of workers manning the storage at the time of the incident, there were no occupational impacts. The negligible activity concentrations measured outside of the store proved the lack of off-site impacts. However, the unplanned spread of contaminants, the improper closure of the overpack of the second package (a noncompliance with the general rules of conduct of the facility) and the late detection of the incident, led ASN to classify this incident at level 1 of the INES.

7.24.3 Lessons learned

The typical (years- or even decades-long) storage of radioactive materials/waste during a decommissioning project requires that great attention be paid to the manufacturing specifications of storage drums and to the environmental conditions of the stored materials. These challenges in turn require frequent, careful surveys and the due care and maintenance of monitoring systems (Box 7.3).

BOX 7.3 Waste storage violations at decommissioning EURODIF enrichment plant, France (WISE, 2020)

Eurodif's gaseous diffusion plant at Tricastin was closed in 2012. In the subsequent cleaning phase 2013–16, chlorine trifluoride was injected into the plant systems

Continued

Box 7.3 Waste storage violations at decommissioning EURODIF enrichment plant, France (WISE, 2020)—cont'd

to remove residual uranium. In addition to recovering large amounts of uranium, this flushing process helped reduce the overall radiation levels. Under the decree of February 5, 2020, dismantling operations of the Eurodif plant were officially authorized. The decree specifies that these activities must be completed no later than December 31, 2051, a 30-year project. During an inspection held at the decommissioning EURODIF enrichment plant on December 13, 2017, ASN discovered numerous violations in connection with the storage of waste: In particular, waste was found to be stored in various locations not designated for this purpose. In a letter dated January 11, 2018, ASN demands the company to rectify the situation within 1 month—indicating a certain sense of urgency. The letter includes a lot of detailed prescriptions.

7.25 High uranium groundwater concentration discovered at former fuel fabrication plant (WISE, 2020)

7.25.1 Decommissioning process; problem encountered; impacts

The Industrial Nuclear Fuel Company (in French, Société Industrielle de Combustible Nucléaire, SICN) plant is located at Veurey-Voroize, 12 km from Grenoble on the banks of the river Isère, and occupies a land of some 11 ha. The plant, comprising facility No. 65 (nuclear fuel fabrication plant) and 90 (fuel pellet workshop), was responsible, until the mid-1980s, for the development and manufacturing of nuclear fuels, based on metallic uranium or uranium oxide, in particular for graphite gas, heavy water, and fast neutron reactors. From the late 1980s to the early 2000s, activities moved toward the production of prototypes, special uranium fuels, the assembly and the instrumentation of experimental devices for reactor research, and development of equipment for manufacturing of fuels based on mixed uranium and plutonium oxide (MOX). The operations of facilities No. 65 and 90 were stopped, respectively, at the end of 2001 and in June 2002.

The operating organization then proceeded, until 2005, to remove uranium-bearing materials and operational waste as part of the permanent cessation of operations. Final dismantling operations of these two facilities began after the decommissioning license in 2006 and are now in progress.

In a report issued on September 24, 2008, the French Radioprotection and Nuclear Safety Institute (in French, Institut de Radioprotection et de Sûreté Nucléaire, IRSN) announced that noteworthy uranium concentrations were discovered in groundwater at the decommissioning COGEMA/SICN nuclear fuel fabrication site located at Veurey-Voroize. Two contamination areas were identified. During a period of 3 years, average concentrations went down more than three times. The contaminated zone stretches from the nuclear site to beyond the site borders (IRSN, 2008).

As for other case studies mentioned in this book (for example, the Cintichem case, Section 7.17), it is difficult to establish whether the environmental contamination originated or worsened during the decommissioning period or (more likely) during operation. In either case, groundwater contamination stayed on-site during and after the decommissioning phase, and post-decommissioning environmental remediation may be required.

The radioactive materials used at facilities No. 65 and 90 were based on uranium, which could be depleted, natural, or enriched in isotope 235 up to 5%, 10%, or 93%.

Discharge limits for Veurey-Voroize plant are fixed in terms of maximum concentrations in the receiving environment after discharge. In the context of shutdown and decommissioning activities, radioactive liquid effluents (aqueous effluents contaminated with uranium) can come from nuclear buildings and originate from soil washing, sinks and showers, the analytical laboratory, and decontamination of solid waste. These effluents are treated in the plant's water treatment station before being discharged into the river Isère. Over the past few years, around a few MBq of uranium has been released annually in liquid form.

7.25.2 Lessons learned

Experience has shown that groundwater contamination can result from operations (and to a minor extent) D&D of nuclear facilities. It is important that the gradual disappearance of barriers during D&D does not cause the diffusion of contamination below ground (especially in liquid form) and imperil the environment.

7.26 Tank W-1A project, Oak Ridge National Laboratory, Tennessee, United States (Knoxblogs, 2012)

7.26.1 Decommissioning process; problem encountered; impacts

Tank W-1A was an underground waste tank that had radioactively polluted for decades the groundwater in ORNL historic central campus. The 15 m^3 stainless steel tank started operation in 1951 and was used for collection and storage of liquid radioactive waste from ORNL's research and production. The tank was removed from service and emptied of most of its contents in the early 1990s after high activity levels were discovered in the surroundings, chased, and attributed to the vessel. However, there were indications of radioactively contaminated groundwater in the vicinity even earlier. The soil became contaminated through a leak at the inlet to Tank W-1A and other leaks from the piping. The tank and the radioactive soils around it—also known as

Corehole No. 8, after the name of a monitoring well where the contamination was first found—were indicated to be the major source of radiological groundwater contamination in Bethel Valley (which encompasses the main ORNL area). From the tank, the plume migrates toward First Creek. After the source was localized, there have been a series of aborted attempts to excavate the radioactive soils and get rid of the old waste tank.

The decades-long cleanup project, which URS/CH2M Oak Ridge (UCOR) inherited from Bechtel Jacobs Co. (BJC) when it took over as DOE's environmental manager in 2011, and satisfactorily completed, has proven to be terribly difficult.

In 2001, a cleanup project was stopped after workers bumped into unexpectedly high radiation fields (60 mSv/h) when they began to excavate the soil around W-1A.

Another attempt was conducted by BJC: They began an Operational Readiness Review (ORR) in April 2011, but it was halted and then closed due to many negative findings making success uncertain. During that review, "it became apparent that the project team was not sufficiently prepared to successfully complete the effort. As a result, BJC terminated the ORR" (Knoxblogs, 2012). Following the initial failure, more staff was added, several organizational changes were made, and corporate oversight was enhanced. BJC prepared a new ORR and submitted in July 2011 to DOE for approval. Field work at the ORNL site began shortly after DOE's approval. The removal of the underground waste tank and the highly radioactive soils around it was subcontracted to DEMCO Inc.

In 2011 and 2012, two bedrock wells were installed to extract contaminated groundwater and route it to the on-site wastewater treatment system. Older system components were replaced, and new wells and a refurbished pumping system began operation in March 2012. Monitoring in First Creek proves that the plume has been contained. Bechtel Jacobs has been supervising the $25 million project, but DEMCO was responsible for providing the site superintendent, equipment, equipment technicians, and workers. The project includes the removing of the old tank and the concrete pad beneath the tank. The radioactively contaminated soils were excavated and transported to the Nevada National Security Site (NNSS) for disposal. Some of the soils were not expected to meet the criteria for disposal at NNSS, and those were containerized and transferred to an Oak Ridge storage facility. After the excavation, workers backfilled the area in ORNL's central campus and returned the site to the pristine grade.

Excavation of the tank began in September 2011; the tank was successfully removed in January 2012 and was sent to be cut up and disposed off-site. Soil and secondary waste disposals were completed in May 2012 (OREM, 2019).

During the tank removal and excavation of radioactive soil, the cleanup workers received significant, but acceptable radiation doses. There were no skin contamination incidents or internal exposures during these removal activities. The average total external exposure received by the 19 workers who worked full-time in the weather enclosure was approximately 3.50 mSv; the highest individual exposure was 5.36 mSv. In total, 316 shielded containers of contaminated soil (well over 500 m^3 in volume) were sent to the NNSS.

7.26.2 Lessons learned

This case study has been categorized among those which are relevant to environmental contamination (although this event was initiated before decommissioning, it got worse during decommissioning). In fact, the case is also relevant to occupational impacts during decommissioning. It highlights that old underground facilities present serious challenges to their decommissioning. First, due to their poor accessibility, there are significant issues in physical and radiological characterization, deployment of decontamination techniques, and physical dismantling and removal. Second, these types of components are present in a large number of nuclear installations. However, early nuclear design and construction practices often did not incorporate decommissioning requirements. Third, contamination of underground components is typically complicated by contamination of surrounding soils, which extends a D&D project to an environmental remediation one. A comprehensive review of underground decommissioning projects, and safety and practical guidance is given in IAEA (2006b).

7.27 Oak Ridge National Laboratory's Building 3026 decommissioning (Knoxblogs, 2010)

7.27.1 Decommissioning process; problem encountered; impacts

To date, the ultimate demolition of old hot cells at ORNL remains a serious challenge. Constructed in 1943–45, Building 3026—one of the lab's initial—is still one of its biggest hazards. Building 3026 was originally built to support the development of the atomic bomb in World War II. It later was used for a range of processing and research operations with radionuclides. Operations ceased in Building 3026 in the mid-1980s. Access to the facility had been restricted since 1998, and the facility had fallen into severe disrepair. Building 3026 C/D is a hot cell and laboratory facility including two adjacent structures divided by a firewall with separate ventilation systems. The main structure in Building 3026-D is a two-story, 1.5-m-thick concrete cell block, which is split into two radioactive material handling cells. The two hot cell banks, which were formerly used for storage, treatment, conversion, and disposition of radionuclides, are located in the Building 3026-C area. The contractor UT-Battelle demolished the wooden superstructure covering the hot cells in early 2010. This was a time-critical action as the deteriorating structure interfered with the fire protection system and impaired the containment of the hot cells. Although Building 3026's wooden superstructure has been demolished, the exposed hot cells require action to avoid release of radiological contamination into the environment, potentially impacting a large number of employees in vicinity of the cells. DOE had planned the demolition of the hot cells soon after the superstructure was removed. The project objective was to remove any remaining legacy materials, abate hazardous materials, characterize hot cells and process equipment, decontaminate as needed, remove the hot cell concrete to slab or grade, and decontaminate the slab. However, due to the legal conflicts regarding the assignment of this demolition order, DOE was unable to begin the demolition. As an interim measure, UT-Battelle sealed the hot cells using an epoxy coating (Fig. 7.15). However, this work missed the goal of stopping the leakage. The coating started to peel off and contamination was released from inside the cells onto the concrete pad, which required cleanup of contamination around the hot cells. The entire area had to be classified as a contamination area, which is subject to constant surveillance and maintenance. Environmental agents, including uncommonly heavy rains, contributed to further ruin the integrity of the coating. The worsening conditions of the structure were posing a significant risk that continuing leaks from the interior of the exposed hot cells will produce a spread of radiological contamination. Routine surveys indicated a number of spots where the coating has developed pits or lost adherence, and liquid could be observed percolating through in those deteriorated locations. These areas were being decontaminated as contamination was being discovered; however, with time and prevailing humid weather, the contamination tended to reappear. At least, six

FIGURE 7.15 Workers adding a protective coating to hot cells at Building 3026 at ORNL; the coating was intended to prevent hazardous releases until removal of the hot cells. ORNL, Oak Ridge National Laboratory. *(Credit to DOE, Wiki Commons).*

decontamination campaigns were required. The major concern to ORNL and the public was the possibility of a major rainfall in concurrence with a contamination event that would spread the contamination beyond the control area and onto Central Avenue and surrounding facilities. About 4500 people work daily within a 1.6 km distance from the facilities.

Finally in April 2012, four cells were demolished and removed from the "C" side of the building. The remaining two on the "D" side were more contaminated than originally assumed and are being addressed at the ongoing demolition stage.

At the Building 3026 area, employees recently finished pumping grout into a 15-m-long underground transfer tunnel formerly used to load radioactive material into a hot cell. This critical task ensures there is no pathway for the contamination to spread during demolition of the final two hot cells.

Demolition of these hot cells is beginning soon (DOE, 2019a).

In November 2019, workers at the Building 3026 hot cells started installing a protective tent to cover the area during demolition. Using a 175-t crane, teams set a foundation of 92 8-t blocks for the protective tent. Workers then began using the crane to construct 20 steel trusses to install the frame. To complete the structure, nearly 1200 m^2 of fabric is being added in two sections (Fig. 7.16).

7.27.2 Lessons learned

As stated in other case studies, old nuclear facilities are prone to leakage, often resulting in contamination of external areas. In planning for decommissioning of these facilities and soil remediation, care should be taken (1) to fully characterize the extent of contamination and (2) not to spread contamination to other areas. Constant monitoring of the worksite and surroundings is required.

FIGURE 7.16 Crews are manually adding 1200 m^2 of fabric to the trusses to complete the cover for the protective tent at the Building 3026. *(Credit to DOE).*

7.28 Alpha-5 the "worst of the worst," Oak Ridge National Laboratory, Tennessee, United States

7.28.1 Decommissioning process; problem encountered; impacts

Alpha-5 (9201-5) Building adjacent to the Y-12 nuclear weapons plant, ORNL (still an active production facility), is a notorious example of a big problem facing the US DOE. The building is a large four-story structure, steel brace frame including process and manufacturing equipment. Alpha-5 covers around 57,000 m^2 and was constructed in 1946 to house Calutrons (a mass spectrometer based on electromagnets, originally designed and built in the 1940s for separating the isotopes of uranium and producing highly enriched uranium for nuclear weapons). In addition to uranium enrichment, Alpha-5 former operations included several metallurgical and machining processes with uranium and beryllium. Alpha-5 contamination is due to beryllium, enriched uranium, depleted uranium, lithium, asbestos, and PCBs, and has been made worse by the intrusion of water from an old and deteriorating roof. Legacy materials include, but are not limited to, nonprocess equipment, containers, tools, and miscellaneous contaminated office equipment. Since unused buildings are often used as storage for external materials, there are also "orphaned" items—often of uncertain characterization—from other ORNL buildings that may or may not have disposition problems. The volume of material requiring disposition was initially estimated around 20,000 m^3, but this could be a low figure.

Since it ceased operations in 2005, Alpha-5 has suffered considerable degradation. In 2008, it was first noted that the facility was extensively flooded, external piping and related supports were corroded, and reinforced concrete roof panels had lost integrity. The 2008 assessment concluded that the combination of the large facility size, quickly degrading conditions, and huge amounts of items to be removed made Alpha-5 one of the worst liabilities in the DOE's complex.

The basement is a marsh. About 7500 m^3 of water has penetrated the building, allowing contaminants to pollute the groundwater. Heavy moisture inside has made mold proliferate, mandating respirators to anyone entering the building.

The real issue for DOE is that there is a long list of cleanup projects either in planning or being actively implemented, but the money is not enough to speed up the program especially for high-risk, expensive projects. Financial restrictions are additional to the formidable technical problems that some of these facilities (e.g., Alpha-5) pose to their decommissioning. In many cases, there are no final decommissioning plans. The first issue regarding excess facilities is their tremendous number, literally many thousands within the DOE complex.

Various reports have dubbed Alpha-5 as the "worst of the worst."

The Alpha-5 situation is particularly critical because the DOE spent already some 100 million $ on a 2011 project that removed a huge amount of redundant materials and equipment from the main building. Around 5430 containers were sent to Nevada site for disposal, and the project was supposed to be a major step toward full decontamination and demolition.

But the roof failures have caused a never-ending series of waterborne issues—although actions have been taken, e.g., some roof repairs—to prevent the situation from worsening (Knoxville News, 2016).

However, since cleanup efforts began, officials revealed that the facility has degraded at an increasingly alarming rate. In particular, a 2014 NNSA site assessment indicated that roof degradation continues to be widespread throughout the facility with varying levels of severity. This has resulted in significant water intrusion and the spread of radiological and toxicological contamination.

"Additionally, the assessment identified the potential for an explosion or reaction associated with remaining contaminants and personnel safety issues related to the degraded condition as high-risk areas. Overall, the assessment concluded that this facility presents a high risk to the workers and environment and should not be accepted. The assessment noted that demolition remains the only viable risk-accepted standard. Further, it noted that funding will need to be diverted from mission work to prevent the realization of imminent risks and mitigate the consequences of realized risk events" (Knoxblogs, 2015a).

The demolition of Alpha-5 does not pose as many radioactive hazards as some of the DOE's completed demolition projects did, such as the Rocky Flats plutonium facility in Colorado or the K-25 uranium-enrichment plant at ORNL. But the cleanup will be compounded by Alpha-5 sitting next to active production facilities at Y-12. Due to security issues, the location would have to restrict the access and movements of cleanup workers or pose potential hazards to Y-12 workers not involved in cleanup operations (Fig. 7.17).

A significant move toward Alpha-5 remediation is the foreseen construction and operation of the Mercury Treatment Facility at the Y-12 National Security Complex. The construction site was officially opened in November 2017. When the new plant is complete, it will reduce mercury concentrations in the East Fork Poplar Creek and enable large-scale cleanup and demolition at Y-12 complex, including Alpha-5 among others. The plant will reduce and control any mercury releases as demolition workers take down Alpha-5 and other buildings and will also tackle the mercury-contaminated soils. When in full operation, the plant will process over 10 m^3/min of water. It has also a 7500 m^3 storage tank to gather storm water. DOE anticipates the plant will go operational in 2022 (DOE, 2017b). The challenges and risks involved in decommissioning of a mercury-contaminated system at Y-12 are described in DOE (2019c). A case study regarding mercury contamination (albeit centered on specified members of the general public rather than the wide environment) is briefly described in Box 7.4.

FIGURE 7.17 Workers empty Alpha 5 moving the building one step closer to becoming demolition ready *(Credit to DOE, Wiki Commons).*

BOX 7.4 Mercury and other chemicals

Liquid mercury was used as a coolant for early experimental fast reactors. Clementine, the first liquid metal cooled reactor, Los Alamos Scientific Laboratory, was mercury cooled. Mercury was also used at certain prototype reactors such as PFR and DFR at Dounreay. The advantages of mercury as a liquid metal reactor coolant are its low melting point (−38.8°C) and chemical inertness in air and water. Its disadvantages are toxicity, relatively low boiling point (356.6°C), relatively low thermal conductivity, high neutron cross section, and high density making pumping difficult and expensive. Due to these drawbacks, mercury has long fallen out of favor for nuclear applications: However, it can still be found at nuclear facilities in older instruments such as gauges, thermostats, and pressure regulators. See also IAEA (2006c).

Due to its high volatility, the greatest health risk is inhalation of mercury vapor, especially in a confined space. Therefore, spilled mercury may represent a serious hazard. Chronic mercury poisoning can have severe effects on the nervous system and kidneys. Short-term exposure (up to a few weeks) to high concentrations of mercury can cause cough, difficult breathing, angina, nausea and vomiting, diarrhea, fever, and high blood pressure. Longer-term exposure to low concentrations can result in trembling, sleeplessness, irritability, migraines, and amnesia. Suitable PPE should be used with mercury waste.

A case quoted in NYS DH (2012), although it did not occur at a nuclear or radiological facility, is representative of a hazard that could be looming during the dismantling, e.g., of a radiobiological laboratory. One day, a family was surprised to find tiny, silvery beads dripping from a doorpost. They investigated the puzzling drops and found that the source was a container of mercury, one of several abandoned items stored in the top floor. When the container broke, about half a kilogram of mercury oozed through the floor into the room below. A specialist team responded. Two people had to leave their home for 9 days until completion of cleanup.

7.28.2 Lessons learned

Following the final shutdown of a nuclear plant, a hazardous situation can eventually develop if no action is soon taken toward full decommissioning. The so-called "no action" approach is unfortunately common to many shutdown facilities. No action being taken often results from the perception that the risks associated with the shutdown facility can be readily tackled any time, and therefore, there is no urgency. In other cases, no action being taken may be due to a lack of funds for the decommissioning of a given facility or priority given to the decommissioning of other facilities (the Alpha-5 case). Eventually, prolonged inaction may end up with sort of semiabandonment of the plant. The loss of records can be particularly serious. There are various risks associated with this approach (or rather lack of a decommissioning strategy). For example, the knowledge of the design and construction features of the plant and its operating history tends to fade away fast, both due to the dispersal of staff familiar with the plant and the loss of documents. The historical memory of key staff plays a major role in decommissioning, particularly for facilities built and operated in the early times of the nuclear industry, when records were rarely kept and maintained accurately. Besides, reuniting a team of people familiar with the plant may be nearly impossible a few years after its final shutdown, as the former staff rapidly disband and find new jobs or retire. Even when a decommissioning project is managed by contractors, experience has shown that having some former staff in the decommissioning team will be helpful.

A long period of inaction invariably leads to higher decommissioning costs including extra contingencies to cover increased uncertainties. Timely planning and implementation will avoid these difficulties. Other risks arising from "no action" include a lack of maintenance, which causes systems and components to corrode, contaminated fluids to leak, and drainage pumps and sumps to get out of order. Rainwater and groundwater start to leak in and out of the facility, which will result in a spreading of contamination, with associated risks to decommissioning workers, the public, and the environment. Another serious risk is that, owing to reduced surveillance, contaminated material or even radioactive sources are stolen because of their perceived salvage value. Failure to protect radioactive sources has already resulted in fatal accidents worldwide (see the notorious Goyana accident in Brazil). A policy of no action is generally disapproved of by regulators and is not endorsed by the IAEA. Decommissioning means positive management initiative and actions toward a predetermined end state, supported by adequate human, technical, and financial resources. To make a successful decommissioning project, the duration of any period of no action following permanent shutdown of a facility must be minimized and ideally reduced to the drafting of a detailed final work plan (IAEA, 2003).

7.29 Overview of cases at Sellafield, United States (Sellafield, 2019; ONR, 2020)

Find out the cause of this effect, or rather say, the cause of this defect, for this effect defective comes by cause. Shakespeare (1564–1616) Hamlet, II, ii, 101.

7.29.1 Misrouting of samples—October 10, 2019.

During the successful demolition of the First Generation Reprocessing Plant Stack, some samples were removed of its concrete support. These samples were shipped a laboratory for analysis using procedures typical of the civil engineering sector.

This analysis was meant to assist the stack demolition project and to better understand the long-term behavior of aging concrete.

The samples were erroneously shipped using arrangements for materials that are not radiologically classified.

The incident analysis confirmed that these materials posed no radiological risk to the workers, the public, or the environment.

Following the discovery, as an immediate step, Sellafield Ltd immediately issued a notice to the teams in charge of releasing materials from site to stress procedures and objectives: Shipments of this kind were temporarily halted.

7.29.2 Building evacuation—September 4, 2019

Monitoring equipment in a maintenance facility at Sellafield registered elevated levels of airborne activity.

As a precaution, and according to procedure, employees were evacuated from the building.

Operations in the facility were halted while analysis was conducted to identify the cause of the incident. Two small areas with elevated surface contamination levels were discovered within controlled areas. These were decontaminated to below permissible limits.

During the incident, slightly increased stack discharges—though well below permissible limits—were measured, before the building ventilation was stopped. There was no risk to employees or to the public.

7.29.3 Discharge pipeline damage—July 26, 2019

A Sellafield Ltd supervisor localized a leak from a short section of a lagoon discharge pipeline under a concrete block. The pipeline conveys low-level liquid effluents (e.g., cooling water, rain water) to sea.

In response to the incident, a crew assisted by the site emergency services installed an interim reroute from the damaged section of pipe and then placed a temporary repair of the damaged pipe to return to normality. Subsequently, a sturdier pipe was installed while final replacement was being procured.

An investigation disclosed that the pipeline had failed due to internal and external corrosion. No contamination to the ground underneath the discharge line was measured.

7.29.4 Ductwork defect—October 12, 2018—updated May 2019

Improvement works have been conducted on the ventilation systems of two old facilities at Sellafield.

During these works, defects were found on the ventilation ductwork outside the buildings.

The first defect was found during a routine safety assessment of a steel ventilation duct wall belonging to a material handling facility. Also, during maintenance work on a duct running from on-site laboratories, more defects were found.

The ducts extract ventilation air from the buildings, and there was no detectable activity released through the defects. The potential radiological impact on the public and the environment has been evaluated as very low.

The Environment Agency issued an enforcement notice to Sellafield Ltd requiring that the defects be fixed and maintenance upgraded.

Moreover the two organizations concurred that a broader inspection and repair program be undertaken for ventilation systems in other site areas; an independent review of the site's ventilation management should also be conducted. This broader program was expected to take some 2 years.

During the broader site review, minor defects were found at the external ventilation ductwork in two decontamination facilities (February 2019) and the Active Handling Facility (April 2019). The former ductwork conveyed air extracted from the decontamination tanks, which had not been routinely used for more than 2 years.

The initial incident, which triggered a broader review of Sellafield's ventilation systems, was rated at INES level 1 (an anomaly).

7.29.5 Mixed metal shipment—January 14, 2019

A metal recycling contractor for low-level radioactive metals originating from Sellafield received a mixed metal waste batch.

The metals waste contained rain water, which was not declared to the recycling contractor and violated the facility's normal parameters. No safety or environmental consequences were identified.

7.29.6 Water Supply Pipe Failure—March 1, 2018

During a period of very cold weather, an overhead external water (i.e., uncontaminated water) supply pipe failed and water flowed into an active pipe trench and the underground level of a contaminated excess building. This caused the contaminated water from the flooded underground level to flow into a concrete compound on the building side and along an excess trench. The contaminated water discharged to ground at these two places. The remainder of the contaminated water was pumped, by standing systems and the fire brigade, to normal effluent management systems and released to sea via the approved route.

Sellafield Ltd. conducted analyses to estimate the radioactive releases to sea and ground based on the approximate volumes of liquids released and water samplings. The whole volume of leaked water is estimated as 680 and 565 m^3 of this is believed to have been contaminated. ONR accepts that these estimates are on the safe side.

The estimates of radioactive discharges are based on nine water samples taken from the basement of the contaminated building, the sump in the compound on the building side, and the trench. All nine total β counts and three of the nine total α counts were below the detection limits and hence negligible. Six samples showed that the main α emitters (^{241}Am and $^{238,239,\ 240}$Pu) had exceeded the release limits for notification of occurrences by a factor of 4. The radionuclides detected are those expected by Sellafield Ltd., and the radioactive amounts released are considered modest. For example, if all of this contaminated water had been discharged to sea, this would have been than 1/50,000 of the annual authorized releases.

The incident met the "Ministerial Reporting Criterion—Abnormal occurrences leading to a release or suspected release or spread of radioactivity on- or off-site, which requires special action or special investigation by the operator" and was rated level 1 on the INES.

The incident quoted here is one of many due to freezing conditions that caused the rupture of nuclear pipes. See the Dresden 1 case (Section 7.22).

7.29.7 Unexpected contamination found, March 8, 2013

An inventory inspection of Pipe Bridge 39 discovered contamination around its base and steel supports. Nuclide analysis of samples showed that the total activity exceeded the regulatory limits for notification to the regulatory body.

The analysis also indicated—based on nuclide composition—the contamination to be historical. Sellafield's investigation determined that the origin of contamination was a deteriorated, excess ventilation duct formerly connected to an installation that had been broken apart in the 1990s.

ONR agreed with Sellafield Ltd's statement about the origin of contamination and proposed actions. Sellafield sealed the contamination to impede spreading. Sellafield moved quickly toward the removal of the excess ventilation duct. No serious risk due to this event incurred to the public or the workers. This event was rated at level 0 on INES.

7.29.8 Unexpected contamination found, March 8, 2012

During a routine survey in March 2012, surface contamination was found on the paving slab of a building, whose main structure had been taken apart in the 1980s. When contamination was first found, it was considered to be below the notification level. Following in-depth ground sampling in June 2013, Sellafield Ltd determined that the contamination level was fairly higher than that initially reported and informed at once the regulators of the increased levels of contamination.

Sellafield Ltd concluded that the cause of the contamination was historical, based on nuclide composition of soil samples and knowledge of the operations formerly conducted in the thereafter demolished building.

ONR agreed with Sellafield Ltd's statement about the origin of contamination and proposed actions. Sellafield sealed the contamination to impede spreading. Sellafield moved quickly toward the removal of the excess ventilation duct. No serious risk due to this event incurred to the public or workers. This event was rated at level 0 on INES.

7.29.9 Unexpected contamination found, February 2, 2011

During a routine survey, site personnel found a puddle of some 25 mL of a brownish liquid under the lowest point in a ventilation duct of the unused Plutonium Purification Plant (a legacy facility). The puddle was in an area classified as low level contaminated. The plant had been cleared of superfluous equipment pending dismantling. Ventilation is a safety-related system because it provides containment by impeding radioactive substances to be released from highly contaminated areas.

Sellafield Ltd's analysis confirmed that the liquid contained plutonium at a concentration above the statutory limit for reporting by a factor of around 5. The contaminated area was safely cleaned up.

Neither the public nor workers were affected by the event.

Contaminated condensation leaked out through a gasket in a ventilation joint, which was designed decades ago. The event has underlined many inadequacies in the design of the ventilation system, and Sellafield Ltd took action to correct these defects. The event was rated at level 1 on the INES.

A weak point is that the site organization does not plan to complete decommissioning of this facility until 2058: Therefore, a safety case accounting for this and other site events should be developed to guarantee continued safety throughout this long period.

7.29.10 Unexpected contamination found, July 13, 2009

Contamination was discovered on and around excess ductwork from B209 to the ventilation fans at the bottom of B230 stack in the B229 complex. This event met the Ministerial reporting criteria.

The extent of activity released by this event was still being ascertained by the licensee at the time the event was reported. Structural constraints in the area affected by the leak meant that excavation and removal of the contaminated material had to be meticulously planned including considerable preparations. This reflected in the belated estimate of the activity leaked. A preliminary assessment based on surface contamination provided an estimate of α activity of some 1.5 MBq of fixed and mobile contamination (equating 150% of the amount specified for Ministerial reporting.). Later on, improved access allowed the taking of more samples.

The company confirmed the B209 duct work was the origin of the contamination. Various engineering improvements were evaluated to avoid a recurrence. The event was rated level 1 on the INES.

7.29.11 Flooding Incident, July 6, 2001

During a heavy rainfall, localized flooding some 15-cm-deep occurred outside a laboratory building and went beyond an outer door into the building. The water penetrated an old laboratory, which was under D&D. Historic plutonium contamination, possibly from underneath a floor liner, was removed by the flooding and dispersed over the laboratory floor. The area was later decontaminated. There was no activity release outside the building and no radiation exposures incurred by the workers.

The activity removed was estimated at 10 MBq of ^{239}Pu and 200 MBq of ^{241}Pu which is, respectively, 10 and 2 times the legal reporting value. The incident was rated INES 1.

The licensee conducted remedial work on the building to avoid a recurrence of the water flooding. Another flooding incident is described in Section 7.2.

7.29.12 Lessons learned

The minor incidents described in this section are representative of situations that may occur at old shutdown facilities, especially when surveillance and maintenance programs have been somehow neglected for years. The reader should consider that these events are not particularly relevant for their direct causes, but they could highlight root causes potentially affecting other plant systems and causing more serious impacts. A similar event that occurred at Dounreay is illustrated in Box 7.5.

BOX 7.5 Discovery of contamination, Dounreay, Scotland, United Kingdom (ONR, 2020)

On May 20, 2007, licensee's staff conducted a radiological survey inside a manhole that provided maintenance for a then unused drain line. The incident

Box 7.5 Discovery of contamination, Dounreay, Scotland, United Kingdom (ONR, 2020)—cont'd

occurred in a part of the site where historical contamination was suspected. A contamination smear sample from within the manhole detected around 1.3 kBq of α activity, which was later found to include 60% plutonium/americium. The manhole was sealed, and the area designated as radioactive contamination high.

As part of their investigation, the licensee installed an interim enclosure over the manhole as an extra containment measure for any plutonium present and to ease control over workers carrying out more radiological checks. Surveys inside the manhole were carried out on July 3, 2007, through the interim enclosure by staff wearing ventilated suits. No more contamination was discovered inside the manhole. However, the origin of the plutonium contamination from the smear sample remained undetermined.

The α contamination found on June 20, 2007, on a filter treating water from a spring just outside the site boundary, but close to the contaminated manhole was assessed accurately. The assessment showed that the initial finding was invalid.

The event was rated of no safety significance on the INES.

7.30 Expect the unexpected when preparing facilities for decommissioning (DOE, 2007)

7.30.1 Decommissioning process; problem encountered; impacts

The High Flux Beam Reactor (HFBR) at the Brookhaven National Laboratory (BNL) was a heavy-water cooled and moderated reactor that went critical in 1965. It operated at 40 MWt, in 1982 upgraded to 60 MWt. Following a reanalysis of the safety impact of a loss of coolant accident, the reactor power was later reduced to 30 MWt. The HFBR was shut down in December 1996 for routine maintenance and refueling. During that shutdown, a leak of tritium-contaminated water was disclosed by routine sampling of ground water wells next to the reactor's spent fuel pond. The reactor remained closed for 3 years pending safety and environmental evaluations. In November 1999, the US DOE decided to close down the HFBR for good. The D&D of the HFBR complex, including many structures and systems to run and maintain the reactor, was over in 2009; all the control rod blades were also taken away during the decommissioning. The emptied and decontaminated HFBR containment, with the reactor vessel still inside, is currently kept under surveillance (Fig. 7.18).

On September 19, 2006, the HFBR building was subject to extensive flooding due to the break of an air conditioning unit cooling water pump seal. Some 100 m^3 of water coming from the BNL domestic water ductwork spread on the equipment floor. The originally spilled water had no hazardous or

FIGURE 7.18 Brookhaven HFBR (the domed building in the foreground). HFBR, High Flux Beam Reactor. *(Credit to DOE, Wiki Commons).*

radioactive contaminants; however, samples taken from the builtup water within the building exhibited high tritium concentrations. The clean air conditioning water that flooded in the building mixed with residual tritium from within the HFBR. As a result, all 100 m^3 of spilled water had to be processed as radioactive waste.

7.30.2 Analysis; solutions

An expert panel conducted an independent investigation and causal analysis of this incident. The causal analysis pinpointed the following root causes:

- Supervisory alarms were "deactivated," which eliminated remote alarm functions.
- HFBR decommissioned status caused reduced maintenance of alarm panel.
- HFBR decommissioned status caused reduced maintenance of air conditioning cooling towers—Insufficient management control for the reductions in surveillance and maintenance (S&M) of remaining systems while the HFBR was getting ready for further decommissioning work.

The following recommendations were intended to avoid recurrence of this incident:

- Conduct an analysis of the remaining active systems and establish their necessity throughout all phases of remaining decommissioning work. Based on this analysis, deactivate unnecessary systems and alter other systems to perform the expected functions.

- Place a leak detection system adjacent to the remaining liquid systems.
- Revise the S&M manual to make sure the HFBR is maintained in a safe, alert, and compliant configuration.

7.30.3 Lessons learned

- When discontinuing operations and reducing building attendance in preparation for decommissioning, a rigorous evaluation of the need for and operations of still active systems should eliminate unnecessary systems. Timely deactivation of unnecessary systems will reduce the risks from system component failure and minimize S&M.
- Operations procedures should be revised to take account of the changed functions of systems operating in support of decommissioning work. This includes the need for regular system surveillance and operability checks of alarm panels.
- Facilities that are being deactivated in view of decommissioning are sensitive to a loss of continuity in S&M. This may happen due to key staff formerly responsible for S&M leaving the plant.
- When deactivating facilities in view of decommissioning, there could be harmful "system" effects if components are in disrepair. Tolerating a system to work with a component in disrepair (e.g., cooling tower cooling fins) can result in the rupture of an important system component (e.g., pump seal break from cooling tower debris).

7.31 The unique case of Bevatron, Berkeley, California, United States (Laraia, 2017)

7.31.1 Decommissioning process; problem encountered; impacts

The Bevatron was a huge particle accelerator (proton synchrotron) located at Lawrence Berkeley National Laboratory (LBNL), United States. Its name derived from it being able to inject energies of billions of eV (**B**illions of **eV** Synchro**tron**) to protons aimed at a stationary target. Bevatron started its operations in 1954. The antiproton was discovered there, and this historic achievement was rewarded with the 1959 Nobel Prize in physics to Emilio Segrè and Owen Chamberlain. The Bevatron site occupied an area of 12,000 m^2.

When the machine was built, the technology was incapable of tightening a particle beam into a narrow aperture; thus the beam space was about 4000 cm^2 in both directions. The combination of a wide beam aperture and enormous energies demanded a 12,000-t iron magnet to control the beam and a huge vacuum system. The circular accelerator had a diameter of some 60 m. There was also a concrete shielding of 13,500 t. Originally, the Bevatron was constructed with no upper shielding, but shielding had to be installed later when the air scattering effect was detected.

In addition to the “primary” protons, the machine generated secondary beams of protons, neutrons, pions (a pion is a particle having a mass approximately 270 times that of an electron), “strange particles,” and many others particles. Given that Bevatron was a new, experimental apparatus, the nature and energy of generated particles were generally unknown a priori: so were their interactions with surrounding materials as the decommissioning project showed later (see the following).

The Bevatron used a large amount of energy, which was very expensive and eventually led to its demise. During its lifetime, the Bevatron shifted from protons to a heavy ion accelerator for high-energy physics trials; later on, it was converted to a nuclear medicine treatment center.

While Bevatron was a “weak-focusing” accelerator, the next generation of accelerators used “strong-focusing” particles and thus required much tighter apertures and less expensive magnets. The CERN Proton Synchrotron, 1959, and the BNL Alternating Gradient Synchrotron, 1960, were the first next-generation machines, with an aperture approximately an order of magnitude narrower than Bevatron in cross section, and reaching 30 GeV proton energy, yet with a less massive magnet ring. Over time, these and other upgraded features made Bevatron redundant.

Bevatron was permanently shut down in 1993. When the Bevatron was shut down, the scientists, engineers, and other operations staff moved to other positions within LBNL or left the lab altogether. No decommissioning plan was ready at that time. Except for a few offices, the building sat unused until 2008. A deliberation of Berkeley City Council reads: “The Building 51 structure housing the Bevatron is deteriorating and consumes disproportionate maintenance resources. It does not meet current building codes, the roof leaks in several locations, and portions of the structure do not comply with current seismic design standards … The structure is seismically unsafe. Its demolition would provide a future safe working environment for an as—yet unidentified activity at Berkeley Lab” (Berkeley City Council, 2005). The demolition of the Bevatron began in 2009 and ended in early 2012.

7.31.2 Analysis; solutions

Fig. 7.19 shows the interior of the Bevatron building in 2010. At that time, the decommissioning project was approaching the end, and the magnet ring had already been removed, leaving only the structural framework of the installation.

Much more radioactive material was found than expected initially. In detail, the project incurred several surprises:

FIGURE 7.19 Interior of Bevatron building during its demolition. *(Credit to Daniel Parks, Wiki Commons).*

- It had anticipated to free release around 60% of shielding blocks; the actual free release of shielding blocks was 20%;
- It had anticipated to free release around 75% of slabs and foundations; the actual slab and foundation free release stayed at 20%.

These surprises were mostly due to DOE and the D&D organization not having specified minimum detection concentrations in a consistent, timely, and clear manner and not having closely supervised the measurements. These discrepancies affected the decommissioning process of blocks, slabs, and foundations and cost some 10 million $ more than expected.

Tritium was a special case. Prejob estimates included:

- activation mostly generated prior to 1974 (three half-lives of tritium);
- minor amounts of tritium;
- tritium expected to be located near other activation products; and
- tritium expected to be detected in soil and groundwater under thin floors.

Instead tritium was unexpectedly found

- under thick foundation slabs;
- not under slabs with the highest activation;
- at concentrations over 150 Bq/L (more than anticipated) and inconsistent with expected equilibrium conditions.

Possible (but unconfirmed) tritium sources include

- accelerator cooling water leak;
- migration from high-activation to low-activation areas may have been caused by groundwater flow.

Clearly, the environmental contamination (including outside the accelerator building) was quite different from that assumed in the decommissioning plan.

The demolition of the facility generated some 29,000 t of material that was then shipped in over 1420 transports to the Nevada National Security Site (NNSS) for disposal as LLW and mixed LLW (see Glossary for mixed waste).

The rainy season provided an additional challenge. During the last phases of the project, rainfall was heavy. Consequently, almost all pieces of metal scrap and concrete fragments had to be packaged wet. Large amounts of absorbent material had to be embedded together with the demolition waste to meet the NNSS no-free-liquids disposal requirements.

7.31.3 Lessons learned

Research accelerators are unique installations. Due to their unique construction and operating features, the generation, environmental distribution, and matter interactions of primary and secondary particles are unpredictable to a large extent. Therefore the predecommissioning characterization of accelerator buildings and on-site and off-site land areas is a *priori* widely unknown. This implies that an accelerator successful decommissioning/environmental remediation project needs to spend more characterization resources than ordinary reactor projects before physical dismantling is undertaken.

As for other case studies presented in this book, the reader should also note that in general a prolonged postshutdown "no-action" period is inimical to safe and smooth decommissioning (the Bevatron idle period was 15 years). Firstly, it causes the loss of knowledge (the "tacit knowledge" embedded in staff memories and experience) that would be helpful in decommissioning planning and implementation. Secondly, it increases deterioration of SSC, which will be more expensive to remedy later.

7.32 Decommissioning-related experience at US irradiators (NRC, 1991a)

7.32.1 Decommissioning process; problem encountered; impacts

The following highlights several case histories in the United States of design and operational flaws that have affected wet irradiators and ultimately their decommissioning. A schematic of wet irradiator is given in Fig. 7.20. The pool water has shielding purposes. It can be seen that there are multiple barriers between the radioactive source and the environment consisting of a series of (from the source outward) (1) the source cladding, (2) the pool water, and (3) the pool structure. Therefore, incidents giving rise to environmental contamination are due to a combination of leaking sources, water contamination, and leaking pools.

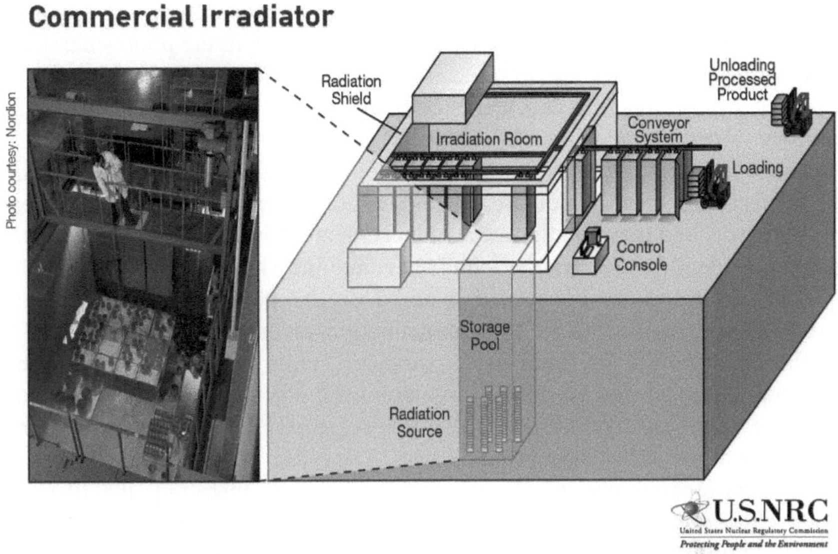

FIGURE 7.20 Commercial irradiator. *(Courtesy of NRC, Wiki Commons).*

RTI, Incorporated (RTI) was licensed in 1970 to operate a pool irradiator near Rockaway, New Jersey. The irradiator and the surrounding buildings cover a 2-ha site. RTI also owned approximately 81 ha of land, directly across the road encircling the licensed site. The irradiator used up to 1.1×10^5 TBq ^{60}Co sealed sources for the sterilization of medical supplies. RTI had also a modest production of irradiated cosmetics, spices, food packages, etc. The irradiation took place in a room above the pool; when the ^{60}Co sources were not in use, they were housed in racks in the pool. The irradiator at the RTI plant is enclosed by a concrete biological shield.

In the 1970s, the licensee disposed of solid radioactive waste by on-site burial. Records of such burials were poorly kept. The only recorded burials had taken place in the northeast corner of the site, and the spent resins employed in pool cleanup were also disposed of there. The boundaries of this disposal area were unknown. These burials produced ^{60}Co soil contamination in the areas around the burials.

In the late 1970s, a material commonly known as "redwood" (an abbreviation for radiation-hardened wood) apparently struck one of the fixed source racks as the container with the radwood was descending into the pool, damaging one of the sealed sources. Leakage from the damaged source contaminated in the pool water. The contamination was not detected soon. Flocculent and other agents used in cleaning the (believed uncontaminated) pool of dirt and algae were dumped onto the ground south of the irradiator building. The ^{60}Co contamination in the pool was discovered in 1986 and mainly eliminated by ion exchange resin and by adsorption on activated charcoal and diatomaceous earth. As the contamination in the pool water went down, it became acceptable to regenerate the resins. The resin backwash was disposed of by pumping the water onto the ground south of the irradiator building. Over several years, this produced a significant buildup of ^{60}Co in the area south of the building. ^{60}Co-contaminated soil was also discovered outside the licensed site in a wooded area, clearly caused by water runoff in that area.

In response to these findings, the management initiated a series of remedial actions to identify, characterize, and remediate the site. Finally, in 1990, the USNRC informed the licensee that based on the possession limits for the irradiator sources and the ^{60}Co contamination recorded, another $150 000 would be required for financial assurance. While the licensee provided financial assurance for the irradiator sources, he objected that the possession limit for ^{60}Co contamination should be decreased based on surveys and remediation operations already carried out. The NRC, however, kept the point: Radioactive contamination was present in the pool; therefore, the licensee could choose to either provide adequate financial assurance or demonstrate that the site met unrestricted release criteria. In 1997, based on the findings of the licensee's remedial operations, the review of records, radiological surveys, and the results of independent measurements, the USNRC confirmed that the decommissioning had met the unrestricted release criteria in the formerly contaminated areas (NRC, 1997).

There have been no irradiator incidents in the United States that caused groundwater contamination, but there have been cases of soil contamination (NRC, 2020) additional to aforementioned RTI case. One such incident occurred in 1988 in Decatur, Georgia, due to a leaking source of ^{137}Cs in the form of Cs chloride, which is highly soluble in water. The leakage extensively contaminated the facility and the soil around the building. Extensive cleanup work was needed, at considerable cost to the licensee as well as the US DOE, which had provided the source. No members of the public were exposed.

At Decatur, the leaking ^{137}Cs source stayed undiscovered for a long time, because the licensee made no use of the online pool water monitoring system coupled with the demineralizer. The contamination was eventually detected when water samples were taken and the pool was radiologically surveyed: These actions were prompted by the activation of the radiation monitoring system, which had automatically blocked the sources in their storage location due to high local radiation levels (NRC, 1991a).

If the demineralizer had been in routine use, the water contamination could have been discovered earlier and mitigation actions started.

A similar incident occurred at Dover, New Jersey, in 1982. This incident involved a damaged ^{60}Co source and resulted in the contamination of water that spread to the building floor and the soil around the plant. As in the Decatur case, a large-scale cleanup was needed, but there was no groundwater contamination and no public exposures.

The decommissioning-related incident described in the following occurred at the Defence Nuclear Agency's Armed Forces Radiobiology Research Institute in Bethesda, Maryland. The licensee measured for a few weeks a loss of pool water that was around three times higher than expected from evaporation only. The licensee carried out tests to better characterize the water loss and took daily samples of the pool water to check compliance with unrestricted release limits. The licensee also inspected the pool liner and discovered caustic stress corrosion on several welds. It was found that welds made during the plant construction were not compliant with industry standards and were vulnerable to caustic stress corrosion: This eventually caused the water losses.

In another case also described by NRC (1991a), while a licensee was attempting to decontaminate pool water further to a leaking source event, a hose on a filtration system got broken. Contaminated water was inadvertently spread onto the facility floor and the soil nearby. The licensee did not report the incident to USNRC and tried to cover it up.

7.32.2 Lessons learned

- Equipment employed during D&D activities must be adequately maintained, and the workers should be trained to comply with specified procedures.
- In several cases, pool contamination caused tough issues for decommissioning. When contamination migrates to the environment, the issue gets worse. Careful monitoring and maintenance of pool water chemistry is vital to avoid such difficulties.
- Financial assurance must be sufficient to deal with all of the liabilities at the time of decommissioning. In the RTI case, the old practice of disposing of solid radioactive waste on-site, including radiation sources, which was hardly recorded, as well as the dumping of contaminated pool water on the land outside of the building prompted the regulator to require considerable upgrades of the financial assurance for decommissioning.
- In the last example, the licensee did not comply with the required notification of incidents to the regulator and even tried to conceal the incident. This serious regulatory breach highlighted the lack of safety culture and poor management on site.
- Good record keeping is imperative to support eventual decommissioning.

7.33 Radioactive waste found in city sewage (Knoxblogs, 2015b)

7.33.1 Decommissioning process; problem encountered; impacts

K-25 was the code name given by the Manhattan Project to the program to produce enriched uranium for atomic bombs using gaseous diffusion. Overtime, the name came to refer to the production plant located at ORNL,

FIGURE 7.21 K-25 demolition, Oak Ridge, 2013. *(Credit to DOE, Wiki Commons).*

Tennessee, the main gaseous diffusion building, and ultimately the site. When it was built in 1944, the four-story K-25 gaseous diffusion plant was the world's largest building. Production of enriched uranium ceased in 1964, and gaseous diffusion ended in 1985. The Oak Ridge Gaseous Diffusion Plant was renamed the Oak Ridge K-25 Site in 1989 and the East Tennessee Technology Park (ETTP) in 1996. Demolition of all ORNL gaseous diffusion plants was completed in 2017 (Fig. 7.21).

In early 2014, radioactive contamination was found in the city of Oak Ridge's sewage treatment facility (the Rarity Ridge Treatment Plant, RRTP).

This unpleasant surprise was attributed to ^{99}Tc, which had migrated from a decommissioning project at the ORNL K-25 uranium-enrichment site now known as ETTP. ^{99}Tc is a low-energy β-emitter that was present in the gaseous diffusion process used to enrich uranium. The radionuclide, which can be mobile in the environment, had entered pipelines, leading to the sewer system.

Although the radioactive contamination was considered not to pose a health hazard to workers at K-25 or to local drinking water, the event prompted several cleanup activities—including removal of technetium from the RRTP.

In 2014 and 2015, some 280 m^3 of radioactive sludge was removed from the facility and dispatched out of the city for treatment and disposal. The sludge, which is about 3%–4% solid, comes from the RRTP digester.

"Overall ^{99}Tc levels continue to decline within, but it is premature to estimate the exact number of shipments (of sludge) or even the estimated time frame that will be needed to bring the facility back to normal operating parameters," said a spokesperson for UCOR (URS-CH2M Oak Ridge), the DOE's cleanup manager in Oak Ridge (Knoxblogs, 2015b). At the highest point of contamination (the RRTP digester), ^{99}Tc concentrations dropped from some 20,000 Bq/L in February 2014 to around 6300 Bq/L in November 2015.

On November 5, 2015, UCOR made the 15th shipment—each of them nearly 20 m^3—from the RRTP to a Perma-Fix Environmental treatment facility in Richland, Wash.

It was noted over time that heavy rainfalls tend to increase activity levels at the RRTP and affect the number of sludge transports. This is due to a sort of cleansing mechanism, which carries away more debris and increases the suspended solids. The ^{99}Tc concentrates in the solids, thus increasing ^{99}Tc levels in the sludge and in the truck. Maximizing the radioactive material removed from the RRTP reduces the number of sludge transports.

Another mitigation measure consisted of isolating the pipelines near the K-25 demolition so that they could not reach the lines leading to the RRTP. Isolation was put into effect by pouring grout plugs into manholes and within sections of the piping. Secondly, UCOR took steps to contain the radioactive and other hazardous contaminants embedded in the demolition rubble at cleanup sites. And thirdly, for oncoming decommissioning work, UCOR planned to either remove piping contaminated with ^{99}Tc before demolition or seal the equipment with foam to keep the contamination fixed in situ as the building is demolished.

Most of the demolition rubble was sent to a local landfill that was designated for Superfund cleanup waste. Superfund sites are listed locations in the United States requiring environmental remediation. However, some of the equipment could not be disposed there because of their high activity levels. Demolition debris was given shipment priority to minimize the time they are in the field and exposed to rainfall.

It is assumed that there will be no long-term ^{99}Tc impacts that would require replacing sewage lines, basins, or equipment at the RRTP. Apparently, the Oak Ridge RRTP contamination case is unique.

7.33.2 Lessons learned

The spread of radioactive contamination washed away from a decommissioning site is not an unrealistic scenario, especially if contaminated materials are left—albeit temporarily—in the field. From the decommissioning site, contamination can spread to water courses and other sources of drinking water.

7.34 Inactive water heater spread contamination, Hanford site, Washington, United States (IAEA, 2018)

7.34.1 Decommissioning process; problem encountered; impacts

By 2013, Hanford Building 327, also known as the Post-Irradiation Testing Laboratory, had been demolished (Fig. 7.22), the associated contaminated soil removed, and the building site reclassified. This achievement eradicated the potential for such incidents as the one described in the following.

On September 2, 2003, water was seen streaming out of Building 327, through a loading hatch, and pouring onto the ground below. The building had remained unattended since early 2003. When the water leak was noticed,

FIGURE 7.22 Workers are demolishing the 327 Building at the Hanford site's 300 area. *(credit to DOE).*

workers went into the building and found a 20 L electric water heater in a cubicle had broken apart and had torn the cubicle door from its supports. Nearly 10 cm of water was found in sectors of the main building floor and basement in contaminated areas, and a few meters of water in the elevator shaft. Samples from the water that had accumulated outside of the building had no α-contamination and 0.074 Bq/mL of β-contamination.

Although nobody was contaminated and the cleanup expenses were small, the impact could have been much worse had the circumstances been different.

7.34.2 Analysis; solutions

Plant staff even ignored that this water heater was there. It was not indicated on the plant drawings. When the personnel vacated the facility, neither electricity nor the water connection to the heater was cut off. Witnesses confirmed that water may not have been used from the water heater for many years.

Initial suspicions pointed to a broken thermostatic controller coupled with a faulty temperature relief valve. The valve was checked and found to be operable. The controller check was inconclusive. The controller thermostat contacts were open, but the mechanism seemed to correctly open at 60°C. The overheating contacts were closed initially: They opened at 82°C but would not reset. The heater element had electrically failed, and the covering was clearly broken in three positions.

Inspection of the tank by a certified welding engineer found no evident pitting or general corrosion and no weaknesses on the tank welds.

Another initiating event for the failure could have been the blast of a bubble of hydrogen originated by water dissociation or corrosion. This condition is well known and is described in manuals. However, no burning signs were visible nor was there an ignition source within the heater. There was a sign of water line at the height of the water inlet (the low part of the tank). The water line could point to a gas bubble.

7.34.3 Lessons learned

All installations should make sure that unused water heaters are fully emptied and deenergized. If no use of water from a water heater tank is planned for over 2 weeks, the heater should be deenergized in accordance with the owner's manual instructions. Locking valves on the water supply and venting should be considered.

Personnel manning a given installation must know in detail what systems and equipment are present in their installations to adequately maintain operations equipment and deactivating systems when they become redundant.

Pressure relief valves on water heaters should be tested as prescribed by manufacturer's instructions. At least, the "try lever test" should be conducted yearly to confirm that the relief valve is operable (BOA, 1998).

If a water heater tank is discovered in an energized condition but no water had been used for several weeks or more, a safety hazard should be considered likely. The system should be deenergized and cautiously vented as any gases inside may include hydrogen and form explosive mixtures.

Box 7.6 reports a laboratory event, which also highlights the need for continuing surveillance and maintenance in a postshutdown period. Box 7.7 describes another event associated with the malfunction of a water heater.

BOX 7.6 Lack of maintenance in a shutdown lab (IAEA, 2018)

Decommissioning process; problem encountered; impacts:
Little maintenance was provided to a laboratory during the shutdown phase. This complicated the decommissioning. For instance, a broken pipe and a broken window allowed water from the pipe and rainwater entering the window to solubilize and move ^{137}Cs to several spots and in depth. The door frames being in poor and bad condition posed no restraints to the cesium migration.

Analysis; solutions
During a survey, the γ spectrometric analysis disclosed contamination under door frames, which prompted extensive sampling at various depths. The floor around and under the door frames was excavated, and material filling up to 50 cm depth was removed. This work caused a delay of 2 weeks in the schedule.

Lessons learned
During construction and operations of a radiological facility, specific attention should be paid to areas where contamination can build up. Even after the shutdown, during the operation-to-decommissioning transition, maintenance of the facility, as needed, should not be neglected.

BOX 7.7 Water heater sets fire to plastic tank, Hanford Tank Farm Site; June 22, 2018 (DOE, 2018b)

An emergency shower water tank was being sanitized before starting operation. A Tank Farm's stationary operating engineer (SOE) was sent to power up the shower prior to Mission Support Alliance (MSA) filling the tank with water, taking a sample and confirming the water was hygienic. The SOE powered the breakers in the power panel and then left to attend routine activities. The MSA team then went away to look for hoses long enough to fill a polymerized plastic (poly) tank with water. When the MSA workers came back to the emergency shower, they saw smoke and flames spreading from the bottom of the poly tank. The Hanford Fire Department (HFD) was called and put out the fire with a manual extinguisher within a few minutes. The HFD established that the cause of the fire was the heater elements; as they heated, they deformed the bottom of the tank where they were lying and set it on fire. The shower was removed from service.

This event, although it caused no immediate risks to the workers or the environments, has been quoted here as representative of the impacts that extremely high temperatures can produce during decommissioning projects.

7.35 Unusual case of environmental contamination (Webb, 2006)

7.35.1 Decommissioning process; problem encountered; impacts

Contaminated feral pigeons were found by personnel of the Royal Society for the Prevention of Cruelty to Animals (RSPCA) at a private property in Seascale, a small town close to Sellafield after a few birds were culled and analyzed. Pigeons had been regularly fed in the property garden for some 20 years. By 1998, the number of pigeons had grown to such unbearable levels that neighbors complained and a cull were arranged by the RSPCA.

The investigation by the Committee on Medical Aspects of Radiation in the Environment (COMARE) and the Radioactive Waste Management Advisory Committee (RWMAC) determined that pigeons became contaminated through entering the buildings in the old Sellafield fuel separation area. This area has many buildings constructed in the 1950s and 1960s with wide openings, through which pigeons entered the facility. Effective doses to the two people living in the property who went in contact with the birds were estimated 0.6 mSv/y and to other members of the public around 0.006 mSv or less.

The property was cleaned up, pigeons on the Sellafield site were drastically culled, and provisions were made to impede birds accessing contaminated buildings. This event was rated by the then owner (BNFL) at level 1 of the INES scale: The evaluation was based on the fact that the transfer of off-normal activity off-site highlighted loss of on-site control. Although the transfer of radioactive material by the birds had been detected for a while, the birds had been left going in and out of the old buildings and getting contaminated over many years.

It should be noted that, depending on case-specific circumstances, contaminated animals can be a public or an occupational hazard.

7.35.2 Lessons learned

Many cases of animals found contaminated inside or outside nuclear facilities (birds, rodents, spiders, and other species) are quoted in the technical literature. Sections 5.14 and 7.28 document cases where the presence of biohazards (molds, etc.) was a factor contributing to the decommissioning strategy. While nuclear activities should not endanger the public and the environment at large, relatively less focus is given to animals as contamination bearers. This scenario should be given special attention when facilities (such as some at Sellafield) have been disused and in disrepair for years. In addition, facilities designed and built in the early days of the nuclear industry tended to have many and large openings, which facilitated the direct spread of contamination to the environment in an accident or the indirect transfer of contamination by animals.

Needless to say the decommissioning of medical or biological facilities includes biohazards as inherent to the project itself. The interested reader can usefully consult WHO (2019) and TEXAS A&M (2012).

7.36 Transportation of excess heat exchangers, Sizewell a Nuclear Power Plant, United Kingdom (ONR, 2020)

7.36.1 Decommissioning process; problem encountered; impacts

In July 2013, the Sizewell licensee shipped eight excess pond heat exchangers to the Siempelkamp factory in Germany (via Winfrith) for smelting and later disposal of residual activity. In September 2014, Siempelkamp started segmenting the heat exchangers for smelting; in doing so, however, about 500 L of slightly contaminated water (remaining from flushing operations) was released, contaminating Siempelkamp staff (though not to the point of exceeding or even nearing any legal limits). The water remained within a containment area. Sizewell A was notified the incident on September 25, 2014. The event implied noncompliance with the Carriage of Dangerous Goods regulations and also with license condition 34, dealing with leakage and inadvertent release of radioactive waste.

Sizewell investigation determined that the root cause of the incident was that the transfer of crucial information was based on verbal communication; in particular the requirement that the heat exchangers be empty of water was forwarded with ambiguous wording, inducing the project team to believe that the mere draining of the heat exchangers would be sufficient (draining was meant to tolerate some water retention). Following its investigation, the Sellafield licensee established and took engagement to several corrective actions regarding improvements in communications, training, and briefing.

ONR concurred with the licensee's conclusions and declared itself happy with the licensee's response. ONR's investigation also confirmed that changes already applied to the licensee's operations (for reasons independent of this incident) should, as an additional benefit, reduce also the probability of a recurrence.

Concerning the violation of the Carriage of Dangerous Good Regulations, ONR applied its enforcement management model and concluded that a verbal warning was sufficient.

The incident met the "Ministerial Reporting Criterion—Abnormal occurrences leading to a release or suspected release or spread of radioactivity, on- or off-site, which requires special action or special investigation by the operator" and had the INES rating 0 (not relevant).

7.36.2 Lessons learned

The reader should note that procedural mistakes in the off-site shipping for release of materials originating from a nuclear site may impact sites at a significant distance from the nuclear site itself and should therefore be assessed with special care: The potential damage of the image of the shipping organization is not a minor concern. In this case study the impact on international relations could have further complicated the matter. Other examples in this regard are given in this book in Sections 7.10, 7.23, and 7.29.

7.37 Leakage and spreading of solutions during decontamination, Ignalina Nuclear Power Plant Unit 1, Lithuania (Bellona, 2010)

7.37.1 Decommissioning process; problem encountered; impacts

The two 1500-MWe Soviet-type RBMK reactors (a model similar to Chernobyl) at Ignalina, Lithuania, were commissioned, respectively, in 1982 and 1983. Following the Chernobyl accident in 1986, Ignalina NPP (INPP) was subjected to many international safety assessments, and it is likely that—with some upgrades—its overall level of nuclear safety was comparable with the safety standards of modern Western reactors. However, unlike other modern NPPs, RBMK reactors have no protective building that might contain radioactive substances in case of a severe accident. Therefore, Western organizations determined that RBMK intrinsic risks could by no means be reduced to the point of allowing continued safe operation. For this reason, one of the requirements for Lithuania to join the European Union (EU) was the permanent shutdown of Ignalina's two units. INPP Unit 1 was shut down at the end of December of 2004, and Unit 2 was closed on December 31, 2009. Most of the financial burden associated with the immediate—though spread over several decades—dismantling of INPP was borne by the EU.

The following quotations are extracted from the Commission report investigating the accident that occurred at INPP 1 on October 5, 2010.

"... during implementation of the scheduled works foreseen in Project B12 'Decontamination of internal surfaces of [State Enterprise] Ignalina NPP Main Circulation Circuit, Blowing, Cooling and Water Purification Systems' – Unit 1 circuit decontamination – seal failure has occurred in one of the components which resulted in the flowing out of the chemical agents (nitrogen acid (1%) and potassium permanganate) used for decontamination outside of the circuit boundaries. Since the strict safety requirements are applied for implementation of this project, the works are performed in the leak-tight rooms, the chemical agents and substances contaminated by radioactive nuclides have drained through the specific drainage systems." The statement remarked elsewhere that no chemicals or contaminated substances spread beyond the limits of the NPP's controlled area. Blaming the Project Contractor, the Commission stated:

"The Contractor of the Project has proposed decontamination technologies based on the guideline documents of the operating company concern "Rosenergoatom" RBMK-1000 type reactors whereby [main circulation circuit] decontamination is performed using the solution which comprise of nitrogen acid ratio 75 [to] 100 [milligrams per cubic decimetre]. In the technology proposed by the Contractor the nitrogen acid concentration in solutions was 1% or 10,000 [milligrams per cubic decimetre]." However, "If they were decontaminating the circuit, all leaks there would be inside sealed premises, and no problems are supposed to arise with any radiation release. The leaks were going into the sump and back into the floor drain tanks. The direct causes of the incident were summarized by the Commission as follows:

> " the sealing failure of one of the components of [the main circulation circuit] during decontamination could be conditioned by the following factors:

- Chemical corrosion of [the main circulation circuit's] internal surfaces which occurred due to the impact of the chemical agents used for decontamination.
- Erosion of [the main circulation circuit's] internal surfaces which was caused by the chemical agent used for decontamination and [consisted] of insoluble sludge particles having an intensive impact.
- Electrochemical corrosion in electrolyte medium of various metals contact areas.
- While preparing the program, the total impact of all factors (solution composition, decontamination temperature, flows rate, time of impact, corrosive elements concentration) on [the main circulation circuit] equipment was not taken into account by the Contractor. The Commission concluded:

- [The main circulation circuit's] sealing failure [...] occurred during Unit 1 Main Circulation Circuit decontamination due to the shortcomings of the decontamination technology proposed by the Contractor.
- The technology proposed by the Contractor was in fact not tested in nuclear industry enterprises before. It contradicts the Technical Specification regulations of the Public Procurement Tender for the implementation of this project."

7.37.2 Lessons learned

During decommissioning, mishaps and errors are inevitable although engineering measures and the careful drafting of technical/administrative procedures should strive to minimize the likelihood that incidents occur. However, failing prevention, mitigating provisions should be in place to minimize the impacts: First and foremost, containment of any radioactive material generated by the incidents should be ensured to contain contamination within controlled areas. This factor makes the difference between contained and uncontained incidents such as those indicated in Table 4.4 of Section 4.9 Loss of Containment (5.44, 6.18, 6.34, 7.1, 7.2, 7.24, 7.27, 7.39, Box 7.8, 7.42).

7.38 Overview of leakages at German installations in permanent shutdown or decommissioning state (BASE, 2020)

All eight NPP units finally shut down in 2011 as the result of a political decision and Grafenrheinfeld shut down in 2015 have submitted their requests for a decommissioning license to the regulatory body; the request was also submitted for Gundremmingen B, which shut down in December 2017. Of these, Brunsbüttel, Isar 1, Biblis A and B as well as Neckarwestheim 1 (GKN 1)

BOX 7.8 Loss of ventilation at Dounreay

In May 1998, an excavator damaged an electrical supply cable, which supplies a complex of nuclear facilities at the decommissioning Dounreay site, Scotland, United Kingdom; as alternative electrical sources did not activate as foreseen, the ventilation fans went out of service. Facilities were evacuated. The fans were put back to service after 15 h. When the facilities were reentered, they were found free of contamination. The environment on- and off-site was surveyed, and no radioactive release was discovered. On the basis of "significant failure in safety provisions with sufficient defense in depth remaining to cope with additional factors," this event was rated at level 2 of the INES.

have been awarded the decommissioning license in 2017. Most NPPs are still in the postshutdown phase or are just beginning with active decommissioning. The German licensees currently face several hindrances to complete the decommissioning process in a timely manner without increasing costs dramatically. For example, at the time of writing, the number of transport and storage casks could be inadequate to allow timely defueling and removal of all spent fuel assemblies

7.38.1 Leakage on a pipeline not yet approved for dismantling, Philipsburg 1 Nuclear Power Plant; January 14, 2020

The plant was finally shut down in 2011. Philipsburg 1 (KKP 1) is owned by EnBW Kernkraft GmbH. Among decommissioning-related events, one could quote the 2015 award of a contract to Westinghouse Electric Germany GmbH to dismantle the reactor pressure vessel (RPV) and its internals. The contract scope includes the planning, equipment manufacture and on-site segmentation of the RPV, its internals, and peripheral structures. This contract will be executed by a consortium including NUKEM Technologies Engineering Services GmbH (NTES) and GNS Gesellschaft für Nuklear-Service GmbH under Westinghouse lead.

The Ministry of the Environment, Climate Protection and the Energy Sector as the competent nuclear authority in Baden-Württemberg granted the decommissioning license on April 7, 2017.

EnBW expects the dismantling of the Philipsburg 1 to take around 15 years.

During dismantling work, a pipe that had not yet been removed was being sawn. Contaminated water leaked into the surrounding area within the interior of the reactor building and resulted in a low level of surface contamination. The removal process of the pipe was stopped, and the leakage was corrected using temporary means. The sawn-off pipe section was then cut out, and the pipe end sealed. Although the pipe was not due to be released, the manager in charge accepted it as de facto released. The reason for the erroneous step was under investigation at the time of reporting. The leak did not lead to any personnel contamination, release, carryover, or increased release of radioactive substances into the environment.

7.38.2 Leakage in the waste water collection system, Mülheim-Kärlich (KMK) Nuclear Power Plant, October 30, 2019

The Mülheim-Kärlich nuclear power plant (KMK), located around 10 km northwest of the city of Koblenz, was built between 1975 and 1986 as a two-loop pressurized water reactor with a capacity of 1300 MWe. In August 1987, commercial operation of the plant started. After only 1 year, the plant was taken out of service for legal reasons. After a long legal dispute, it was decided in the framework of the so-called "Atomkonsens" (nuclear consensus) of 2000 to shut down the NPP. The decommissioning license for the first phase of dismantling was granted in 2004, and since then, dismantling of the plant has been continued. The fuel had already been removed in 2002 during the postoperational phase. The activation of structures in the reactor core region and thus also the radioactive inventory were less than in comparable plants due to the short operating life. Besides, the long shutdown period considerably decreased the activity inventory by radioactive decay.

Orano, together with partner EWN (Entsorgungswerk für Nuklearanlagen GmbH), has been awarded two contracts by German utility RWE to carry out strategic dismantling operations for the Mülheim-Kärlich and Biblis NPPs. RWE awarded the two partners an overall site contract for Mülheim-Kärlich.

The agreements cover segmentation and packaging of the heart of the Mülheim-Kärlich nuclear plant, the RPV, as well as critical equipment such as the RPV-head and the RPV-internals. The operations will be conducted from 2020 to 2022. The work will be performed mainly under water, using remote-operated technology.

A section of the wastewater collection system was taken off-service and separated from the system section remaining in operation with a plug-in disk in a pipe tee. This insert was removed after the project work was completed. When pumping radioactive wastewater from a tank in the wastewater collection system, wastewater overflowed into the off-service section through the open pipe section and then leaked through open drains/vents into the auxiliary building (a controlled area). The wastewater was returned to the wastewater collection system via the floor drains. The open drains/vents were closed, thereby preventing more leakage. The cause of the leak was the premature removal of the plug-in disk from the off-service section of the wastewater collection system. The system for collecting radioactive waste water is one of the KMK systems that are still of safety-related rank with regard to the retention of radioactive materials even in D&D operations. The leak caused a slight contamination in the building. The leak had no effect on the safe decommissioning operation or on the personnel.

7.38.3 Drip leakage in the suction line of the separator of the wastewater treatment plant, Philipsburg 1 Nuclear Power Plant; June 18, 2019

During a routine inspection of the centrifuge system of the wastewater treatment plant, drip leakage was found on a piping section. The treatment of wastewater with the centrifuge system was then interrupted. The drip leak was stopped. The task of wastewater treatment system is to collect the radioactive water and process it in such a way that it can be returned to the nuclear power plant's circuits or released in a controlled manner. The leak had no effect on the shutdown operation, people or the environment. The surface contamination due to the leaked wastewater (within the control area) was below 0.5 Bq/cm^2. The affected pipe section was replaced.

7.38.4 Pipe leakage from wastewater collection system, AVR HTR; November 22, 2018

AVR was a prototype pebble-bed reactor located at Research Centre Jülich; it was first connected to the national power grid in 1967. During a severe water accident in 1978, ^{90}Sr leaked and widely dispersed. Soil and groundwater contamination below the reactor was discovered in 1999. At the end of 1988, AVR was shut down and transferred to a safe enclosure state before the license for dismantling was granted in 2009. AVR decommissioning has been extremely difficult, lengthy, and expensive. Government organizations took over the dismantling and disposal. In 2003, the reactor and its nuclear waste became government property. A milestone in the dismantling process was the intact removal and transport of the reactor vessel in May 2015. To this end, a material lock equipped with a crane was constructed next to the reactor building (Fig. 7.23) to make space for the vessel handling and positioning. The vessel was filled in 2008 with 2000 t of lightweight concrete to fix radioactive substances inside and to stabilize the vessel structures. After lifting out of the building, the reactor vessel was turned horizontally in the material lock to transport it to the purposely built on-site store. Currently, no dismantling method for the AVR vessel exists. It is planned to develop a process during the next several decades and to begin with vessel dismantling at the end of the 21st century. After the reactor building is dismantled, soil and groundwater will be decontaminated. Fuel removal was also difficult and lasted 4 years. During that time, it became evident that the AVR bottom reflector was broken; about 200 fuel pebbles remain wedged in it.

FIGURE 7.23 AVR reactor. *(Photo by M. Laraia).*

Due to a fault message signaled in the control room via the hazard alarm system, a water accumulation was detected in the pump sump located in the room of the waste water collection tank. The sump pump was switched on to return the water to the tank, after which the fault message was deleted. An initial inspection of the room showed no evidence of a leak. It was only in the following days that drip leakage from a pipe connected to the tank could be discovered after the tank filling had reached the height of the pipe. The cause of component damage is attributed to age-related corrosion. As an immediate measure, the leak was sealed with a cuff. Due to the advanced dismantling of AVR, the damaged pipeline was no longer required and was therefore removed as part of a dismantling project. Sampling of the leaked water was of little radiological concern.

7.38.5 Pipe leakage in an intermediate heat exchanger of the operations cooling circuit, Brunsbüttel Nuclear Power Plant; October 24, 2018

Brunsbüttel (KKB) started operation in 1976. It had an electrical capacity of 800 MWe. In line with the general phase-out of German NPPs, it was legally taken out of service in 2011. However, the plant had been idle since 2007 following a grid-facilitated trip. The decommissioning license was granted in 2017. Also in 2017, the owner Vattenfall awarded a contract to dismantle and package its RPV internals to the Areva-EWN Consortium.

Following an indication of internal leaks in one of the two redundant intermediate heat exchangers for the cooling circuit II (BKK-II), the affected cooler was taken out of operation. During the inspection of the cooler, which was opened on the Elbe river side, one leak and previous damage to 11 of the 1148 cooler pipes were found. The diagnosed pipes were closed. According to the licensee, the findings revealed a damage pattern matching the known damage mechanism at KKB due to the wear of the coolers exposed to the Elbe water. To prevent recurrence, plastic protective sleeves were placed in the rolled-in area of the 11 closed tubes and also on the remaining tubes of this cooler. The leakage found had no effect on the function of the heat exchanger and the availability of the BKK-II. The second redundant cooler was used for cooling during the period of the inspection and refurbishment. With the two coolers and their associated cooling lines, the heat from the cooling points of all functionally relevant components (such as fans, pumps, engine coolers) is dissipated to the river Elbe.

7.38.6 Findings on the drainage line of a control area laboratory, Brunsbüttel Nuclear Power Plant, January 31, 2018

The plant is shut down and defueled. During a plant inspection, a small leak was found on a plastic pipe for laboratory water in a service building. This line is part of the wastewater treatment and serves to return initially weakly radioactive wastewater from the chemical laboratory to the control area after treatment. The line is in a casing that has a monitored outlet for any leaks. The radiological evaluation revealed a slight contamination of the drip pan; no violations of warning or limit values were incurred. Examination of the defective plastic line revealed that the area of a weld seam had a striking manufacturing defect. Three additional plastic pipes were identified, which also run from the control area back to the control area via the service building. It was found that a leak detection probe was out of order in a section of the waste water piping. The damaged line was since repaired.

7.38.7 Unavailability of an emergency power diesel due to a leak in the high-temperature cooling water circuit, Krümmel Nuclear Power Plant; March 29, 2017

The Krümmel NPP—a single 1260 MWe boiling water reactor—is jointly owned by Vattenfall and E.ON. The reactor was linked to the electrical grid in 1983, but has been off-service for several years. It had a transformer fire in 2007 and, following repair, was restarted in June 2009 but taken offline again a month later due to a transformer fault. Krümmel was one of eight older power reactors that had their operating licenses abrogated by the Government shortly after the Fukushima accident in March 2011.

During a test on an emergency diesel generator, drip leakage occurred in a cooling water pipe in the area of the diesel engine cooling circuit. A crack in the cooling water pipe was identified as the damage. The damaged pipeline was replaced. The cause of the damage was determined by an examination of the replaced component and attributed to vibration crack corrosion. The vibration level in the affected pipeline area is to be reduced by additional supports. In the present case, there were no effects on the shutdown activities. In the current state of the plant, it is required that at least two emergency power diesel units are available for emergency power supply. Additional emergency power diesel generators were fully available during the repair of the diesel concerned.

7.38.8 Inspection findings when checking the pipes of the waste treatment systems, Brunsbüttel Nuclear Power Plant; November 23, 2016

Due to earlier events of corrosion findings in the waste treatment systems, a special inspection program including piping wall thickness measurements was launched at KKB with a focus to the corrosion of partly unused piping of the following systems: wastewater treatment, concentrate treatment system, wastewater collection system, and drainage of the reactor building. During the investigation of the ferritic pipelines through which water flows intermittently and which are otherwise empty, piping wall weakening up to complete rusting was found in the reactor building sump system. During further inspection in the wastewater collection system, local external damage due to mechanical impacts (presumably during previous maintenance work) was found on a small pipe. Corrosion findings were also found on a horizontal sump line of the drainage system and related transfer lines, as well as in the drying section. A refurbishment program for the affected pipelines was started at KKB, and administrative measures were taken to avoid generation and spreading of contamination.

7.38.9 Evaporator leakage, Biblis Nuclear Power Plant; November 28, 2016

Biblis NPP was finally shut down in 2011. On March 30, 2017, the licensee RWE Power received legal approval to decommission and dismantle units A and B of Biblis NPP. On June 1, 2017, the decommissioning process began to be implemented, and the plant is now in a state of a slow phase-out. The NPP is planned to be dismantled and released from regulatory control within the next 15 years. The focus of the initial phase of dismantling work is on systems that are no longer needed. In October 2019, Orano, together with partner EWN (Entsorgungswerk für Nuklearanlagen GmbH), was awarded a contract by RWE to carry out dismantling at the Biblis NPP. The contract included two lots for the twin units at the Biblis NPP. The scope at the Biblis work includes the RPV internals of Units A and B. The operations will be conducted from 2021 to 2023. The work will be performed mainly underwater, using remote-operated technology.

An activity transfer from the coolant preparation system to the auxiliary steam condensate system was signaled via a process computer message. The auxiliary steam condensate system and the evaporator systems for the coolant treatment were taken out of operation. The transfer of activity occurred via corrosion-induced leaks within the evaporator in the coolant preparation system. The evaporator was separated and deactivated, and the contaminated auxiliary steam condensate was discharged into the wastewater treatment plant. No activity entry into other systems and no release into the environment occurred. The system for coolant storage and preparation is used to separate coolant fractions. Heating of the evaporator is provided by the auxiliary steam supply. The damaged evaporator is no longer required for decommissioning; therefore, it remains permanently out of service.

7.38.10 Findings in the wastewater discharge line, Brunsbüttel Nuclear Power Plant; November 2, 2016

During a survey of the turbine building, a leak was found on a discharge line of the wastewater treatment system. The finding point is located in the system section behind the discharge tanks. Only processed and measured water is stored here that does not exceed the limit values for Elbe river discharges as specified by the operating license. The water that leaked was contained in the sump of the turbine building. No radioactive contamination was found during the sampling. Subsequent to the first finding, a further small leak was found in the reactor building during a pressure test of the discharge line. The cause of the leaks was local corrosion attacks in the ferritic material of the pipeline, starting from the inner surface. The leaks were temporarily sealed with a clamp. A new pipeline was soon ordered for the remaining operations.

7.38.11 Small leakage on a secondary cooling water line, Brunsbüttel Nuclear Power Plant; September 2, 2016

During a survey of the turbine building, a small leak was found on a vent pipe of one of the two cooling water lines of the secondary cooling water system (using Elbe river water) (BKK II). The damaged pipe was initially sealed with a clamp and has now been repaired by removing the old connector and welding in a new, identical one. The cause analysis conducted during the repair work revealed a local flaw. In the present case, the licensee assumes the same damage mechanism (local pitting corrosion) as in the case of an earlier, extensively investigated event in 2005. With the two cooling lines of the BKK II system, the heat is dissipated from the cooling points of the three emergency power diesel units. The required amount of cooling water was not affected by the leakage, and the affected cooling line was available without restriction.

7.38.12 Leakage at the weld seams of a ventilation pipe of the active wastewater treatment plant, Muelheim-Kaerlich (KMK); April 4, 2016

The plant is in the decommissioning/dismantling phase. During an inspection, leakage points due to corrosion were found on the ventilation line of a tank in the active wastewater treatment system. The leakage points were initially sealed with pipe clamps and then replaced with pipe pieces made of the same material. The metallographic examinations of the removed pipeline sections revealed pitting corrosion from the inside as the cause of leakage. However, the direct cause of the damage could no longer be determined due to the decontamination activities already completed. It is believed that concrete sludge from demolition work from 2008 to 2011 led to deposits that caused the corrosion. The leaks had no effect on the function of the system and caused no inadmissible contamination in the affected area. At KMK, active wastewater treatment is one of the systems that are still required for decommissioning and dismantling. Building floor drainage, decontamination wastewater (from the decontamination of systems or components), and system wastewater (from the emptying of tanks and components) are collected in the wastewater treatment system and decontaminated by various processes.

7.38.13 Crack in a pipeline of the active wastewater treatment plant, Muelheim-Kaerlich; November 25, 2015

When checking the evaporator system of the active wastewater treatment, a crack was found on a component of a pipeline. The crack had no radiological impacts. The defective component was replaced by an identical component and was subjected to metallographic examination. The investigations revealed vibration corrosion as the cause of damage. Checks of the same component on the parallel evaporator system did not show any adverse effects. The licensee assumes that a single error was the likely cause of the event. At KMK, active wastewater treatment is one of the systems that are still required for decommissioning and dismantling.

7.38.14 Leakage in a reserve water system, Philipsburg 1 Nuclear Power Plant; June 23, 2015

During a routine test, a leak was found on a pipeline of the reserve water system for the Independent Sabotage and Accident Protection System (the German acronym is USUS). The leak was sealed. The cause of the leak was local corrosion attack progressing from the inside outward. The USUS system was used during the plant operations to protect it from external impacts, e.g., plane crash or explosion. At the time of the event, the USUS auxiliary cooling water was only used to cool important cooling points (e.g., diesel coolers, room air cooling in the USUS building). The reserve water system is no longer in operation.

7.38.15 Leakage on a measuring line in the coolant preparation, Biblis Nuclear Power Plant; May 18, 2015

During an inspection, a boron deposit was found on a measuring line in the auxiliary system building, from which a leak emanated. The affected pipe section was replaced and subjected to metallographic examination. The cause of the leak was found to be various corrosion mechanisms, which can be attributed to the concentration of chemicals used in decontamination during the decommissioning process. The mode of operation of the coolant treatment system was then changed so that it will no longer be exposed to such corrosive media. The affected pipe section is part of the boric acid measuring circuit on the evaporator column, which is used to separate the coolant from the system drainage into a deionized fraction and boric acid. At the time the leak was detected, the affected system area was unlocked for inspection work.

7.38.16 Small leakage during pressure test on a nuclear heat exchanger, Biblis Unit B Nuclear Power Plant; March 25, 2015

During a pressure test, drip leakages were found on two heat exchanger tubes. The cooling tubes were closed with reversible plugs. The cause of the leakage was manufacturing errors in the retubing of the coolers in the area of tube rolling. A new pressure test of the heat exchanger was carried out. The second pressure test revealed no leaks in the cooler tubes. In the 2015 state of the plant, it was mandatory to remove the heat from the fuel pool. The availability of the cooling line was not affected by the leak.

7.38.17 Leakage in pipelines in the nuclear auxiliary cooling water system, Biblis Unit B Nuclear Power Plant; February 18, 2015

During plant surveys, leaks were found on four lines in the nuclear auxiliary cooling water system. All leaking areas have been repaired. To identify the cause of the incident, metallographic examinations were carried out on all affected lines. Corrosion damage was identified as the cause of the damage. The leaks did not restrict the flow of cooling water to safety-related components.

7.38.18 Leakage in a line of the wastewater treatment system, Philipsburg 1 Nuclear Power Plant; December 13, 2014

During a survey of the plant, a leak with a diameter of about 3 mm was found in a pipe in the wastewater treatment system. The affected pipe area was temporarily sealed with a sealing clamp. In the investigation program subsequently carried out for the area concerned, further traces of leakage were found in the same pipeline area and in an adjacent line. Leakages were caused by corrosion penetrating the pipe wall. The affected sections of the discharge line were replaced. The task of wastewater treatment is to collect the radioactive water and to process it in such a way that it can be returned to the NPP cycle or released in a controlled manner. The leakages were located in the reactor building in the discharge line to the outlet channel. The leaked liquid was treated water. An increase in activity was not found either at the activity measuring point in the discharge line or in the chimney exhaust air. No contamination was found at the leakage points. The leaks therefore had no impact on workers, the reactor system, or the environment.

7.38.19 Small leakage on the ventilation line of the circulating air cooler of a secondary cooling water line, Krümmel NPP; December 11, 2014

During an inspection in the cooling water pump house, a small leak was found on a pipe bend in the ventilation pipe of the circulating air cooler in the area of one of the four auxiliary cooling water pumps. The subsequent examination revealed that the pipe bend was corroded due to local damage to the inner coating. In the present case, the licensee assumed a random error as the cause. As an immediate measure, the leak was sealed with a bandage. The final repair was carried out by replacing the damaged pipe bend. According to the licensee, the leakage had no effect on the function of the fan (it serves to cool the air in the area of the auxiliary cooling water pump) and on the redundancy of the four-line auxiliary cooling water system. The secondary cooling water system has the task of dissipating the heat from the connected unit cooling systems via the intermediate cooler to the Elbe river water.

7.38.20 Leakage in a drain line of the USUS auxiliary cooling water system, Philipsburg 1 Nuclear Power Plant; March 28, 2014

When the pump house of the Independent Sabotage and Accident Protection System (the German acronym is USUS) was inspected, a leak was found on a drain line connecting to the main line of the USUS auxiliary cooling water system. The leak was temporarily sealed. The affected pipe section was later replaced. The cause of the leak was a corrosion attack originating from the inside of the pipeline. USUS was designed to tackle external impacts, e.g., plane crash or explosion. In the current state of the plant, the USUS auxiliary cooling water system no longer has any safety functions.

7.38.21 Damage to an overflow pipe on the waste water collection tank of the radioactive waste water treatment system, Biblis-A Nuclear Power Plant; April 3, 2014 (Fig. 7.24)

During a survey of the plant, the shift personnel discovered a leak in an overflow pipe of a waste water tank in the system for storage and treatment of radioactive wastewater. The cause of the leak was damage to the pipeline during scaffolding work for the replacement of piping. The affected section of the system was blocked off and taken out of service, and was replaced later. The pipe concerned is a plastic pipe (polypropylene), which in this case contained deionized water. Since only deionized water leaked out, there was no contamination of the area. The system for the treatment of radioactive waste water is one of the systems that are still required after the plant has been shut down. Within the wastewater treatment system, the radioactive wastewater accumulating in the control area is collected, processed, and released after being measured.

FIGURE 7.24 Biblis-A NPP. *(Credit to Peter Stehlik, Wiki Commons).*

7.38.22 Leakage in a drainage line during system decontamination, Philipsburg 1 Nuclear Power Plant; December 11, 2013

During decontamination work, a small leak was discovered in a pipe in the reactor emergency shutdown system (SCRAM). The leak caused a slight contamination in the immediately surrounding area. There were no increased values of indoor air activity or activity releases into the environment. The affected area was cleaned, and the leakage point was blocked off. The cause of the leak was material erosion due to the decontamination agents during system decontamination. With the current plant status, there are no remaining requirements for the SCRAM system. The decontamination work serves to reduce the activity and to prepare the system for later dismantling.

7.38.23 Leakage in the housing of a valve in the nuclear cooling system during reactor pressure vessel decontamination, Philipsburg 1 Nuclear Power Plant; November 24, 2013

During decontamination work, two small leakage points were discovered in the body of a valve of the nuclear cooling system, which led to drip leakage. The leak caused a slight contamination in the immediately surrounding area. There were no increased values of indoor air activity or activity releases into the environment. The affected area was blocked off. The leak was due to a corrosion attack by the decontamination agents used in the decontamination of the nuclear cooling system. The corrosion attack was limited to the affected area on the valve body, where was a surface inhomogeneity in the cast structure. The affected after-cooling line was used for core cooling in the previous power operation and is now deactivated. There are no longer any process requirements. The decontamination work serves to reduce the activity within the system and to prepare the system for subsequent dismantling.

7.38.24 Leakage in a vent line of the reactor water purification system during system decontamination, Philipsburg 1 Nuclear Power Plant; November 18, 2013

During routine radiation protection measurements, a leak was found in a pipe of the reactor water purification system. The leak next to a weld was sealed off. The room was cleaned. The leak was not associated with increased indoor air activity or increased activity releases through the exhaust air chimney. The cause of the leak was material erosion due to the decontamination agents used during the decontamination of the reactor water purification system. In the present state of the plant, there are no more process requirements for the reactor water purification system, which is therefore slated for full dismantling. At the time of the event, the affected line was used for ventilation during the decontamination of the reactor water purification system. The decontamination work serves to reduce the activity within this system and thus to minimize the radiation exposure of the personnel during future dismantling work.

7.38.25 Leak in the sewage pipeline to river Havel, Rheinsberg Nuclear Power Plant, September 11, 2013

The legacy fleet of the former German Democratic Republic (GDR) including Rheinsberg NPP and the five reactor units of Greifswald are being decommissioned by EWN Entsorgungswerk für Nuklearanlagen (EWN), a state-owned company under control of the Federal Ministry of Finances. The Rheinsberg plant has been in decommissioning since 1995. Its RPV was transported to Greifswald's centralized interim storage facility (Zwischenlager Nord), also operated by EWN.

During the routine tightness check of the pipeline to the Havel river (a 7042-m-long waste water pressurized pipe), an inadmissible pressure drop was found. Due to a suspected leak, the tightness control was interrupted, and measures were taken to localize the leak and then repair the damaged section of the line. Several leaks were found during the repairs, which lasted over a month. The line was replaced for a length of 291 m in the affected section. Initial investigations showed that the leaks were caused by foreign bodies in the line. Further material tests were carried out by the manufacturer of the pipe. The affected pipeline is the approved pathway for the discharge of weakly radioactive and chemically treated water, such as from the laundry and showers. The soil contamination values measured in the leakage areas were below detection and the regulatory exemption limits.

7.38.26 Small leakage on a ventilation pipe of the nuclear auxiliary cooling water system, Neckarwestheim 1 Nuclear Power Plant; September 19, 2013

Licensee EnBW formally applied on May 6, 2013 to the Baden-Württemberg Government to be authorized to decommission and demolish Neckarwestheim 1 Nuclear Power Plant (GKN 1). The unit was among the eight oldest German reactors taken out of service shortly after the Fukushima Daiichi accident in March 2011. Baden-Württemberg's ministry of environment awarded the GKN 1 decommissioning license on February 3, 2017. EnBW expected to begin dismantling activities in the same year. The plan is to start with the dismantling of the primary coolant pipelines. Preparations are under way for the subsequent dismantling within the RPV. It is tentatively anticipated that the plant decommissioning project will be fully completed in 10–15 years.

During a routine check, a minor leak was detected in the area of a weld seam of a vent pipe of one of the four nuclear secondary cooling water systems. The affected area was sealed with a clamp. It was subsequently reported that a reduction in the pipe wall thickness was detected in the radiographic tests carried out in other comparable weld seams. Due to the combination of materials present in the affected welds, the damage mechanism is attributed to electrochemical corrosion. As part of an ad hoc test program, other comparable pipes were randomly examined with the help of radiographic tests. According to the licensee, all affected pipeline sections have been replaced. The functions of the nuclear auxiliary cooling water system were not impaired by the leak.

7.38.27 Leakage at a beam pipe, Geesthacht 1 research reactor; September 22, 2013

The reactor was finally shut down on June 28, 2010, and was in the post-operation phase at the time of the incident. The reactor is free of fuel elements. In one of the beam pipes, through which the neutrons came from the reactor core into the test hall for specific experiments during the operations of the research reactor, the collimator of the beam pipe was to be flooded due to a planned modification. A leak in the pipe to the pool was found, and the work was stopped immediately. The leak could not be located. In the present case, there was no impact on people or the environment.

7.38.28 Small leakage on a secondary cooling water pipe in the turbine building, Brunsbüttel Nuclear Power Plant; January 15, 2013

During a test in the secondary cooling water system, a leak was detected on a pipe in one of the two cooling water lines. The damaged area of the pipeline was temporarily sealed with a clamp. The final report submitted by the licensee stated that there was local aging-related corrosion pitting, which was caused by underrusting and damage to the coating. The repair was carried out by buildup welding. In the present case, there was no impact on the post-operation status of the plant. Due to the small leakage, the cooling capacity required for cooling auxiliary systems and emergency diesel generators was not impaired.

7.38.29 Discovery on a waste water discharge line, Neckarwestheim 1; August 16, 2012

During a routine inspection, initial corrosion and a slight leak were discovered in the weld seam zone of the nuclear sewage discharge line. The affected pipeline has been repaired. In the final report submitted by the licensee, it is declared that the causes of the damage are widespread corrosion and pitting corrosion. According to the report, the damage found is fairly normal in a component used for around 35 years. The wastewater system is used to collect and decontaminate the radioactive wastewater generated during power plant operation and also when the plant is out of service. The event had no environmental impact. The wipe tests carried out in the discovery zone gave contamination levels below the maximum permissible values.

7.38.30 Small leak in a pipeline of the concentrate processing system, Brunsbüttel Nuclear Power Plant; August 28, 2012

During a plant inspection, a small leak was found on a pipe of the concentrate processing system. The findings revealed damage that penetrated the pipe wall in the area of the pipe bend. In the final report submitted by the operating organization, it is stated that the laboratory examination of the removed pipe bend showed abrasive wear (due to filter residues transported by the liquid flow). According to the licensee, there was no evidence of corrosion attacks in the affected area. The task of the concentrate preparation system is to process the radioactive concentrates from the filters and evaporators in the control area until they are ready for transport and storage. In the present case, the leakage had no direct impact on the system and workers. The area in which the leak was found is for restricted access. The leaked quantities did not lead to an increase in the local dose rate or to any radiological limit violations.

7.38.31 Leakage in a drainage line in the auxiliary cooling water system, Philipsburg 1 Nuclear Power Plant; March 10, 2012

During a routine inspection in the cooling water pump house, a pencil-size leak was found. The leak was first temporarily sealed, and the affected pipe area was later replaced. In the investigation of the wider pipeline area, the cause of the leak was found to be corrosion pits, whose formation is attributed to such boundary factors as stagnant water and Rhine water. The leaked cooling water was collected in the building sump and pumped out via the system drainage. In the current plant state (decommissioning), the availability of one of the two secondary cooling water lines is required. The second secondary cooling water line was available at all times with no problem.

7.38.32 Corrosion damage to a vent line in the active wastewater treatment plant, Muelheim-Kaerlich; February 11, 2011

During an inspection of the active wastewater treatment system, a small leakage due to microbiologically induced corrosion was found on the vent line of the wastewater collection tank. The leak had no radiological effects. The affected pipe section was replaced by a new pipe section made of the same material. The active wastewater treatment will continue to operate even with the plant in a shutdown status. The safety function is the retention of the radioactive substances collected and processed in the system. The gaseous radioactive substances accumulating in the collection tanks are discharged to the exhaust air system of the control area via the ventilation lines. The exhaust air is filtered and then released via the exhaust air chimney.

7.38.33 Lessons learned

All the aforementioned events were announced under category N (normal message). All of them were classified by the operating organization at level 0 of the INES rating scale (no or very little safety-related/radiological significance). In simple words, all these events were minor, and the impacts from liquid leaks were contained within controlled areas.

7.39 Soil contamination at Greifswald Unit 1 Nuclear Power Plant, Germany (BASE, 2020)

7.39.1 Decommissioning process

Greifswald nuclear power station was the largest NPP operating in the former East Germany before it was closed shortly after the German reunification. The nuclear units were Soviet-type VVER-440 reactors. As upgrading to the modern safety standards was not economically feasible and against public sentiment, the new unified German government decided in early 1991 to decommission the four active units, close Unit 5, which was being commissioned at the time, and stop construction of Units 6–8 (Fig. 7.25). Decommissioning of Units 1 through 5 began in 1995. At the time of writing, decommissioning of the whole Greifswald site is nearing completion. The end use could be restricted or unrestricted.

FIGURE 7.25 Greifswald NPP. *(Photo by M. Laraia).*

7.39.2 Problem encountered 1; impacts

7.39.2.1 Soil contamination west of Unit 1

During radiological control measurements on June 10, 2010, an increased contamination in the soil was detected. According to the licensee, this increased contamination refers to a 1974 event. In this event, 400 L of contaminated water leaked out of a room in building 1 and entered the ground through the external outside wall of the building. The contaminated soil (approximately 5 m^3) was disposed of at that time, but some contamination remained. The affected area was closed off, and the remaining contaminated soil was disposed of. The surface contamination determined in the present case and the overall activity met the criteria for reporting the event in category E (urgent report). The operating organization has classified it at level 0 of the INES rating scale (no or very little safety-related/radiological significance).

7.39.3 Problem encountered 2; impacts

7.39.3.1 Increased contamination in the monitored area of Unit 1

A radiological control on June 10, 2010, revealed increased contamination in the soil below a pipe duct in the monitored area (approximately 2 m^2). The affected area was cordoned off, and the contaminated soil was disposed of. Subsequent investigations and analyses showed an overall activity of 8E + 8 Bq and a maximum surface contamination of 7.54 E + 3 Bq/cm^2 over an area of approximately 24 m^2. There was no impact on people or the environment. In the final report submitted by the licensee, it is stated that the contamination (^{137}Cs, ^{60}Co) measured is due to a 1987 event in which liquid leaked out of an evaporator system. The surface contamination values determined in the present case exceeded those legally permitted for these radionuclides by more than 100 times and the total activity by more than 10 times the permitted values. The criteria for reporting the event under category E (urgent report) were thus met. The operating organization classified the event at level 0 of the INES rating scale (no or very low safety/radiological significance).

7.39.4 Lessons learned

A cursory comparison between the German leaks described in this section and in Section 7.38 clearly shows that these two events originated from a past era that used less rigorous safety criteria. In particular, the lack of containment that should have retained the leakage within the controlled area is striking. This difference explains why these two events were given a reporting category E (urgent report), whereas the cases described in Section 7.38 were given category N (normal report). In all cases, however, the safety and radiological impacts were insignificant. According to German regulations, category E events should be officially notified by telephone and in writing at the latest 24 h after the event has been known. In contrast, category N events should be notified at the latest on the fifth working day after the event has been known using the notification form.

7.40 Actual or potential fire-related events at shutdown or decommissioning installations in Germany (BASE, 2020)

7.40.1 Blocked fire door in the access building of a storage facility, Würgassen Nuclear Power Plant; December 5, 2018

Würgassen NPP (KWW) was commissioned in 1971 and shut down for economic reasons in 1994. Nuclear D&D of the entire plant was successfully completed in 2014. This achievement makes KWW the first NPP decommissioning project in Germany to be fully completed. The plant is currently being operated as a radioactive waste storage site. Two storage facilities for radioactive waste remain at the site. Only low- and medium-level radioactive waste from the dismantling and operation of the plant is stored here.

During an inspection, it was found that the fire door of the radiation protection office in the access building to a waste storage facility was blocked by a wedge in the open position. An internal inquiry revealed that two employees who had been waiting in the radiation protection office for a lock operation had blocked the door in the open position. After completing the work, they forgot to remove the wedge. Appropriate instructions were then given to all persons involved in the waste storage activities. The on-site tools that could be used for blocking fire doors in the open position were removed. The fire protection functions were inadmissibly impaired due to the unlocked fire door. In the event of a fire in the radiation protection office of the access building, the secure exit route from the waste store could have been blocked.

7.40.2 Failure to close a fire damper during a functional test, Biblis-B Nuclear Power Plant; November 15, 2018

During the periodic inspection of fire dampers, one damper did not fully reach the *closed* position. The reason for the failure to close was friction due to stuck contamination. The fire damper was cleaned and serviced immediately. The test was then successfully completed. The affected fire damper is located in the auxiliary system building and serves as a fire protection separation between a waste storage room and an adjacent traffic route. Due to the failure of the fire damper to close, the fire protection separation was temporarily not available.

7.40.3 Inadequate reactivation of a fire alarm system, Brunsbüttel Nuclear Power Plant (KKB); December 14, 2018

By the end of 2018, the existing Integrated Services Digital Network (ISDN), telephone connections to the fire detection system at KKB should be replaced by ALL-IP (note: An ALL-IP Voice Network uses internet protocol (IP) techniques to link telephone calls and establish communication, in a similar way as computers use internet to exchange information. As of early 2014, no more ISDN connections have been assigned by Deutsche Telekom, and notifications of enforced termination had already been sent out for existing connections). During the preparatory work to this transition, the telephone line to the fire brigade control center in Elmshorn was taken out of operation. As a result, it was temporarily impossible to forward alarms via manual call points in the KKB (so-called push button detectors) to the control center of the public fire service. However, the fire detection and fire report within the KKB was fully available. The first firefighting response by the KKB fire department was not affected either. After completion of the work by a specialist company, the reporting facility in KKB was not reactivated. The error of not reregistering was recognized by the shift crew on the following day due to a signal in the area of the fire alarm system's user interface and corrected immediately. The investigations initiated by the operating organization to clarify the cause with the specialist company that carried out the work were under way at the time this notice was issued. In the present case, there was no impact on fire detection and fire reporting in KKB. During the deregistration of the reporting facility due to the work, the alert of the fire brigade control center could have been activated if required by ordinary telephone line or via a dedicated radio service for authorities and organizations tasked with security issues.

7.40.4 Malfunction of a smoke aspiration system in part of the waste storage facility during routine tests, Würgassen Nuclear Power Plant; December 5, 2017

Two storage facilities for radioactive waste remain at the KWW site. One of these (UNS) resulted from the 2005 conversion of a former emergency building (Fig. 7.26). Low- and medium-level radioactive waste from the dismantling and operation of the plant is stored here. During a routine test, it was found that the smoke detector in the smoke aspiration system (RAS is the German acronym) of the recirculating air system in the waste storage facility was not functioning properly. No fire alarm and no fault message were triggered. The subsequent causal investigation revealed that a power supply was defective. A single fault is assumed to be the cause. After replacing the defective power supply with a new power supply, the test could be carried out successfully. In the present case, there was no impact on the waste storage operations. The RAS is a component of the technical fire protection and has the task of monitoring smoke in the ambient air extracted from the waste storage facility.

7.40.5 Partial failure of the smoke detection system in the area of the biological shield, AVR HTR Nuclear Power Plant; April 17, 2018

During a routine test of the smoke alarm system in the area of the biological shield, it was found that one of the two lines being tested did not trigger an

FIGURE 7.26 Würgassen NPP independent emergency building (UNS) converted into waste storage facility. *(Photo by M. Laraia).*

alarm. During the subsequent investigation conducted by checking individual subsections, a leak was found at a suction point; contamination was also found at the suction point in another area of the pipe system. In a third area, the problem could not be clearly determined due to inaccessibility. The system could be made fully functional again by reinstallation. The cause of the failure is attributed to the dismantling activities under way. As part of the precautions against recurrence, preventive maintenance measures were taken, including a more frequent review of the functionality of the systems, especially if they can be affected by dismantling activities.

7.40.6 Malfunction of a remote control valve in a sprinkler system, Brunsbüttel Nuclear Power Plant; May 17, 2017

In a routine test of sprinkler systems, a remote control valve should be activated by the power station control using a trigger module. Due to a malfunction, the remote control valve did not activate as per design. However, the valve could still be operated using manual control. In the two plant rooms concerned, firefighting using portable fire extinguishers would still have been possible. A reserved valve was substituted for the faulty valve. To clarify the cause, the defective remote control valve and its activation device were further investigated. In August 2016, a similar failure of a remote control valve of a spray water extinguishing system was reported at KKB. Since the 2017 case raised a recurrent issue, the responsible supervisory authority required that all fire protection devices of the NPP be checked. The fire protection measures at KKB are intended to fight internal fires and are therefore important for safety. The deluge systems in KKB are fixed; sprinkler-based fire extinguishing systems have the task of extinguishing initial fires in areas with large unprotected fire loads (e.g., areas with difficult accessibility, rapid fire propagation, and insufficient smoke and heat dissipation).

7.40.7 Open fire protection doors, Biblis Unit A Nuclear Power Plant; December 7, 2016

During routine inspections, several open fire protection doors were found in the reactor building and in the auxiliary plant building, which were blocked by different objects. The doors themselves were intact. The improper blockages were removed, and the fire doors closed. Fire protection doors are part of the structural fire protection measures. During these events, the fire protection function was incapacitated due to the open position of the fire protection doors.

7.40.8 Fire produced during treatment of filter material, Philipsburg 1 Nuclear Power Plant; September 16, 2016

After the drying process of two drums loaded with filter material in the drying chamber (a part of the treatment of solid radioactive waste treatment), a burning fire was discovered in one of the two drums. The drying chamber is classified as a restricted area. The blaze was successfully fought with CO_2 hand extinguishers. The cause of the fire was an excessively high drying temperature in combination with an excessively short cooling and dwell time after the drying process. When drying filter material in future, a lower drying temperature will be selected, and the cooling and dwell time will be extended. The staff was then trained accordingly. The measurements of air activity concentration, local dose rates, and additional wipe tests did not reveal any noticeable values in the drying chamber, so it can be safely assumed that no release of airborne activity occurred into the plant rooms or the surrounding areas.

7.40.9 Blocked fire door, Biblis Unit A Nuclear Power Plant; December 9, 2015

When entering a room in the annex building, a fire door was found in a blocked open position. The fire door connects the two rooms where the emergency diesel generators are installed. The cause of the blocking was a door bearing broken due to material fatigue, which led to the lowering of the fire protection door and thus to jamming of the door. How the door was opened could not be clarified. By jamming the fire door in the open position, the fire protection separation of the two rooms was not possible. The door bearing was replaced, and measures were taken to control the opening angle of the door and thus prevent it from getting stuck again.

7.40.10 Fire in a control cabinet of the hazard alarm system, Karlsruhe KNK II Nuclear Power Plant; October 31, 2013

The compact sodium-cooled Karlsruhe KNK II was run as a prototype fast breeder reactor with a capacity of 17MWe for 14 years until its final shutdown in August 1991. KNK II was subjected to early dismantling operations beginning in 1993, and by mid-1994, all fuel had been removed. A number of preparatory, decommissioning tasks were carried out, but it was not until 2003 that dismantling work began in the reactor containment building. The unrestricted release of the site is planned to be reached soon.

During a survey of the plant, a fire was noticed in the power supply cabinet of the alarm system. The fire was extinguished by the plant fire. The power supply cabinet of the alarm system was repaired and put back into operation, so that the alarm system was available again a day later. The cause of the fire was an insulation fault on a cable. The fire had no effect on the facility. Until the control cabinet was repaired, the systems were monitored in situ by the operations managers. The affected power supply cabinet and two other control cabinets of the alarm system were replaced by new cabinets.

7.40.11 Fault in a valve in the fire extinguishing system, Unterweser Nuclear Power Plant; December 11, 2011

The 1400-MWe Unterweser NPP (KKU) was in operation from 1978 to 2011. The licensee Preussen Elektra applied for permission to decommission and dismantle KKU in May 2012. The decommissioning license was granted in February 2018. The last fuel rods were removed in February 2019. The dismantling work inside the controlled area is expected to take some 13 years.

During a routine test, it was found that one of two fittings for the fire water supply to the containment building did not open fully. The direct cause was damage to the sealing rings inside the valve. In the report submitted by the licensee after completion of the material technical tests, it is stated that the damage to the sealing rings was due to material wear. The valve was repaired using new sealing sets. The permissible repair time results from the requirements for the operational readiness of the fire extinguishing system. In this case, it was 14 days.

7.40.12 Smoldering fire from residues in a waste container within the drying system, Isar 1 Nuclear Power Plant; September 30, 2011

Together with several German NPPs, Isar 1 (KKI 1) was permanently shut down in 2011. A first step toward decommissioning was taken in January 2017, when Preussen Elektra was granted the decommissioning license (phase 1) for Isar 1. This license permits the dismantling of plant parts, which have not been specified for dismantling phase 2 (reactor pressure vessels and biological shield). It is planned that phase 1 will be completed by 2023 and phase 2 by 2026. A waste processing center will be operated on the Isar 1 site and subsequently used for the dismantling of Isar 2 (planned for final shutdown in 2022). The final dismantling of Isar 1 can only be completed when the processing of waste from the dismantling of Isar 2 has been completed. The sites of Isar 1 and 2 are therefore only expected to be released from regulatory control in around 2040.

After a 200 L waste drum was dried within the drum drying system and the heating was switched off, a smoldering fire occurred in the solid mixed waste, consisting of coating material (hard rubber). The smoldering fire was extinguished by the personnel using the existing CO_2 system as well as foam and water. The cause of the smoldering fire after self-ignition was the lowering of the ignition temperature of the mixed waste due to its consistency (powders). The drying process has been changed since so that self-ignition will no longer possible. There was no release of radioactive substances into the room exhaust air and no surface contamination from the extinguishing work.

7.40.13 Lessons learned

The risk of fire and related impacts on the workers, the public, and the environment is probably the major concern for most deferred dismantling projects. To mitigate this risk is possible and necessary, but considerable amounts of human, technical, and financial resources should be allocated to this end.

A general reference addressing all aspects of fire protection for deferred dismantling projects—based on US DOE requirements—is given in WYATT (1994).

To determine if firefighting objectives are met for a permanently shutdown plant, some organizations (e.g., US DOE) require a fire hazards analysis specific to that plant. The risk of fire in a safe enclosure (SE) can be mitigated by removing combustible materials and ignition sources as far as possible during the transitioning phase to SE. However, this will not completely cancel the risk of fire. Therefore, a firefighting plan and related equipment must be in order. The existing firefighting plan for the operations phase will in most cases be revised as ongoing and planned activities, building layout, water supply, access routes, etc. will in general change in the operations-to-decommissioning transition. In some plants, due to the limited fire risk, a fire brigade on site is no longer required for SEs while in others firefighting requirements can even be increased due to the lack of the workers on site who can at least issue an early warning. For multireactor sites, the SE can benefit from the on-site firefighting team. For single-unit SEs, the local fire brigade may provide assistance with fires. Prior to moving to a decommissioning phase new firefighting plans must be developed in cooperation by the licensee and the local fire brigade and agreed to by the regulatory body.

7.41 Defective measurements at the discharge of radioactive effluents, shutdown or decommissioning nuclear power plants in Germany (BASE, 2020)

7.41.1 Nonrepresentative values measured at a sampling point for activity in the outlet channel, Philipsburg 1 Nuclear Power Plant; September 19, 2019: problem encountered; impacts

For environmental monitoring purposes, sufficient mixing of the plant's water discharges should be ensured at the sampling point: This should take into account the decommissioning status of Philipsburg 1 Nuclear Power Plant (KKP 1), the ongoing operation at KKP 2, and the final shutdown of KKP 2 in 2022. Therefore, both the location of the point of discharge from the KKP 1 wastewater treatment plant and that of the sampling point at the end of the main cooling water channel into the river Rhine changed. It was later determined that the new locations of the sampling point and the discharge point did not allow a representative measurement of the activity of the wastewater discharged from KKP 1 into the Rhine. The cause had not yet been clarified when the preliminary event report was issued. The sampling point for activity measurement and the discharge point were moved back to their original locations. Subsequent tests showed no violations of the discharge limit values. Furthermore, the radiological monitoring of the wastewater from KKP 1 was ensured by an additional activity measuring point in the discharge line, which automatically stops the discharge process when limit values are exceeded.

7.41.2 Failures in the exhaust air monitoring, AVR HTR Nuclear Power Plant; July 26, 2016; problem encountered; impacts

Malfunctions of the two fans of the reactor building exhaust air monitoring system were first signaled in the plant control room during the night. On the following day, on-site examination of these early notices confirmed that both fans were out of operation. The malfunctions were promptly eliminated. The interruption of the exhaust air monitoring lasted about 4 h. According to the licensee, the cause of the event can be referred to a faulty power supply to the two blowers. The clarification cast light on an incorrect setting of the tripping limit values for the voltage monitoring relays. The setting values were corrected accordingly. The exhaust air system has the task of continuously monitoring the discharge of radioactive substances in the exhaust air of the AVR. If the exhaust air monitoring fails, dismantling work must be interrupted immediately. No dismantling activities took place during the unavailability of the exhaust air monitoring.

7.41.3 Incorrectly set parameters for concentration estimates after changing the filter at aerosol measuring points, Philipsburg 1; February 2, 2015; problem encountered; impacts

During the weekly filter change at the aerosol activity measuring points of the chimney exhaust air monitoring, an excessively high value for the volume flow limit was determined. This setting error could not have triggered the alarm messages in the control room if the specified aerosol activities were exceeded. This affected the measurement of aerosol activity concentration of the chimney exhaust air and the monitoring of the turbine building exhaust air. The reason for the incorrect setting of the volume flow limit was an incorrect understanding of the meaning of this parameter and an incorrect change of the corresponding test instructions. This error was noticed shortly after a filter change in 2012. Incorrect settings made at several measuring points were subsequently reversed, but not at the aerosol activity measuring points affected in the present case. This was only noticed when the filter was changed in 2015 and then corrected. The related messages (alarms) were not available for the period of the incorrect setting. However, other measurements of aerosol activity were still available, such as the pulse rate measurement, which can also be used to determine an increase in the activity concentration, as well as the accounting of the integrated activity emissions by means of the weekly measurements of the aerosol filter. These later reevaluations showed that the permissible discharge values were not exceeded over the period of the incorrect setting.

7.41.4 Failure of an aerosol measuring point, Karlsruhe KNK II NPP; March 30, 2013

Following an error message in the alarm center, it was found that the filter belt on a monitor for measuring radioactive aerosols in the building exhaust air had been damaged. A replacement measuring point was then put into operation in accordance with the operational procedures. After using a new filter belt, the old measuring point could be put into operation again. The accounting of the radioactive discharges via the exhaust air was not affected.

7.41.5 Lessons learned

Airborne and liquid releases during the decommissioning phases are generally much lower than during previous operations. However, especially in conjunction with large-scale decontamination and dismantling activities the generation—and ultimately release—of significant amounts of airborne particulates or contaminated solutions is to be expected. Therefore, monitoring systems should be kept operational and accurate during the entire decommissioning process.

7.42 Incidents at solid waste treatment systems in shutdown or decommissioning installations in Germany (BASE, 2000)

7.42.1 Mishap when operating the powder suction system, Brunsbüttel Nuclear Power Plant; May 25, 2018

For a conditioning campaign, the dried filter concentrates (powder resin) are sucked out of 200-L barrels into a cast container with the help of a powder suction system (whose German acronym is PUSA) located in the area of solid waste storage. As early as February 2018, a suction port with a sampling connection caused a slight leakage of suctioned material, as identified by PUSA operations and monitoring personnel. A similar incident occurred 3 months later. The leak led to a slight contamination on the surface of the cast container. The conditioning campaign was suspended. According to the licensee's assessment, the affected intake manifold, which is considered subject to wearing out, failed unexpectedly and quickly. The solid waste storage at KKB is part of the control area. According to the licensee, there was no contamination spread, and warning or limit values were not reached either. There was never any danger to individuals.

7.42.2 Deflagration during the cold test of a dismantling device for reactor core components, Greifswald Unit 4 Nuclear Power Plant; April 26, 2018

In Unit 4 of the Greifswald NPP, the device in question was to be cold-tested using a dummy to dismantle a core component. This component is an activated item from the Rheinsberg NPP (KKR), which was stored in the shaft storage of Greifswald NPP Unit 3 and 4 in 1982 and had to be dismantled. After about 80% of the first cut had been made by the cutting device, the workers heard a bang on site. The cutting device was immediately taken out of operation, and the testing was stopped. No damage occurred to the cutting device, dummy, or transfer container. The bang was a deflagration of waste gases from thermal decomposition, which had concentrated to an ignitable mixture at the lower opening of the transfer container. The flushing device used for the discharge of the gases proved not to be effective enough to prevent the gas mixture from accumulating. As a precaution against recurrence, the licensee made some procedural improvements for the discharge of the exhaust gases.

7.42.3 Rainwater leaked into the area of solid waste storage, Brunsbüttel Nuclear Power Plant; June 22, 2017

During a heavy shower, a leakage of rainwater occurred in one of the rooms of the solid waste storage area (a restricted-access area). Several barrels of compacted, low-level radioactive raw waste were wetted. The water was drained through the building drainage system for treatment. The damaged pipeline was temporarily sealed, and the room dried. The rainwater pipe that previously ran from the roof through the building was closed and shut down after the external roof drainage system was rebuilt. There was no release, carryover, or discharge of radioactive substances. Wiping tests carried out on the accessible barrel and floor surfaces showed no higher activity values than those normal in this storage room.

7.42.4 Incidents in storing and handling solid radioactive residues, Brunsbüttel Nuclear Power Plant; August 13, 2014

Due to abnormal events in the filter concentrate disposal campaign that came to light in 2012 (very strong corrosion was found in an already emptied drum and subsequent examinations revealed more drums with significant signs of corrosion), KKB started a comprehensive review of the accumulation and storage of radioactive solids and their handling and conditioning activities. The focus of this review was the inspection of the so-called caverns (six enclosures in the solid storage area within the KKB control area). The six caverns contained low- and medium-level radioactive waste from reactor operations. The waste consists of basically filter resins, evaporator concentrates, and miscellaneous waste. In the follow-up report submitted by the licensee, it was stated that 164 out of 573 barrels examined showed significant damage. There were in total 632 barrels in the caverns. The previous investigations have not shown any signs of damage to the caverns. According to the licensee, radiation protection measures have been defined for the inspections.

7.43 Wildfire hazards

7.43.1 Decommissioning process; problem encountered; impacts (BBC News, 2020)

Chernobyl NPP reactor No. 4 in Ukraine became on April 26, 1986, the scene of a major nuclear disaster station (INES level 7, the highest), which heavily contaminated the plant, the nuclear site, and the environment. The radioactive gases produced by the accidents contaminated vast areas in Ukraine, Belarus, and Russia and impacted in many ways other European countries. An exclusion zone was established in the aftermath of the accident: It covers an area of around 2600 km^2 surrounding the plant site where radioactive contamination from nuclear fallout is highest and public access and residence are restricted.

Chernobyl NPP went on generating nuclear power until the plant's last operational reactor was permanently shut down in 2000. A huge shield built to long-term protect the reactor against environmental agents (named NEW Safe Confinement, aka Arch due to its form) was installed on-site in November 2016, replacing a deteriorating "sarcophagus" hurriedly built in 1986 to seal the burning reactor (Fig. 7.27). The construction was completed in 2018.

A blaze spread in April 2020 in the surroundings of the nuclear plant and raged for many days burning the contaminated forests in the exclusion zone and presumably releasing radioactive particles in air. There had been concerns the blaze could threaten the site of the 1986 catastrophe. According to various

FIGURE 7.27 New safe confinement during construction, 2013, Chernobyl NPP, Ukraine. *(Credit to Tim Porter, Wiki Commons).*

statements, the huge fire went as near as 1–2 km from the plant site. Hundreds of firefighters as well as planes and helicopters were sent to extinguish the fire. Aircraft dropped 538 tons of water on the blaze. Eventually the fire was contained, although some "smoldering" parts of the forest floor remained.

This event highlights a form of external hazard that decommissioning plants can be exposed to and the need for preventive measures especially in sensitive areas (e.g., forests, dry vegetation).

As an example of wider concerns on this hazard, the US DOE issued in 2000 an Environmental Assessment that evaluated the potential environmental impacts of wildland fires that threaten operations at the Los Alamos National Laboratory (LANL) and strategies for addressing such events.

In 2019, the DOE issued a supplemental study (DOE/EA, 2019) covering new strategies that reflect conditions that have changed at LANL since the 2000 report was published, including longer fire seasons, changes in vegetation, and global climate change.

7.43.2 Lessons learned (DOE, 2019b)

A wildfire or wildland fire is an unplanned, uncontrolled fire burning in a vegetated natural environment (forest, grassland, or prairie). It can be classified according to the origin of ignition, physical properties, combustible materials, and weather factors.

Wildfire evolution and severity depend on multiple factors such as combustibles, topographical features, prevailing wind, temperature, and moisture.

It has been estimated that 90% of wildfires in the United States originate from human-made activities, including:

- Burning debris, yard waste
- Unattended campfires
- Unextinguished cigarette butts
- Fireworks
- Arson (a person was indicted for the aforementioned Chernobyl case)
- Equipment failure, transformer sparks, powerline arcing
- Engine sparks, car crashes, train sparks

It can be seen that some of these factors are pertinent to industrial activities (e.g., nuclear decommissioning). For example, the Mendocino Complex, California, fire which extended over 1860 km^2, destroyed 280 structures and caused a fatality, originated from a spark or hot metal fragment produced by a hammer that was pushing a metal rod into the ground. 25 wildfire events occurred at the nuclear sites (many of them under active decommissioning) listed in Table 7.1 (DOE, 2019b).

At Sandia National Laboratory (SNL) the boom of a truck came within arc flash restricted area boundary distance and cut an aluminum cabling line that initiated a fire in dry grass. At Moab Project Site grass fire stemmed from welding a metal plate onto a railcar.

TABLE 7.1 Wildfires occurring at DOE nuclear sites; January 1, 2013, to June 10, 2019.

Hanford	**2 (a 2000 event is described in detail in Section7.47)**
INL	1
LANL	3
LBNL	1
LLNL	2
Moab	1
NNSS	3
National Renewable Energy Laboratory (NREL)	2
Paducah	2
Pantex	2
Sandia National Laboratories (SNL)	3
SRS	1
Strategic Petroleum Reserve (SPR)	1
WIPP	1

Naturally occurring fires in 2018 occurred at Nevada National Security Site (NNSS, caused by lightning) and Paducah Gaseous Diffusion Plant (power line fell).

Wildfires belong to the broader category of external events that can impact decommissioning (and operating) nuclear installations. Box 7.9 is a brief note about earthquakes.

BOX 7.9 Earthquakes potentially impacting decommissioning installations

As well known, nuclear facilities are designed and built to withstand earthquakes. I have found no published reference to earthquakes damaging decommissioning facilities, but minor damage may have happened since a number of retired facilities are located in seismic areas, e.g., Humboldt Bay and San Onofre in California or many Japanese reactors. Of course, the risk from an earthquake is much greater for operating facilities, but the structural deterioration for facilities left in a dormant state for many decades creates an increased seismic concern that should be faced.

7.44 Unexpected discovery of subsidence, Hanford site, Washington, United States; February 10, 2020 (DOE, 2020a)

Common sense is not so common Voltaire (1694–1778), Dictionnaire Philosophique (1764) Self-Love.

7.44.1 Problem encountered; impacts

At the Waste Information Data System (WIDS) Site 600, which is identified as a nonradiological waste disposal landfill, a team was removing tumbleweeds inside the site fence line. The area is known to have obstacles as well as many spots of surface subsidence. The impediments were marked with "bike flags" as warnings for future work to remove tumbleweeds with a tractor-mounted long-toothed rake. The workers were wearing goggles, thick gloves, and gaiters to protect from the thorns on the tumbleweeds. The Lead D&D Worker (LW) was in charge of surveillance by slowly driving a small truck with low sides, intersecting the area, and planting bike flags on unmarked obstacles. The front tires of the truck got stuck in a subsidence so deep that the LW was unable to back out. The driver suffered no injuries, and the truck remained undamaged. Following the event notification, a tractor was sent to the site. A recovery rope removed the truck from the subsidence. Two subsidence spots were identified at Site 600: One was around 1.5 m × 3 m and 1.5 m deep, and the other was 1.5 m × 1.5 m and 1 m deep. Senior management restricted access to WIDS sites until the sites hazards could be reviewed.

7.44.2 Lessons learned

The Cause Evaluation Team identified two root causes (1 and 2) and one contributing to the event (3).

(1) The Central Plateau Risk Management (CPRM) had taken a non-interfering approach with WIDS work activities, basically allowing workers to operate unsupervised. After the LW got the truck stuck, they posted warning flags around the subsidence. The LW did not view the case as an "event." They determined the stuck truck case on the basis of past practice, not by following the prescribed procedure. The removal of the stuck truck was viewed as ordinary work. When the LW called the field work supervisor (FWS), the LW did not describe the whole field condition; therefore, the FWS only instructed the LW to suspend tumbleweed removal. The FWS did not try to get more information, and therefore, did not recognize the situation as an "event" and considered that freeing the truck was the priority. The FWS was most interested in getting the workers and equipment out of the area prior to the end of shift.

(2) The LW did not notice the subsidence until the truck got stuck. Potential subsidence spots were not addressed in detail during the prejob brief because the site in question was not a documented subsidence area. This particular WIDS Site had not been considered as a subsidence because it is a nonradioactive site (a wrong decision as the facts showed).

(3) A visual survey of site in question was carried out by the FWS manager, but a comprehensive physical walk down was not carried out. Resources required to perform a walk-down prior to driving on the site were not allocated or evaluated. A specific procedure states: "Prior to driving any vehicle on (specified) WIDS waste site, walk downs should be performed." However, the procedure does not require a thorough walk-down or what the walk-down should achieve (e.g., the identification of subsidence areas and related hazards).

7.45 Natural gas hazards at Fort St. Vrain Nuclear Power Plant (NRC, 1991b)

7.45.1 Decommissioning process; problem encountered; impacts

When Fort St. Vrain (FSV) NPP was licensed for operation in 1973, no natural gas pipelines were located nearby; thus, neither the licensee's final safety analysis report nor the NRC's safety review addressed the hazards associated with natural gas pipelines in the area. In 1974, a 40-cm low-pressure gas collection pipeline was installed in the vicinity of the plant. This pipeline crossed a corner of the licensee's site about 1.5 km from the reactor building. In the period 1981–83, 12 natural gas wells were constructed within 1.6 km of FSV reactor building; seven of them on land formerly owned by the FSV licensee, which had been acquired by the gas company. Some of these wells were located within the exclusion area boundary, but all were outside of the protected area. Nine of these wells were connected to the 40-cm collection pipeline by a 15-cm pipeline. The 15-cm pipe passed within 400 m of the reactor building. The gas company and the FSV licensee determined that possible gas incidents would not cause adverse effects further than 100 m from the wellheads. The licensee did not prepare any safety analyses for the wells or pipelines.

In 1987, the FSV licensee agreed to the drilling of a gas well within 360 m of the reactor building, which was within the exclusion area boundary and 100 m from the protected area fence. The pipeline linked with this well passed within 170 m of the NPP switchgear area. The licensee's safety assessment concluded that in case of a well fire caused by a blowout, the area affected by the fire including also the equipment needed to control the fire would not extend beyond the existing drilling site. The licensee also determined that the air temperature farther than 120 m from the fire location would not be higher than the ambient temperature, and therefore, there was no additional hazard to the NPP and no unreviewed safety question. The safety assessment did not consider the impacts of a rupture of either the 15-cm or 40-cm pipeline and did not assume the release of a cloud of natural gas, which might drift toward safety-related SSC and ignite and burn (or explode).

On August 29, 1989, the NPP licensee notified the NRC that the plant would be shut down for good. In November 1990, the licensee submitted a decommissioning plan for NRC review. While reviewing the decommissioning plan, the NRC became aware that plans to introduce natural gas at FSV as part of a proposed gas-fueled repowering of the plant could lead to an unreviewed accident.

During this review, the NRC realized that the licensee had not taken due account of the existing gas hazard. The licensee took corrective actions to limit the amount of natural gas that could be released from a large rupture of the 15-cm collection pipeline. These actions consisted of closing a 15-cm valve and opening a 4-cm bypass valve in the line that conveyed gas from the wells to the 40-cm collection pipeline. This arrangement would limit the gas leakage from a rupture in the 15-cm pipeline by limiting the flow of gas back from the 40-cm pipeline.

The NRC concluded that, with the corrective actions implemented, potential gas accidents on site would not impact the safe storage of spent fuel or other radioactive components. Eventually the NRC issued the FSV possession only license (POL) amendment on May 21, 1991. This clause consisted of a license condition requiring the licensee to seek NRC's approval before making changes to the gas collection system that would involve the opening of the 15-cm valve.

Following reactor decommissioning, FSV was converted to a gas-fired plant. The first natural gas turbine was commissioned in 1996. Two more turbines were constructed by 2001. Heat recovery steam generators allow the plant to operate in combined cycle: In this operational mode, waste heat extracted from combustion-turbine exhaust gases produces secondary steam, which operates the plant's original turbine and power generator. As of 2011, the nominal power of the plant was 965 MW.

According to Lochbaum (2011), there is a similar situation at the three-unit Indian Point NPP, with 65-cm and 75-cm diameter gas pipelines some 120 m from the reactor buildings.

In April 2019, Holtec and Indian Point licensee Entergy agreed that Entergy would transfer Indian Point license to a Holtec International subsidiary for the purposes of decommissioning. Holtec submitted a license transfer application to the NRC in November 2019, and the Commission is currently reviewing the application. Holtec plans to move all spent fuel from Units 2 and 3 into dry cask storage by 2024. As Unit 1 was shut down in 1974 all of its spent fuel is already in dry cask storage. Unit 2 was shut down on April 30, 2020. Unit 3 is expected to shut down by April 30, 2021. According to Holtec's plan, the demolition of all three units should be completed by 2032, with site remediation completed by 2033. In September 2020 the NRC stated that the gas pipelines is not a danger to the plant (Mid Hudson, 2020).

7.45.2 Lessons learned

While not being a concrete event, the aforementioned FSV case study highlights the explosive and other hazards associated with natural gas wells and pipelines on or near a nuclear site. The NRC positions clearly show that these hazards remain even for a facility in a shutdown or decommissioning state.

Like gas wells and pipelines, the presence of oil or other flammable petroleum products including depots, terminals, pumping stations, and so on causes significant risks to a nuclear plant nearby (even in a shutdown/decommissioning state). There is a vast database of fire and explosion accidents that have occurred frequently in oil depots worldwide, which led to many casualties, severe environmental pollution, and heavy economic losses. Case studies of 435 fire and explosion accidents of oil depots in China in the period 1951–2013 are evaluated in Zhou (2016). From the analysis of various influencing factors, it is statistically apparent that the most dangerous plant zone is the loading/unloading operation area, and the most sensitive facilities are the storage tanks. The vapor cloud explosion could be the most common accident mechanism in oil depots, and management appears to be responsible in all the accident causes.

7.46 Large spread of radioactive materials, CEA-Marcoule vitrification facility, France; July 23, 2009 (IRSN, 2012)

7.46.1 Decommissioning process; problem encountered; impacts

The Marcoule vitrification facility (MVF) is located within Marcoule classified installation. The event in question produced a large spread of active liquids inside the facility due to a chemical reaction and poorly managed facility configuration. Although this event occurred at a classified (military) installation, the French Regulatory Body presented it to the general public because many generic lessons learned were applicable to the civilian facilities: For example, the chemical process that originated the incident is commonly used at any nuclear facility for cleaning up contaminated equipment.

The mission of the MVF was to accept, house, and vitrify HLW solutions. It also stores and monitors the vitrified containers. The event described in the following is referred to the facility's fission products storage section. Its tanks have been used to store concentrated solutions of fission products before vitrification since the UP1 spent fuel treatment plant started operation in 1958. After this plant was decommissioned in 1997, part of the radioactive inventory of the tanks was removed, and the tanks were rinsed with nitric acid (HNO_3). To continue the tank decontamination in preparation for their dismantling, the responsible organization conducted further rinsing with specific reagents, in particular sodium carbonate solutions. The products of sodium carbonate rinses need to be acidified before being concentrated and vitrified in the MVF itself.

The abnormal event presented here involved the acidification of carbonate solutions by mixing them with concentrated HNO_3. This exothermic reaction is complete, provided the reactants are constantly stirred and homogenized. The kinetics of the reaction is fast, and large amounts of carbon dioxide gas are produced. Measures must therefore be taken to avoid overpressure.

For the day of the accident, the plant management had scheduled to acidify 30 m^3 of carbonate solutions by moving them into a tank in five subsequent batches. During the fourth batch, a sudden release of carbon dioxide caused the tank pressure to increase and induced an "airlift" mechanism whereby the solution was lifted by the pressurized gas and some 900 L were ejected from the tank. The overflowed liquid poured into another tank via overflow pipes and into the room environment by overflowing the tank via an experimental compartment adjusted to this tank. This compartment had been designed for the insertion and retrieval of corrosion coupons that could be positioned in contact with the tank contents.

The solution spread from the room to other rooms and then to collector dishes and floor drains from where it ultimately ran into two recovery tanks. Around 60 L of solution spread beyond the containment systems and contaminated surrounding rooms. Although this event significantly contaminated the MVF, it had no direct impacts on the workforce or the environment. This event was rated as INES level 1.

7.46.2 Analysis; solutions

Following investigation, the responsible organization determined that the event was caused by an unnoticed and uncontrolled accumulation of two reagents (carbonate solution and concentrated HNO_3) that had not reacted. The stratification of the two reagents was enabled by their different density and the density of the reaction solution which stayed between the two, combined with inadequate stirring of the tank. When the two reagents finally mixed, the reaction occurred massively and quickly, producing carbon dioxide faster than the tank ventilation could absorb. The organization had not considered this hazard in the risk assessment. Following this event, operations involving acidification of carbonate solutions were suspended.

On the basis of a detailed analysis of the event, measures were taken to correct the failures observed, with a view to restarting safe acidification activities. A series of tests was carried out to prove the safety of the planned modifications. All these measures were reviewed by the Regulatory Body (IRSN).

7.46.3 Lessons learned

Firstly, the investigations showed that the tank's stirring system was not suitable for the objective. This system is designed to flush out sediments and prevent them from building up at the bottom of the tank but does not provide effective constant mixing of two solutions, especially when there is a significant density difference between them. Moreover, introducing the carbonate solution over the acid solution favored their stratification.

The massive spread of liquid throughout the MVF was due to some of it flowing out via the experimental compartment located in the upper tank. This compartment was not sealed and represented a flaw in the containment barrier that had not been noticed in the safety assessment. The affected tank was the only one of the four acidification tanks that was equipped with such feature. The compartment was not identified in the fission products storage safety report and was not indicated on the tank schematic.

This event was therefore caused by inadequate planning of the acidification process. According to IRSN, the event feedback proves the need to specifically verify that the equipment being proposed is fit for other uses than it was designed for, especially when decontamination is introduced. This confirmation may require ad hoc tests. The event stresses the importance of checking the current state of systems and equipment prior to use: This is especially relevant in older plants where records could be missing or incomplete.

The event assessment also identified the inadequate monitoring of the acidification process. Increased surveillance of relevant parameters (especially the solution density) could have signaled an accumulation of carbonate solutions in the tank. This inadequate monitoring results from of the failure to identify reactant stratification in the prejob risk analysis.

The event feedback is a reminder that all chemical reactions that could have a safety impact must be monitored using relevant parameters to ensure that the reactions proceed as planned and identify any anomalies promptly.

7.47 Wildland fire, Hanford site, Washington, United States; June–July 2000 (DOE, 2000)

It is your concern, when your neighbor's wall is on fire (Horace 65-8 BC), Epistles.

7.47.1 Decommissioning process; problem encountered; impacts

Unlike the Chernobyl case (Section 6.3.43 based on the information available at the time of writing), the direct cause of this event was identified with certainty.

As described elsewhere in this book, Hanford played since the 1940s a key role in US national defense for plutonium production purposes. Today, Hanford site has a new mission—environmental cleanup. Under contract to DOE, many private companies work at Hanford to safely decontaminate and demolish the site's old facilities and manage the legacy waste.

On June 27, 2000, a car and a tractor-trailer crashed on Washington State Route 24 near the DOE's Hanford site. The fire generated by the vehicle accident rapidly reached the vegetation on either side of the road. The local vegetation is typical of an arid shrub-steppe environment and is prone to catching fire. Plentiful natural fuel and hot, dry weather conditions pushed the fire quickly across the Ecology Reserve southwest of the central Hanford site and part of the Hanford Reach National Monument. On the following day, the fire passed State Route 240, directly threatening the central Hanford installations. Over 4 days, the wildland fire consumed nearly 800 km^2 of public and private lands. On average, the fire spread at an average speed of some 20 km/h.

When the emergency responders arrived, they found that a significant number of vehicles had blocked both lanes of the road, complicating the firefighters' efforts. Defensive barriers were cut along major roads on the Hanford site and the firefighters (first from Hanford Fire Department (HFD) and then provided by off-site resources) managed to prevent the fire from reaching the Hanford installations. Air tankers and helicopters contributed to blaze containment by dropping enormous amounts of water and fire retardant. At the event peak, the Washington State Governor declared an Emergency, which mobilized all available resources. Toward the closure of the accident, a radiological survey was conducted to assess unused laboratory facilities that had been overrun by fire. The main facilities were intact, although a nearby trailer and metal store had been destroyed. Neither structure stored any radioactive or hazardous substances. These were the only structural damaged sustained on the Hanford site. A detailed chronology of the wildfire is given in DOE (2000).

Wildland fires occasionally occur on or near the Hanford site, many due to lightning bolts. The arid vegetation, jointly with the area's variable wind patterns, can produce wildfires that expand quickly. Since the 1984 large wildfire, the site has been subject to many small and a few large wildfires (Fig. 7.28).

7.47.2 Analysis; solutions

Hanford site managers have taken planning actions to minimize the potential for, and impacts of, a fire on the site, including, among others:

- Fire department pumper units and grass units are maintained in case of fires.
- Wildfire periodic training is imparted to firefighters and heavy equipment operatives.
- Chemical herbicides are sprayed in selected areas.
- Radiological control provisions for fighting fire in certain controlled areas.

FIGURE 7.28 Dried grasses and vegetation can carry wildfire quickly through the landscape. *(Joshua Tree National Park, Wiki Commons).*

- Maintenance of defensible areas around the installations. Defensible areas include green grass and concrete, tarmac, or stone areas clear of vegetation and other combustibles.
- Control of deep rooted vegetation on radioactive disposal sites and eradication of tumbleweeds by controlled burning.
- Agreements have been stipulated with off-site organizations and authorities to request and obtain assistance as needed.
- Radio repeaters are available for an emergency, and telephone lists are available for heavy equipment and weather information providers.
- Fire danger signs are regularly updated.
- Inventory of natural combustibles is kept up-to-date.

The Investigation Board concluded that the HFD response to the event was effective and timely. The fire was a direct and unavoidable result of a vehicle accident. The HFD management was soon aware of the fire severity and gathered all local resources. Later on, when the dimensions of the fire grew, HFD was quick to alert and get assisted by off-site parties and state resources.

The Board also regarded the emergency response of other site staff as timely. The early release of nonessential staff from Hanford facilities was proactive, allowing decrease the health effects to the workforce and provide less crowded access to firefighters and other emergency resources.

Robust and proactive fire planning, including fire-safe designs, vegetation control, and fire barriers, helped to successfully protect Hanford installations. Vegetation management on Hanford waste sites and radiologically controlled areas minimized the release of airborne activity during the fire.

7.47.3 Lessons learned

Regardless the overall satisfactory appraisal of the wildfire management, the Investigation Board identified a number a weaknesses worth improvement as follows (author's selection):

- The existing emergency response processes related to Hanford events impacting the Washington State and national systems, as well as state and national events impacting Hanford systems, did not all perform as expected. Interagency coordination was not ideal.
- Agreements with off-site organizations (authorities and companies) and non-DOE site tenants did not always exist. Existing agreements were not sufficiently detailed, which raised issues that had to be solved during the event.
- The lack of maintenance of defensible firebreaks along state roadways running through the Hanford site allowed the fire to expand rapidly. Hanford organizations should liaise with clean air authorities, state regulators, the DOE-HQ Office of Environmental Health, and the Washington State Department of Transportation to select the most effective ways of installing and maintaining defensible areas along state road shoulders.
- A helicopter fire suppression assistance that had been previously used was not available for this event, and it was not incorporated into the overall fire management because no agreement had been defined with the party providing the assistance.
- The process for collection and review of radiological data during the fire was not formalized, which caused inadequacies of the environmental monitoring and public dose assessment. No process was established for continuing radiological monitoring after the source of airborne radioactivity has been arrested and for coordination and analysis of radiological data.
- There was no formal process to request and use off-site personnel during emergency activities.
- HFD had not addressed in detail the needed resources for large wildfires, and additional resources had to be obtained through the National Interagency Fire Centre.
- Hanford planning did not provide enough data to outside agencies prior to the fire. Information on radiological and chemical hazards had to be forwarded during the fire. Planning did not evaluate that such hazards would require air support.
- The process for communicating basic personnel information was not effective during this event.
- A clear process to obtain off-site specialists and internal volunteers was not in place.

- Site's emergency procedures did not consider how to utilize subject matter experts (e.g., in air operations and fire protection). The emergency management process did not characterize the event for associated hazards, access the need for technical support, obtain resources, or reassess evolving issues.
- The process for bringing equipment in from off-site for emergency response was not formalized.
- The review and approval process for news releases was not in place.
- Traffic control (both onsite and offsite) was not well coordinated. Likewise, onlooker control was not well coordinated.
- Procedures for closure of airspace were not flexible enough to enforce or remove interim flight limitations at a specific time.
- Some corrective actions resulting from a former (1984) Hanford fire proved inadequate or were not performed.

References

Anagnostopoulos, H., March 1995. Cold weather effects on Dresden Unit 1. In: Proceedings of the Third International Workshop on the Implementation of ALARA at Nuclear Power Plants, Long Island, NY (United States); 8-11 May 1994, NUREG/CP 0143. https://www.osti.gov/servlets/purl/54640.

Argonne National Laboratory, 1998. Decontamination and Dismantlement of the Building 200/205 Pneumatic Transfer Tube at Argonne National Laboratory-East Project Final Report, ANL/D&D/98-5. https://www.osti.gov/servlets/purl/12331.

BBC News, April 14, 2020. Chernobyl Fire under Control, Ukraine Officials Say. https://www.bbc.com/news/world-europe-52277414.

BBC News, July 6, 2015. Dounreay Operator Resumes Work at Scene of Fire. https://www.bbc.com/news/uk-scotland-highlands-islands-33408238.

Berkeley City Council, 2005. Demolition of B51 and the Bevatron at Lawrence Berkeley National Laboratory. (LBNL), Milvia St, Berkeley, CA 94704, US, p. 2180.

Building Operators Association (BOA) of Canada, Boiler Room Routines (Part 2), 1998. https://www.boacalgary.com/artboilerroom4.html.

CH2M Hill, 2018a. Plutonium Finishing Plant Work Resumption Plan, CHPRC-03689. https://www.emcbc.doe.gov/SEB/CPCC/Documents/Document%20Library/3rd_Posting/Other/Plutonium%20Finishing%20Plant%20Work%20Resumption%20Plan%20-%20(CHPRC-03689,%20Rev.%200).pdf.

CH2M Hill, July 12, 2018b. Plutonium Finishing Plant Resumption of Work. https://www.hanford.gov/files.cfm/July_2018_PFP_Briefing.pdf.

CH2M Hill, March 5, 2018c. Discovery of Contamination Spread at the Plutonium Finishing Plant during Demolition Activities, EM-RL–CPRC-PFP-2017-0018, CR-2018-0022 (DRAFT Rev. 2), Root Cause Evaluation Report. https://www.hanford.gov/files.cfm/Attachment_1_-_CR-2018-0022__Report_Final_Draft_R2_03-05-18.pdf.

Germany's Federal Office for the Safety of Nuclear Waste Management, Monthly reports on reportable events https://www.base.bund.de/EN/ns/events/reports/monthly/monthly_node.html (in German, Bundesamt for die Sichereit der nuklearen Entsorgung, Monatsberichte zu meldepflichtigen Ereignissen. www.base.bund.de/DE/themen/kt/stoerfallmeldestelle/berichte/monatsberichte/monatsberichte_node.html). Please note that the monthly reports are only in German.

International Atomic Energy Agency, 2003. Decommissioning of Small Medical, Industrial and Research Facilities, Technical Reports Series No 414. IAEA, Vienna.

International Atomic Energy Agency, 2006a. Decommissioning of Research Reactors: Evolution, State of the Art, Open Issues, Technical Reports Series No 446. IAEA, Vienna.

International Atomic Energy Agency, 2006b. Decommissioning of Underground Structures, Systems and Components, Technical Reports Series No 439. IAEA, Vienna.

International Atomic Energy Agency, 2006c. Management of Problematic Waste and Material Generated during the Decommissioning of Nuclear Facilities, Technical Reports Series No. 441. IAEA, Vienna.

International Atomic Energy Agency, 2009. Integrated Approach to Planning the Remediation of Sites Undergoing Decommissioning, Nuclear Energy Series No. NW-T-3.3. IAEA, Vienna.

International Atomic Energy Agency, 2016. Managing the Unexpected in Decommissioning, Nuclear Energy Series No NW-T-2.8. IAEA, Vienna.

International Atomic Energy Agency, 2018. Lessons Learned from Deferred Decommissioning of Nuclear Facilities, Nuclear Energy Series NW-T-2.11 IAEA. Vienna.

International Atomic Energy Agency, News, November 12, 2019. Loss of liquor from the Magnox Swarf Storage Silo original building primary containment. https://www-news.iaea.org/ErfView.aspx?mId=f35a2320-44dc-49d7-a425-ed7714b77f69.

Institute for Radiological Protection and Nuclear Safety (in French, Institut de Radioprotection et Sûreté Nucléaire, ISRN), 2008. Environmental Surveillance and Groundwater Mapping of Ground and Surface Water around Nuclear Sites and Old Radioactive Waste Repositories (In French, Etat de la Surveillance Environnementale et Bilan du Marquage des Nappes Phréatiques et des Cours d'eau autour des Sites Nucléaires et des Entreposages anciens de Déchets Radioactifs). https://www.irsn.fr/FR/expertise/rapports_expertise/Documents/environnement/IRSN_Rapport-HCTISN_150908_maj131108.pdf.

Institute for Radiological Protection and Nuclear Safety, 2012. (in French, Institut de Radioprotection et Sûreté Nucléaire, ISRN), Safety of Laboratories, Plants, Facilities Being Dismantled, Waste Processing, Interim Storage and Disposal Facilities-Lessons Learned from Events Reported in 2009 and 2010, DSU Report 248. https://www.irsn.fr/EN/publications/technical-publications/Documents/IRSN_Report_LUDD-events_2009-2012_DSU248.pdf.

Joint Convention on the Safety of Spent Fuel Management and on the Safety of Radioactive Waste Management, Second Italian National Report, 2008. https://www.isinucleare.it/sites/default/files/contenuto_redazione_isin/jc_second_italian_national_report2008.pdf.

Knoxblogs Atomic City Underground, 2010. ORNL's Hot Spot: Building 3026. http://knoxblogs.com/atomiccity/2010/08/20/ornl_hot_spot_building_3026/.

Knoxblogs Atomic City Underground, 2012. Tank W-1A is out of the ground! Radioactive saga at ORNL approaches an end. http://knoxblogs.com/atomiccity/2012/01/09/tank_w-1a_is_out_of_the_ground/.

Knoxblogs Atomic City Underground, 2015a. IG identifies More Than 200 High-Risk Facilities That Are Dirty and Degraded; 'Worst of Worst' is at Y-12. http://knoxblogs.com/atomiccity/2015/01/28/ig-identifies-200-high-risk-facilities-dirty-degraded-worst-worst-y-12/.

Knoxblogs- Atomic City Underground, 2015b. 75000 gallons of rad-hot sludge and counting. http://knoxblogs.com/atomiccity/2015/12/30/75000-gallons-of-rad-hot-sludge-and-counting/.

Knoxville News Sentinel, January 18, 2016. Oak Ridge DOE's Big Problem: Hazards Outpace Funding for Old Nuke Buildings. https://eu.knoxnews.com/story/news/local/2016/01/18/oak-ridge-does-big-problem-hazards-outpace-funding-for-old-nuke-buildings/90818728/.

Laraia, M., June 23–25, 2015. Safety Culture, the New Frontier of Decommissioning, International Conference ECED 2015 – Eastern and Central European Decommissioning. Trnava, Slovakia.

Laraia, M., June 2017. In: Laraia, M. (Ed.), Chapter 1 "Introduction" of "Advances and Innovations in Nuclear Decommissioning". Woodhead Publishing Ltd. ISBN: 978-0-08-101122-5 (print).

Laraia, M., 2019. Beyond Decommissioning- the Reuse and Redevelopment of Nuclear Installations. Elsevier. ISBN 978-0-08-102790-5.

Lochbaum, D., May 10, 2011. Fission Stories #39: Ripley Wouldn't Believe it. https://allthingsnuclear.org/dlochbaum/fission-stories-39-ripley-wouldnt-believe-it.

Lochbaum, D., December 13, 2013. Critique of the Analysis of Safety and Environmental Risks Posed by Spent Fuel Pool Leaks in the NRC's Draft Waste Confidence Generic Environmental Impact Statement. https://www.nirs.org/wp-content/uploads/radwaste/exhibitb2013-12-13lochbaumfinaldeclarationrepoolleaks.pdf.

Mcgrath, R., Reid, R., March 18–22, 2018. Management on Hazardous Materials during Decommissioning, WM2018 Conference. Phoenix, Arizona, US. https://www.xcdsystem.com/wmsym/2018/pdfs/FinalPaper_18310_0117074107.pdf.

Mc Intyre, P.J., 2012. Nuclear decommissioning policy, infrastructure, strategies and project planning. In: Laraia, M. (Ed.), Nuclear Decommissioning-Planning, Execution and International Experience. Woodhead Publishing. ISBN 978-0-85709-115-4.

Mid Hudson News, September 23, 2020. NRC says AIM pipeline near Indian Point causes no danger. https://midhudsonnews.com/2020/09/23/nrc-says-aim-pipeline-near-indian-point-causes-no-danger/.

Muckenlhaler, F.J., 1992. Where have the neutrons gone - a history of the tower shielding facility. In: CONF-921102–27 Summary Presented to the American Nuclear Society 1992 Int. Conf., November 15-20. Chicago, IL, US.

New York State Department of Health, November 2012. Mercury Spill Incidents Data and Resources. https://www.health.ny.gov/environmental/chemicals/mercury/docs/spill_incidents.pdf.

Nuclear Decommissioning Authority and Sellafield Ltd, Work Begins on Second Emptying Machine at Sellafield Waste Silo, January 10, 2020. https://www.gov.uk/government/news/work-begins-on-second-emptying-machine-at-sellafield-waste-silo–2.

Oak Ridge Office of Environmental Management, July 2019. Tour of Department of Energy's Environmental Management Projects. https://www.energy.gov/sites/prod/files/2019/07/f65/2019%20OREM%20Tour%20Book%20Print%20Updated%207.18.19.pdf.

Office for Nuclear Regulation, December 2014. Enforcement Action in Response to a Fire in the Dounreay PFR Sodium Tank Farm, Project Assessment Report ONR-DFW-PAR-14-014, Rev. 0, 24. http://www.onr.org.uk/pars/2015/dounreay-14-014.pdf.

Office for Nuclear Regulation, 2016. Events Reported to the Nuclear Safety Regulator in the Period of 1 April 2001 to 31 March 2015. http://www.onr.org.uk/documents/2016/events-reported.pdf.

Office for Nuclear Regulation, 2020. Quarterly Statements of Nuclear Incidents at Nuclear Installations (Updated March 5, 2020). http://www.onr.org.uk/quarterly-stat/index.htm.

Ozharovsky, A. November 22, 2010 https://bellona.org/news/nuclear-issues/accidents-and-incidents/2010-11-comment-a-shut-down-ignalina-npp-no-rip-for-lithuanias-cranky-nuclear-corpse.

Parliamentary Acts- 8th Legislature – Discussions on the Judicial Communication Sent by the Sessa Aurunca (Caserta) Magistrate to the Director of the Garigliano Nuclear Power Plant for Failure to Comply with a 'Nuclear Emergency' Alert, Response Announced in the Session of 7-9-1981, p. 4358) (in Italian: ATTI PARLAMENTARI – VIII LEGISLATURA –Discussioni sulla Comunicazione Giudiziaria Inviata dal Pretore di Sessa Aurunca (Caserta) al Direttore della Centrale Nucleare del Garigliano per Mancata Denunzia di Emergenza Nucleare, Risposta Annunziata nella Seduta del 7/9/1981, pag. 4358) http://legislature.camera.it/_dati/leg08/lavori/stenografici/sed0365/rsi0365.pdf.

Sellafield Ltd Incident Reports and Notices, 2019 (Last updated December 10, 2019). https://www.gov.uk/government/collections/sellafield-ltd-incident-reports-and-notices.

S&P Global Platts, Radioactive Water Leak Detected at UK's Sizewell A Nuclear Plant, September 19, 2011. https://www.spglobal.com/platts/en/market-insights/latest-news/electric-power/091911-radioactive-water-leak-detected-at-uks-sizewell-a-nuclear-plant.

Texas A&M University, 2012. Research Compliance & Biosafety. Laboratory Decommissioning. https://rcb.tamu.edu/biohazards/resources/laboratory-decommissioning.

The Christian Science Monitor, April 10, 1995. Leak at Nuclear Plant Raises Concern. https://www.csmonitor.com/1995/0410/10041.html.

Tri Cities Business News, May 2019. Risks Linger with Plutonium Finishing Plant's Demolition. https://www.tricitiesbusinessnews.com/2019/05/hanford-pfp/?cn-reloaded=1.

United States Government Accountability Office, January 2020. Hanford Cleanup-DOE Should Take Actions to Improve Inspections and Oversight of Contaminated Excess Facilities, GAO-20-161. https://www.gao.gov/assets/710/703959.pdf.

US Department of Energy, U.S, October 2000. Department of Energy Response to the 24 Command Wildland Fire on the Hanford Site — June 27-July 1, 2000. https://www.energy.gov/ehss/downloads/type-b-accident-investigation-response-24-command-wildland-fire-hanford-site-june-27.

US Department of Energy, 2007. Don't Let Your Guard Down When Preparing Facilities for Decommissioning and Decontamination, Lesson ID: 2007-CH-BNL-ERP-001 (Available upon DOE authorization).

US Department of Energy, 2014. Keep One Eye on the Past and Two Eyes on the Present, Lessons Learned Data Base No: RCCC-2014-0013 (Available upon DOE authorization).

US Department of Energy, June 19, 2017a. Office of Environmental Management, Richland Operations Office Focuses on Stabilizing PUREX Tunnel. https://www.energy.gov/em/articles/richland-operations-office-focuses-stabilizing-purex-tunnel.

US Department of Energy, November 21, 2017b. Oak Ridge Breaks Ground on Facility Enabling Large-Scale Y-12 Cleanup. https://www.energy.gov/em/articles/oak-ridge-breaks-ground-facility-enabling-large-scale-y-12-cleanup.

US Department of Energy, 2018a. Plutonium Finishing Plant Demolition Resumption Expert Panel- Final Report. https://www.hanford.gov/files.cfm/Plutonium_Finishing_Plant_Expert_Panel_Final_Report.pdf.

US Department of Energy, 2018b. Water Heating Elements Melt and Ignite Poly Holding Tank, Occurrence Report Number: EM-RP–WRPS-TANKFARM-2018-0007 (Available upon DOE authorization).

US Department of Energy, November 12, 2019a. EM Addresses Risks at ORNL Facilities, Preparing for Next Cleanup Phase. https://www.energy.gov/em/articles/em-addresses-risks-ornl-facilities-preparing-next-cleanup-phase.

US Department of Energy, July 16, 2019b. Operating Experience Committee Webinar, External Operating Experience Presentation: Wildfires, R. Natoli. https://orau.org/ihos/downloads/meetings/support-files/2019/wildfires_opex_DOE.pdf.

US Department of Energy, February 26, 2019c. Lessons Learned on Demolition of a Mercury Contaminated Facility. https://www.dndkm.org/DOEKMDocuments/LessonsLearned/112-LLB_Feb_2019_Demo_Mercury_Facility_02.26.19.pdfrowhead.

US Department of Energy, July 2019. Final Supplemental Environmental Assessment for the Wildfire Hazard Reduction and Forest Health Improvement Program at Los Alamos National Laboratory, Los Alamos, New Mexico, DOE/EA-1329 - S1. https://www.energy.gov/sites/prod/files/2019/07/f65/final-ea-1329-s1-wildfire-hazard-reduction-forest-health-improvement-program-07-2019.pdf.

US Department of Energy, February 11, 2020a. Hanford crews complete demolition of PFP's main processing facility. EM Update 12 (6). https://content.govdelivery.com/accounts/USDOEOEM/bulletins/27a4a09#link_1.

US Department of Energy, February 2020b. Office of Environment, Health, Safety and Security, Operating Experience Summary, ORPS Data Indicates Work Process Deficiencies at Active Deactivation and Decommissioning Projects, OES: 2019-03. https://www.energy.gov/sites/prod/files/2020/03/f72/OES_2020-01.pdf.

US Nuclear Regulatory Commission, March 5, 1991a. NRC Information Notice No. 91-14: Recent Safety-Related Incidents at Large Irradiators. http://pbadupws.nrc.gov/docs/ML0311/ML031190579.pdf.

US Nuclear Regulatory Commission, October 3, 1991b. NRC Information Notice 91-63: Natural Gas Hazards at Fort St. Vrain Nuclear Generating Station. https://www.nrc.gov/docs/ML0311/ML031190130.pdf.

US Nuclear Regulatory Commission, January 24, 1997. Removal of RTI, Inc., from Site Decommissioning Management Plan, SECY-97-019. http://pbadupws.nrc.gov/docs/ML9929/ML992920113.pdf.

US Nuclear Regulatory Commission, 2019a. West Valley Demonstration Project (Last Reviewed/Updated, November 05, 2019). https://www.nrc.gov/info-finder/decommissioning/complex/wv.html.

US Nuclear Regulatory Commission, 2019b. Dresden Unit 1 (Last update November 19, 2019). https://www.nrc.gov/info-finder/decommissioning/power-reactor/dresden-nuclear-power-station-unit-1.html.

US Nuclear Regulatory Commission, 2020. Backgrounder on Commercial Irradiators (Last update February 2020). http://www.nrc.gov/reading-rm/doc-collections/fact-sheets/commercial-irradiators.html.

Washington State Department of Ecology, Hanford's PUREX Tunnels: It's Time for the Next Chapter, April 30, 2019. https://ecology.wa.gov/Blog/Posts/April-2019/Hanford-s-PUREX-tunnels-It-s-time-for-the-next-cha.

Webb, G.A.M., et al., March 2006. Classification of events with an off-site radiological impact at the Sellafield site between 1950 and 2000, using the international nuclear event scale. J. Radiol. Prot. 26 (1), 33–49. https://www.ncbi.nlm.nih.gov/pubmed/16522943.

World Health Organization, 2019. Decommissioning of Medical Devices (WHO Medical Device Technical Series), Licence: CC BY-NC-SA 3.0 IGO, Geneva. https://apps.who.int/iris/bitstream/handle/10665/330095/9789241517041-eng.pdf.

World Information Service on Energy, 2020. Uranium Project, Uranium Enrichment and Fuel Fabrication - Decommissioning Issues (Europe) (Last updated March 25, 2020). https://www.wise-uranium.org/edeur.html.

World Nuclear News, June 12, 2009. Old Nuclear Event in the Open. https://www.world-nuclear-news.org/RS_Old_nuclear_event_in_the_open_1206091.html.

World Nuclear News, June 17, 2013. Fine for Misdirected Waste Shipment. https://world-nuclear-news.org/Articles/Fine-for-misdirected-waste-shipment.

World Nuclear News, May 11, 2017. Hanford Emergency Ends as Collapsed Tunnel Sealed. http://www.world-nuclear-news.org/RS-Hanford-emergency-ends-as-collapsed-tunnel-sealed-1105175.html.

Wyatt, D.M., 1994. Fire protection for inactive contaminated structures, WHC-SA-2288-FP. In: Waste Management Conf., Tucson, Arizona, February 27-March 3. https://www.osti.gov/servlets/purl/10139393.

Zhou, Yi, Zhao, X., Jianyu, Z., Du, C., 2016. Research on fire and explosion accidents of oil depots, AIDIC. Chem. Eng. Trans. 51. https://www.aidic.it/cet/16/51/028.pdf.

Chapter 8

Conclusions

As shown by individual cases reviewed in this book and by conclusions based on experience of large nuclear programs (e.g., Sellafield and other UK nuclear installations), the radiological impacts of decommissioning incidents are generally so low that they do not even qualify as accidents. In INES terms (Section 4.2), decommissioning events range from irrelevant to level 2 max. Quoting from ONR (2016a):

- "Of the 3866 events reported to ONR during the period covered by this report (April 1, 2001, to March 31, 2015), the nuclear safety significance of 3857 (more than 99.7%) was very low (rated at or below level 1—anomaly—provided by the seven-level International Nuclear and Radiological Event Scale [INES]).
- Eight events, only one of which has occurred since 2009, were rated at the next significance level of the scale (INES level 2—incident).
- One event—that took place approximately 10 years ago—was of sufficient significance to merit an INES level 3 rating (serious incident); (*author's note: this event occurred in a still operating installation*).
- None of the more significant events reported had any detrimental effect on public safety or the environment."

In Germany, the officer responsible for radiation protection at Berufsgenossenschaft Energie Textil Elektro Medienerzeugnisse (BG ETEM, in English: Professional Association for Energy, Textiles and Electrical Media Products) stated in an interview: "Under the German Ordinance on Radiation Protection, a radiation accident is an occupational incident which may have exposed a worker to a dose of over 50 mSv (millisievert). To date, no such incident has occurred during decommissioning work" (DGUV, 2019).

A recent Korean study takes into account (as this book does) both radiological and nonradiological accidents in the course of nuclear decommissioning (Park et al., 2019). This study defines four risk classes (Table 8.1, elaboration from Table 6 of Park et al., 2019)

Nuclear Decommissioning Case Studies. https://doi.org/10.1016/B978-0-12-823700-7.00008-4
Copyright © 2021 Elsevier Inc. All rights reserved.

TABLE 8.1 Risk classes in Korean study.

Risk class (IV the lowest, I the highest)	Safety requirements
IV	No safety measure
III	Safety management program by operating organization
II	One independent safety measure
I	Two independent safety measures

Overall, 80 potential accidents were reviewed in this study: It was determined that 62.5% of the potential accidents were classified as the risk class IV, 28.75% as the risk class III, 7.5% as the risk class II, and 1.25 (1 case) as the risk class I. The potential accident with the highest risk level referred to the removal of nonradioactive materials, e.g., fuels, lubricating oil, 1,1,1-trichloroethane, laboratory chemicals, lead, mercury, paints, and battery acid. The dismantling phase of the decommissioning process presents the highest risk not only because it has the only accident ranked in risk class I but also because it includes most events classified in risk class II. Actually, most activities performed in the dismantling phase are subjected to nonradiological hazards. The analysis shows that accidents with high risk are much more probably related to nonradiological hazards than to radiological hazards. This conclusion is supported in full by the 170+ events described in this book, which clearly show that the only events resulting in fatalities or serious injuries are nonradiological events.

An additional confirmation to this statement is given in NRC (2002). This study indicates that the radiological occupational impacts are neither detectable nor destabilizing. A generic conclusion can be drawn that for all decommissioning plants, the potential radiological issues are small. This study has also considered mitigation measures and concludes that no mitigation measures additional to those in standard use today are likely to be sufficiently beneficial to be warranted. No individual radiological impact from any occurrence described in this book has ever reached deterministic effects, and stochastic effects are relatively minor. However, some cases of doses approaching or even exceeding annual limits are given in this book and may have resulted in work restrictions or transfers (see Sections 4.1.2, 4.1.4, 6.25, 6.28, 6.29, 6.30, etc.). Remarkably, most of these events resulted in internal exposures due to alpha emitters (transuranics): Mechanisms were inhalation or ingestion through punctures. The highest exposure mentioned in this book (Section 6.25) was 80 mSv (against an annual limit of 20 mSv), well below deterministic impacts and within a relatively safe zone in terms of stochastic risks.

Regarding accidental impacts to the public the fact that most plants have reduced or canceled offsite emergency plans is a clear indication that expected impacts are relatively minor. Two examples follow.

San Onofre NPP's Permanently Defueled Emergency Plan states that a radiological emergency beyond the site boundary is no longer credible. The plan abrogates the two highest NRC emergency levels and maintains NRC supervision for the two lowest emergency levels. San Onofre retains emergency response staff and communications with off-site emergency responders (NRC, 2015).

In the UK, the Office for Nuclear Regulation (ONR) announced in 2016 that there is no longer a requirement for an off-site emergency planning area in the surroundings of the Harwell nuclear site. The ONR determined that there has been a considerable reduction in the site radiological hazard following the shutdown of the reactors in the 1990s, decommissioning work, and the treatment/removal of radioactive waste. Therefore, a radiological emergency that necessitates an external emergency plan is no longer realistic at Harwell (ONR, 2016b).

Several NRC studies demonstrate that even fuel-related accidents at shutdown NPPs do not pose serious radiological risks to public and the environment (as *an argumentum a fortiori* most events assessed in this book refer to defuelled plants, a much less risky situation).

In summary, work in nuclear facilities appears to be much safer than in general industry. Nuclear facilities cannot afford even minor accidents; they accept self-regulation; they are regulated much more severely; and they have adopted a safety culture that should be a model for other high risk industries such as construction. The Three Mile Island disaster in the United States was a watershed that led to the recognition that a general industry approach to safety was not nearly sufficient for nuclear work. This led to the establishment of INPO, and when the NRC finally adopted a "nuclear safety culture" framework in 2011, the transformation from high risk to exemplary safety was complete (Ringen, 2020).

In general, human health risks for most decommissioning options are expected to be dominated by occupational, nonradiological injuries to workers engaged in activities such as construction, dismantling, and excavation.

Historically, actual injury and fatality rates at nuclear reactor facilities have been lower than the average US industrial rates (NRC, 2002). However, (NRC, 2002) dates back to an era when decommissioning activities are at much lower scale than today, and no segregation was made between various phases of a nuclear facility's lifetime (construction, operations, and decommissioning). Besides, (published) information directly comparing fatality and injury rates in nuclear decommissioning against conventional industrial demolition is rarely available.

A 2011 study for the US Blue Ribbon Commission on America's Nuclear Future tried to answer the question: How safe is work in the nuclear fuel cycle? This study (Stoneturn, 2011) (1) reviewed radiation dosimetry data provided by the NRC from 1979 to 2010 and (2) compared injury and illness data for work in nuclear power facilities for both operations and maintenance with such data reported for fossil fuel, hydro, and alternative power plants, including data provided by contractors who perform maintenance work in both nuclear and nonnuclear facilities. The nuclear facilities had risk levels that were one-fifth of what was reported in the comparison plants (Ringen, 2020).

Another issue is due to the demolition sector being historically part of the overall construction sector. Those trying to collect information pertaining to accidents, incidents, or even person-hours within the demolition industry may find it very difficult to separate such statistics from the rest of the construction industry. As the demolition industry is large enough to stand apart from construction, regulations/recommendations meant to control the hazards of the construction industry do not fully respond to demolition priorities and activities. As one example, OSHA reported an overview of occurrences during the cleanup and recovery of the World Trade Centre after the terroristic attacks of September 11, 2001. OSHA "protected thousands of workers by overseeing the fitting and distribution of more than 130,000 respirators and handing out over 40,000 pieces of personal protective equipment and taking more than 6000 air and bulk samples ... there were no work-related deaths after the collapse of the two buildings and only 35 lost time injuries and a remarkably low rate of 2.3 injuries per 100 workers. Demolition sites normally have 4.3 lost time injuries" (US SENATE, 2002).

In fact, it appears that high-rise formwork and demolition is a high risk group—its workers have lost-time injury (LTI) rates that are two to three times as high as the average for all other construction activities (IHSA, 2011). In recent years, however, companies in the formwork and demolition industry have accomplished major steps toward lowering their injury rates. For example, demolition work has changed from mostly manual operations to a heavily mechanical process and from employment of basically unskilled labor to highly trained workers and supervisors. In parallel with these changes, the LTI frequency for demolition fell from 3.7 (2005) to 2.8 (2010)—yet higher that at most nuclear decommissioning projects (see the following). In any case it is expected that safety in nuclear decommissioning will benefit from the progress being achieved in conventional demolition.

Further comparisons between the nuclear decommissioning and nonnuclear construction, demolition, cleanup, and waste management can be attempted on the basis of sporadic data for individual facilities as it is attempted in the following.

Information from the Maine Yankee (MY) NPP decommissioning project is quoted in the following. MY was a commercial 900 MWe reactor that generated electricity from 1972 through 1997 when it was no longer economically viable to operate. MY can boast "Zero lost time injuries from

May 2002 through the end of decommissioning in October 2005. For every 200 000 hours worked, the decommissioning project had a Lost Time Incident Rate of 0.26 compared with a demolition industry rate of 3.6" (MY undated). Toward its end, "the project had completed over two million safe work hours without a lost time accident. Overall, the project has completed approximately 5.4 million hours with a recordable incident rate of approximately 2.3 per 200 000 hours worked" (EPRI, 2005). In comparison, US data for 2018 indicate for Heavy and Civil Engineering Construction a total recordable incident rate of 2.6 cases and 1.7 cases with days away from work, job restriction, or transfer. As a basis for the incident rate, 200,000 person-hours per year is equivalent to 100 equivalent full-time workers (working 40 h per week, 50 weeks per year): 100 workers can be the average crew at work in the nuclear decommissioning of a large installation (BLS, 2019).

A more accurate depiction of occurrence rates at US DOE sites is given in DOE (2020). Four sites have been selected as representative of major nuclear decommissioning programs. It can be readily seen that occurrence rates are low and in line with those given elsewhere in this section (Figs. 8.1–8.4). Data from the four DOE decommissioning sites also show a downward trend, doubtless due to the progressive implementation of better industrial safety policies. In the graphs in Figs. 8.1–8.4.

TRC Rate is the total recordable cases per 200,000 work-hours (approximately 100 person-years). TRC is the total number of work-related injuries or illnesses that resulted in either death, days away from work, days of restricted work activity, days of job transfer, or other recordable cases.

DART Case Rate is the DART (days away, restricted, or transferred) cases per 200,000 work-hours (approximately 100 person-years). DART Case is an injury or illness case where the most serious outcome of the case resulted in days away from work or days of job restriction or transfer.

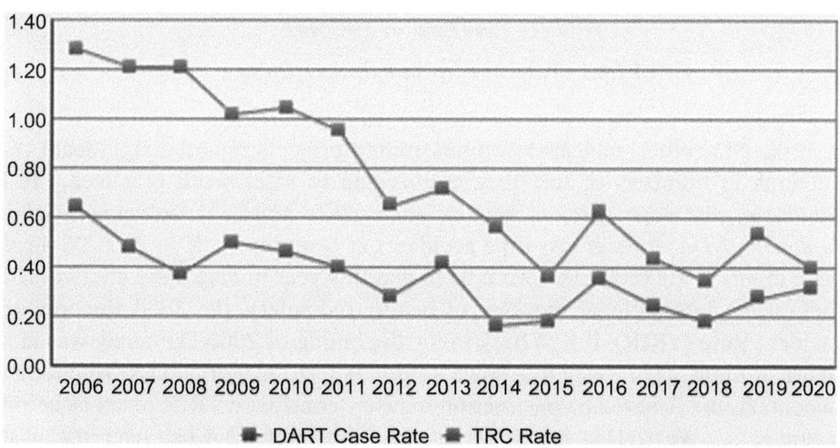

FIGURE 8.1 Hanford safety performance. *(Credit to DOE).*

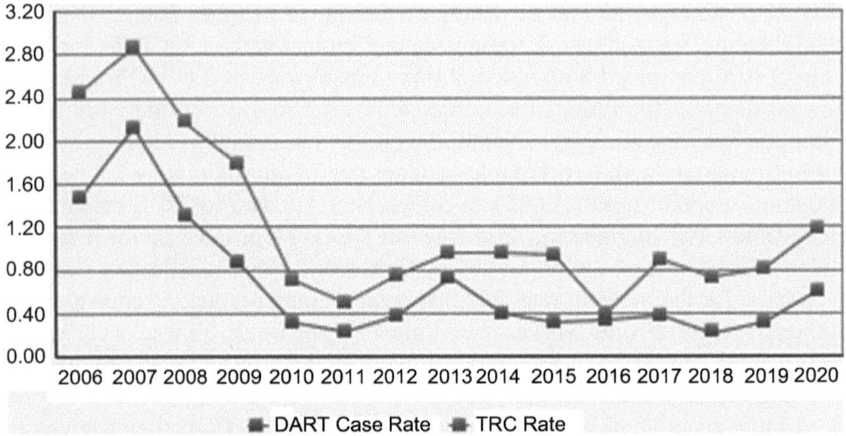

FIGURE 8.2 East Tennessee Technological Park safety performance. *(Credit to DOE).*

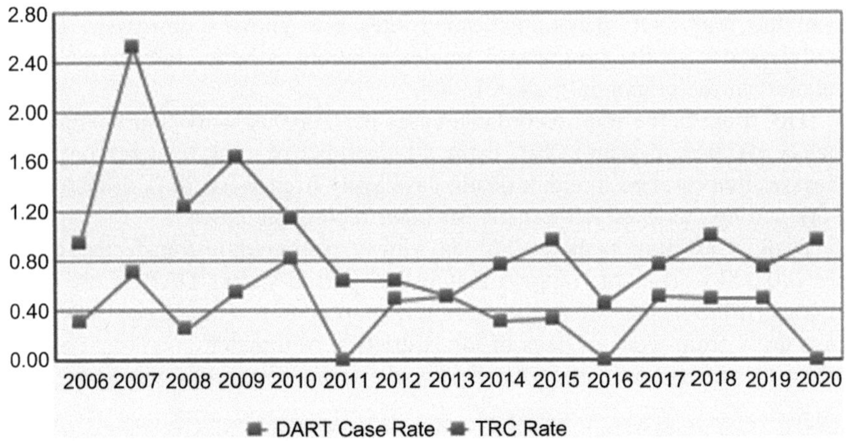

FIGURE 8.3 Paducah safety performance. *(Credit to DOE).*

Like MY, other nuclear decommissioning projects reported significant reductions in numbers of lost time events due to safer work practices. "If I (*author's note: the speaker here is from UK's Nuclear Decommissioning Authority, NDA*) look at loss time accidents at Dounreay, UK in 2005/06 there were eight, in 2006/07 there were three and this year to date there are two. If I look at the international measure of health and safety, the Total Recordable Incident Rate (TRIR), if I go back to the beginning of 2006 Dounreay was at a figure of three, as of today it is down around 0.3; 0.3 is world class. To put that in context, the American construction industry consider a TRIR of six as being average. … We *(NDA)* have five sites at the moment who, over the past

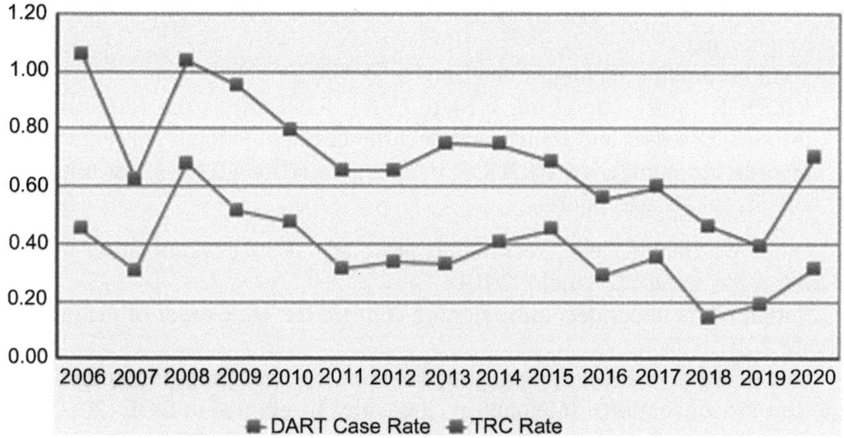

FIGURE 8.4 Savannah River safety performance. *(Credit to DOE).*

12 months, on a rolling 12 month basis, have a TRIR of zero. So there has been a profound change" (HC, 2008). For comparison, UK annual data for the construction industry in the same period indicate around 3000 nonfatal work-related incidents per 100,000 workers (or an incident rate of 3 for a crew of 100 workers—a good approximation for a decommissioning site) (HSE, 2019).

As well known, Sellafield, UK, is a "difficult" site due to congestion and old age of its many nuclear facilities, most of them being decommissioned. And yet even at Sellafield the reduction of significant radiological and non-radiological events is apparent over the period 2011−2018 (Sellafield, 2018). However, a couple of indicators have shown an increase in 2019. A selected list of indicator changes from 2011/12 to 2018/19 is given in the following (except where otherwise noted), namely (Sellafield, 2019):

- Nuclear Sellafield Incident Reports (from 6 to 0)
- Environmental Sellafield Incident Reports (SIR) (from 7 in 2011/12 to 1 in 2017/18; 20 in 2018/19). The increase in numbers of environmental SIRs in 2018/19 are in part due to changes in reporting criteria and to greater focus on visualizing potential impacts. None of the events in 2018/19 resulted in any remarkable impact to the environment. During an extensive site condition review, minor defects have been identified including a defect of the external ventilation ductwork in one of the decontamination facilities (February 2019). This ductwork conveyed the extract from the decontamination tanks, which have not been used for more than 2 years. This event, which prompted a full review of Sellafield's ventilation systems, was assigned an INES level 1 rating.

- Lost-time accident rate (from 0.34 to 0.20); the trend of this parameter is roughly flat.
- Total recordable incident rate (from 0.50 to 0.33).
- RIDDOR injury rate (from 0.14 to 0.05); RIDDOR is the Reporting of Injuries, Diseases and Dangerous Occurrences Regulations. This parameter records the number of RIDDOR events greater than 7 days lost-time accidents and major injuries.

Other parameters, e.g., recordable personal (skin) contamination events exhibit a flat trend (Sellafield, 2019).

Italian NPPs under decommissioning confirm the same order of magnitude (Table 8.2).

One of the few direct comparisons between the decommissioning industry and similar conventional (nonnuclear) industries is reported in DOE (2010) for the West Valley cleanup and decommissioning project. See Table 8.3.

The following data 2016–2020 were made available from the BR-3 NPP decommissioning project, Belgium (Massaut, 2020). Earlier data, when most of the decommissioning work was carried out (1990–2007), are hard to find. After 2007, the dismantling of the plant continued but at a somehow lower pace. Moreover, the main components (primary loop and principal auxiliary loops) had been already dismantled. In recent years, the most important targets of decommissioning work concerned the remaining auxiliary loops, the ventilation, the fuel transfer tank, the pools, and the various concrete structures (Table 8.4).

Once again, all these data confirm that the occupational impacts from planned nuclear decommissioning are no greater, and in all likelihood considerably lower, than assimilated nonnuclear industries.

TABLE 8.2 Number of work injuries and max individual doses for Italian NPPs under decommissioning (SOGIN, 2019).

	SOGIN staff (the operating organization)			Contractors			Max individual dose (mSv)[a]
NPP, year	2015	2016	2017	2015	2016	2017	2017
Caorso	1	1	0	2	0	0	0.14
Latina	3	0	2	0	0	0	0.75
Trino	0	1	1	2	1	0	0.45
Garigliano	0	1	1	3	1	1	0.52

[a]*Legal limit 20 mSv/year.*

TABLE 8.3 Injury rates at West Valley Nuclear Services company.

Year	Lost-time injury rates (per 100 full-time workers)
1999	1.14
2000	0.89
2001	1.60
2002	1.69
2003	0.14
2004	0.00
6-year average	0.91
National Average for Waste Management and Remediation Services Industry	3.9
National Average for Industrial Inorganic Chemicals Manufacturing Industry	1.4
National Average for Heavy and Civil Engineering Construction Industry	3.0

In contrast with the data referring to planned decommissioning projects, the number of significant cases in decontamination, and (pre-)decommissioning of the accident-stricken Fukushima Daiichi NPP is striking (FPLB, 2019). The following summary figures refer to the period from January to December 2018. In 2018, the decontamination employers who violated the laws and regulations related to labor standards were more than 60% of the supervised employers. The number of safety and health-related violations amounted to 122 including dose measurements, preliminary surveys of workplaces, operations management, contamination monitoring of exiting persons, contamination monitoring of removed articles, use of protective gear, provision of radiological tests for decontamination workers, nonsubmission of radiological test findings, and others (prevention of contact with construction vehicles, installation of lifts and elevators, etc.). Note, however, that the number of violations does not necessarily tally with the number of events.

It is clear that the radiological and physical conditions at Fukushima Daiichi NPP are prohibitive, which has made (pre-)decommissioning operations incredibly difficult to plan and execute in a timely, orderly manner. It can be also inferred—unsurprisingly—that a number of organizations and workers were ill-prepared to engage with these extreme circumstances.

As a general conclusion of this book, it can be stated that safety-related occurrences during (planned) nuclear decommissioning projects are lower in number (relative to comparable nonnuclear industries) and far from

TABLE 8.4 BR-3 accidental events, 2016–2020.

- 2016: 2 events
  - Internal workers: 2 events
    - Events with absence from work: 2
      - Number of days of absence from work: 101
- 2017: 3 events
  - Internal workers: 3 events
    - Events with absence from work: 2
      - Number of days of absence from work: 6
    - Events without absence from work: 1
- 2018: 3 events
  - Internal workers: 2 events
    - Events without absence from work: 2
      - On-site: 1
      - Road-work event: 1
  - External employee: 1event
    - Event without absence from work: 1
- 2019
  - Internal workers: 2 events
    - Event without absence from work: 1
    - Event with absence from work: 1
      - Road-work event: 1
        - Number of days of absence from work: 1
  - External workers: 3 events
    - Events without absence from work: 3
- 2020 (up to June)
  - Internal workers—event without absence from work: 1
  - External workers—event without absence from work: 1

representing a challenge to the environmental sustainability of nuclear decommissioning. This of course does not exempt the responsible parties to pursue a constant improvement in their health and safety practices.

References

US Bureau of Labor Statistics, 2019. Incidence Rates of Nonfatal Occupational Injuries and Illnesses by Industry and Case Types, 2018 (last update November 2019). https://www.bls.gov/web/osh/summ1_00.htm.

Deutsche Gesetzliche Unfallversicherung (DGUV, in English, The German Social Accident Insurance), 2019. Relatively Low Accident Risk and Severity. https://www.dguv.de/en/mediencenter/background/energie/atomkraft/index.jsp.

US Department of Energy, January 2010. Final Environmental Impact Statement for Decommissioning and/or Long-Term Stewardship at the West Valley Demonstration Project and Western New York Nuclear Service Center, DOE/EIS-0226. https://www.wv.doe.gov/final/EIS-0226_F-Summary.pdf. https://www.wv.doe.gov/final/EIS-0226_F-Chapter03.pdf (Chapter 3) Affected Environment.

US Department of Energy, 2020. Computerized Accident/Incident Reporting System (CAIRS), Injury and Illness Dashboard (update 2020). https://data.doe.gov/MS/asp/Main.aspx?.

Electric Power Research Institute, 2005. Maine Yankee Decommissioning Experience Report Detailed Experiences 1997 – 2004. http://www.maineyankee.com/public/pdfs/epri/my%20epri%20report-2005.pdf.

Fukushima Prefectural Labour Bureau, Ministry of Health, Labour and Welfare, Japan, Results of Supervision and Instruction Activities for Employers of Decommissioning Workers at the TEPCO Fukushima Daiichi Nuclear Power Plant and Employers of Decontamination Workers in Fukushima Prefecture (For 2018), March 29, 2019. https://www.mhlw.go.jp/english/topics/2011eq/workers/ri/gr/gr_190329.pdf.

House of Commons Committee of Public Accounts, July 2008. Nuclear Decommissioning Authority—Taking Forward Decommissioning, Thirty—Eighth Report of Session 2007—08, HC 370, 10. https://publications.parliament.uk/pa/cm200708/cmselect/cmpubacc/370/370.pdf.

Health and Safety Executive, 2019. Construction Statistics in Great Britain. https://www.hse.gov.uk/Statistics/industry/construction.pdf (updated October 2019).

Infrastructure Health and Safety Association (IHSA), 2011. Canada, Formwork and Demolition. http://www.ihsa.ca/pdfs/magazine/volume_11_Issue_4/formwork_and_demolition.pdf.

Massaut, V., 3 July 2020. SCK-CEN, Mol, Belgium, Personal Communication to M. Laraia.

Maine Yankee Decommissioning Overview. http://www.maineyankee.com/overview/default.htm.

US Nuclear Regulatory Commission, November 2002. Generic Environmental Impact Statement on Decommissioning of Nuclear Facilities. NUREG-0586, Supplement 1 Volume 1. https://www.nrc.gov/docs/ML0234/ML023470304.pdf.

US Nuclear Regulatory Commission, News, June 5, 2015. NRC Approves Changes to San Onofre Nuclear Plant's Emergency Planning Requirements. https://www.nrc.gov/docs/ML1515/ML15156A555.pdf.

UK's Office for Nuclear Regulation (ONR), February 2016a. Events Reported to the Nuclear Safety Regulator in the Period of 1 April 2001 to 31 March 2015. http://www.onr.org.uk/documents/2016/events-reported.pdf.

UK's Office for Nuclear Regulation (ONR), 13 December, 2016. Harwell Emergency Off-Site Planning Area Removed. http://news.onr.org.uk/2016/12/harwell-emergency-off-site-planning-area/.

Park, B.I., Kim, J.Y., kim, C.-L., March 2019. Suggestion of risk assessment methodology for decommissioning of nuclear power plant. JNFCWT 17 (1), 95—106. https://e-sciencecentral.org/upload/jnfcwt/pdf/JNFCWT-17-1-95.pdf.

Ringen, K., 6 July 2020. Stoneturn Consultants, Personal Communication to M. Laraia.

Issue 09 Sellafield Magazine, June 2018. https://assets.publishing.service.gov.uk/government/uploads/system/uploads/attachment_data/file/722217/Sellafield_Ltd_Issue_9_Magazine.pdf.

Sellafield Ltd, October 2019. Annual Review of Sustainability 2018/2019. https://assets.publishing.service.gov.uk/government/uploads/system/uploads/attachment_data/file/843409/Sustainability_Report_2018_19_FINAL.pdf.

SOGIN (in Italian: Società Gestione Impianti Nucleari, the Nuclear Plant Management Company), 2019. Sustainability Accounting 2017 (in Italian: Bilancio di Sostenibilità 2017). https://consumerlab.it/wp-content/uploads/2019/02/GruppoSogin.pdf.

Stoneturn Consultants, March 30, 2011. From Three Mile Island to the Future, Improving Worker Safety and Health in the U.S. Nuclear Power Industry, White Paper Prepared for the Blue Ribbon Commission on America's Nuclear Future. Stoneturn Consultants, 1519 3rd Ave, No 701, Seattle, WA 98101, US.

Hearing before the Sub-committee on Employment, Safety and Training of the Committee on Health, Education, Labor and Pensions, United States Senate, One Hundred Seventh Congress, Second Session on Examining Workplace Health and Safety Oversight of MHSA and OSHA Regulation and Enforcement, 2002. US Government Printing Office, Washington D.C, pp. 107–574. July 1, 2002, Y 4.L 11/4:S.HRG.

Further reading

European Commission, 2019. EUROSTAT, Accidents at Work - Statistics by Economic Activity (updated 19 December 2019). https://ec.europa.eu/eurostat/statistics-explained/index.php?title=Accidents_at_work_-_statistics_by_economic_activity#Developments_over_time.

Annex A-1

Design, construction, and operational measures to facilitate eventual decommissioning of a nuclear facility and ensure sustainability

The future enters into us, in order to transform itself in us, long before it happens.

Rainer Maria Rilke (1875–1926)

A-1.1 Introduction

Prior to the commercial development of nuclear applications, and especially of nuclear power, all available experience was derived from nonnuclear plants, for which the design included consideration of siting, construction, and operating aspects, but little focus was given to decommissioning (this has changed nowadays; for example, the postshutdown environmental impacts of oil and chemical installations are widely recognized, partly reflecting the experience in nuclear decommissioning). Similarly, for the first 20–30 years of the nuclear era, the decommissioning of nuclear installations was not a significant design issue; many nuclear installations were designed and constructed with no specific concern given to eventual decommissioning.

In the past, planning and implementation of decommissioning/environmental remediation (D&ER) were only considered once a facility had exhausted its service life, after an accident or when a leak has worsened to a point that immediate repair was required—or final shutdown was the only viable alternative. In other words, D&ER was only considered at the back end of the service life of a site or facility, in some cases when there was no longer an organization "legally" responsible for the cleanup activities. When the organization responsible for the cleanup could no longer be identified or made liable, the liability ended up being transferred over to the State, i.e., to the tax payer. This directly conflicts with the "polluter pays" principle, long established as one of the pillars of sustainability. Early planning for D&ER may help prevent the previously described final "bottleneck."

By the late 1980s, decommissioning experience and studies had proven that the lack of decommissioning considerations at the design and subsequent life phases of nuclear facilities could produce serious safety consequences on the decommissioning management and incur unnecessary costs.

Beginning in the early 1980s, a drastic change in world's priorities has taken place: Growing awareness of ecological long-term problems has been bringing forth a rejection of the previous attitude to dealing with environmental problems only after they have occurred. The goal is now to plan from the beginning for the entire lifecycle of a human activity to avoid unexpected environmental impacts at a later stage (the so-called cradle-to-grave approach). Life cycle management (LCM) aims to regard each stage in the life of a facility or site not as an isolated event, but as just one phase in a seamless life. Thus, the planning not only covers each single stage but is also a continuing activity, taking into account both current circumstances and projected/possible developments (IAEA, 2006).

As an application of this integrated approach, the "Joint Convention on the Safety of Spent Fuel Management and on the Safety of Radioactive Waste Management" (IAEA, 1997) requests the avoidance of "undue burdens on future generations," a direct reflection of the sustainability principle. The concepts of "protection of future generations" and "burden on future generations" are also highlighted in the IAEA Safety Fundamentals (IAEA, 1995). The statement given by OECD-NEA in NEA (2006) illustrates the same point: "Safety of current and future generations is the paramount concern of decommissioning and decommissioning funding." As a consequence, a more prospective management of human activities also became part of the legislation in many Member States including the accumulation of decommissioning funds and the preservation of knowledge. In many countries, there are also regulatory constraints, which require the organizations to establish and maintain up-to-date plans for the LCM of their facilities including the ability of producing records for any legal review. However, the generation and preservation of records is only the basis for knowledge, which is a much broader concept. While it is important to consider the longevity of the records, (IAEA, 2008) the ability of future generations to utilize them is also essential.

Early planning for decommissioning of nuclear facilities and site environmental remediation (D&ER) is a relatively recent innovation. The life cycle of a nuclear facility including design, construction, operation, and back end (waste management, decommissioning, environmental remediation) is well over 100 years. This typically includes several years for the siting and design stage, one or two decades for the construction and commissioning, a period of 60 years' operation (likely to extend to 80 years for newer installations), and a decommissioning period from a minimum of 10–20 years (for the immediate dismantling of a minority of nuclear power plants) to 100 years (for the deferred dismantling of some nuclear power plants). If the disposal of low-level radioactive waste (LLW) resulting from D&ER is included, a time span of

300–500 years (including institutional control) up to release of the disposal site is typical. The active management of high-level waste (HLW) can last for millennia. Given the timescales involved and the complex generation of vast amounts of data and plans required for eventual decommissioning and environmental remediation, it is essential to systematically organize the planning conducive to the successful completion of a nuclear project.

A wide range of materials/waste will arise from nuclear decommissioning. Some will be radioactive; some will have an economic value and/or can be recycled or reused. Others will have little or no economic value, and this is the waste that has to be disposed of, or stored pending availability of an accepted method of disposal. Waste minimization can be considered a good strategy for limiting the generation of these undesirable by-products. Whereas such by-products are inevitable, or mitigatory actions are required to minimize their volumes.

Features intended to facilitate decommissioning vary in ranking. Some will be inherent in modern plant design and would benefit operation and maintenance, whereas others will represent "good practice." Some design and construction features, however, may have negative cost and program consequences or even reduce plant reliability. A selection of important features relevant to decommissioning and related waste management is highlighted in this annex.

A-1.2 International and national guidance on early planning to facilitate decommissioning

The first step toward successful management and completion of a decommissioning project is to develop an early strategy that is acceptable to the regulator and is feasible in terms of program and cost. This must then be followed by appropriate design and construction considerations to specify the expected tasks and to select the most effective approaches to facilitating eventual decommissioning and minimizing the management of decommissioning waste. During subsequent plant operations, the starting conditions, planned outcomes, and the end state of decommissioning must be more clearly formulated, and the plant operations must be carried out with a view to their decommissioning implications. All these activities reflect into an evolving decommissioning plan.

The decommissioning plan should be dealt with as a "living document," which will require additions or amendments during plant operation and beyond. The decommissioning plan should evolve from preliminary to detailed, taking account of factors such as new socioeconomic circumstances, plant modifications, operating experience, and the availability of better technology.

The top-level IAEA Safety Standard on Decommissioning reads (IAEA, 2014)

".....Requirement 10: Planning for decommissioning The licensee shall prepare a decommissioning plan and shall maintain it throughout the lifetime of the facility, in accordance with the requirements of the regulatory body, in order to show that decommissioning can be accomplished safely to meet the defined end state.

.... The regulatory body shall ensure that the licensee takes decommissioning into account in the siting, design, construction, commissioning and operation of the facility, by means which include features to facilitate decommissioning, the maintenance of records of the facility, and consideration of physical and procedural methods to limit contamination and/or activation."

Other international agencies have joined the IAEA in the concept of LCM of decommissioning (WENRA, 2011).

A comprehensive EPRI report in this field is well worth a mention (EPRI, 2001). The purpose of this report is to provide a framework for preplanning (advance planning) of the decommissioning of an NPP with the goal of reducing delays in moving to decommissioning, including premature plant closures. "Preplanning" is used in the EPRI report to describe strategic planning done at the top management level well before permanent shutdown. Indeed, it is the goal of the EPRI report to prove that such "preplanning," if done properly, will provide benefits in time and money saved when the plant reaches the end of its service life. Preplanning is not meant to develop a detailed work plan but is a higher level of planning aimed at addressing basic issues, which, once identified and evaluated, will enable an operating plant to pursue essential tasks while it is still operating and when it is easier and less expensive to do so. The report is complemented by another EPRI report providing practical experience on decommissioning planning (EPRI 2006).

A three-tier structure of decommissioning planning has been promulgated by the IAEA for many years and is still in use. Typically the decommissioning planning process is structured in three categories:

- Preliminary (aka initial) plan
- Ongoing plan
- Final plan

The key principle of planning for decommissioning is continuity. Experience shows that a significant gap in planning will inevitably result in unnecessary delays (e.g., between final shutdown and the start of active decommissioning) and ultimately extra costs. This has been the case for facilities that were shut down as the result of a political decision (e.g., a national or local referendum): inevitably the uncertainties linked to a political decision stopped or slowed down the progress in decommissioning planning.

The reader should note that the concept of LCM in decommissioning planning is now well accepted internationally. Some countries have specified this approach in their legislation: One such country is Finland (Paltemaa, 2015). Another example is Canada (CSA, 2014).

Title 10 of the United States Code of Federal Regulations, Part 20.1406 states (NRC, 2011):

> Minimization of contamination:
>
> "Applicants for licenses, other than renewals.... shall describe in the application how facility design and procedures for operation will minimize, to the extent practicable, contamination of the facility and the environment, facilitate eventual decommissioning, and minimize, to the extent practicable, the generation of radioactive waste".

A-1.3 Considerations about the selection of siting, design, construction, and operations features facilitating decommissioning

As a preamble, it should be clearly stated that a nuclear facility is not conceived and built with the main purpose of being decommissioned. Firstly, a nuclear facility is intended to generate energy or radioisotopes, and second, to do so safely and cost-effectively. Decommissioning scores inevitably lower in the designer's checklist. However, there are design, construction, and operational features that would serve the plant's main objectives as well as facilitate decommissioning. For example, reducing radiation levels would simplify operating inspections and maintenance—and decommissioning. Features that serve multiple purposes should clearly have priority.

There are other features, however, that are merely intended to facilitate decommissioning. In this case, a cost—benefit analysis would be needed (NRC, 1986). It should be questioned first whether the proposed decommissioning-friendly feature is not counterproductive for the safe and smooth operation of the plant and second whether the extra costs paid at the construction stage would be worth the benefits expected at the time of decommissioning—many decades later (and taking due account of the accumulated inflation).

To pursue optimization, the cost of implementing a new design or construction feature facilitating decommissioning such as new construction materials, additional shielding, installation of equipment removal aids, and so on is compared with a reduction of decommissioning dose. Or else, the costs of implementing a new practice are assessed against person power or against expenses needed for management of waste generated.

Given that multiple factors will be involved in the decision-making process, a technique called multiattribute utility analysis (MAUA) is often used. It is based on a scoring system for alternatives, so that the alternative with the highest score is preferred. In the context of design and construction features to facilitate decommissioning, MAUA may be worth considering since factors expressed in incomparable units are taken into account such as extra construction costs and possibly savings at the decommissioning stage, waste amounts and variable waste management requirements, and occupational radiological/nonradiological impacts during plant operations and at the decommissioning stage. Some general considerations that will influence the decision-making about decommissioning-friendly features are given in the following.

The siting of nuclear installations is commonly based on factors such as location within the electrical grid or the availability of cooling water. However, decommissioning considerations may play some role here (not necessarily the predominant role). For example, siting new facilities on existing nuclear sites would allow benefits to be gained from shared services and utilities. Ensuring safe enclosure of idle plants would be made easier by the availability of staff serving operating facilities onsite.

For prolonged periods of safe enclosure, the integrity of plant structures will depend on site-specific meteorological condition, e.g., extreme temperature variations, airborne chloride contents (e.g., in the vicinity of seawater), acidic fumes generated by industrial plants nearby tornadoes, or floods.

The entombment option (aka in situ disposal) consists of the conversion of a nuclear facility into a disposal site. This option implies that a nuclear site is licensed not only for construction and operation but also for waste disposal. Entombment is strongly discouraged by the international organizations (IAEA, 2014) and rejected in many countries; however, this strategy is still implemented in facilities owned by the US Department of Energy, in the Russian Federation, and in a handful of other countries. Entombment generally includes the dismantling of aboveground parts of the facility and the grouting of belowground parts (Fig. A-1.1).

A similar option facilitating decommissioning is the siting of a waste disposal site in the vicinity of the decommissioning facility: This strategy will considerably reduce the cost of decommissioning waste handling and transportation (commonly a significant share of the overall decommissioning costs). This strategy has been pursued in Finland (Fig. A-1.2). The Onkalo (meaning "small cave" or "cavity") spent nuclear fuel repository is a deep geological repository for the final disposal of spent nuclear fuel, the first such repository in the world for high level waste. It is nearing completion and commissioning in the granite bedrock at the Olkiluoto site, a short distance from the power plant. Olkiluoto was selected in 2000 as the site for a long-term underground storage facility for Finland's spent nuclear fuel.

Where a suitable waste repository is available, the waste acceptance criteria of that repository will have an impact on decommissioning-friendly design and construction features. For example, where a repository will accept the intact disposal of large components (e.g., reactor pressure vessel (RPV) or steam generator) and the transport infrastructure (cranes, railways, etc.) will allow the shipment of such large components, design features permitting the intact removal of large components are possible. The strategy of intact removal—rather than in situ segmenting—is generally considered to facilitate decommissioning (Fig. A-1.3).

A factor affecting the selection of decommissioning-friendly design and construction features could be the industrial capability of a given country. For example, natural cobalt concentrations have been decreased considerably in the design of the reactor primary circuit over the past few decades to reduce radiation levels due to cobalt activation. However, a country's metallurgical industry may or may not have the capabilities to incorporate these stricter requirements in the manufacturing of reactor materials. The alternative could be to import the materials from abroad. For this and other aspects, a trade-off of advantages and disadvantages is necessary.

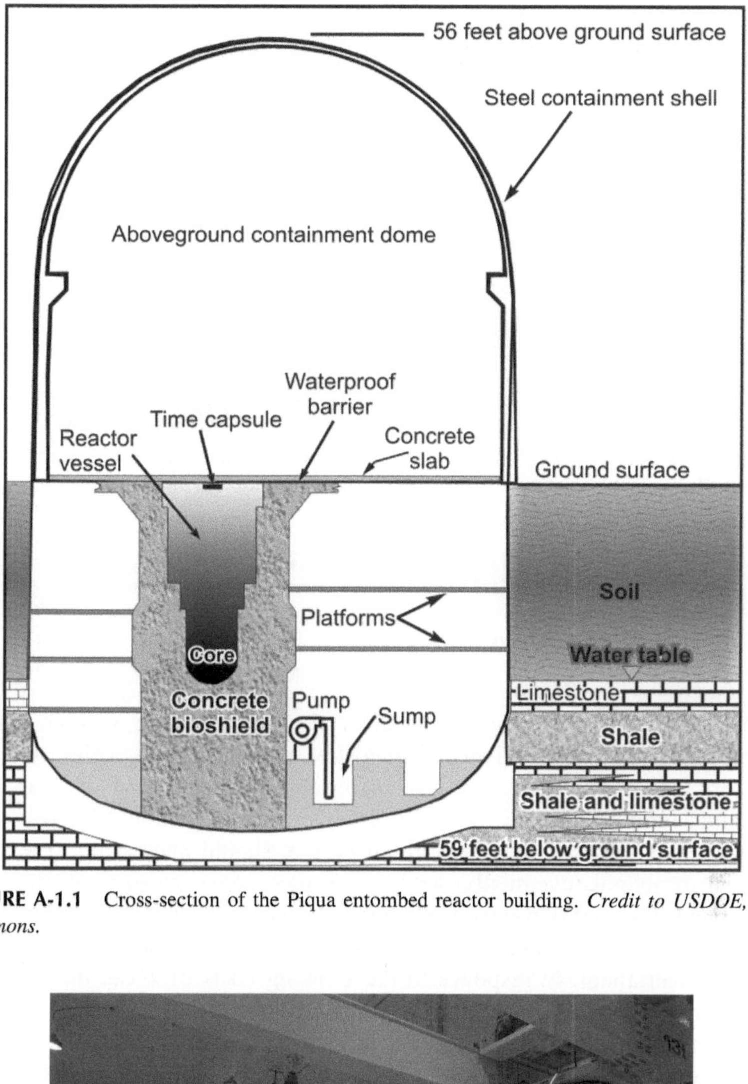

FIGURE A-1.1 Cross-section of the Piqua entombed reactor building. *Credit to USDOE, Wiki Commons.*

FIGURE A-1.2 Olkiluoto, Finland: Onkalo spent nuclear fuel repository. In the photo: Silos for low-level waste. *Credit to Teemu Väisänen, Wiki Commons.*

FIGURE A-1.3 Intact removal of HWCTR steam generator, Savannah River Site, United States. HWCTR, Heavy Water Components Test Reactor. *Credit to USDOE, Wiki Commons.*

A-1.4 Costs and funding of decommissioning

Even though work stops, expenses run on,

Marcus Porcius Cato 234-149 B.C.

In the early 1980s, the cost of nuclear decommissioning was no longer regarded as similar to the dismantling cost of conventional factories. As labor costs rose and radioactive waste disposal costs escalated dramatically, decommissioning costs became a serious concern in the overall life cycle costs of the nuclear industry. Regulators realized that should nuclear utilities suffer financial problems eventually leading to permanent closure of their installations and possibly bankruptcies, the safe shutdown and decommissioning of nuclear installations could become a financial burden also to the taxpayers and the government. In response to the growing costs of waste disposal and other cost items, regulators endorsed the principle that those who had received the benefit (e.g., the plant owners benefitting from the electric power generated) should pay for the decommissioning of their facilities (the "polluter pays" principle). This approach soon led to legislative and regulatory requirements for funding decommissioning; for example, US regulators established minimum funds to be set aside for NPPs, based on thermal power and corrected on account of inflation and other changing parameters.

Decommissioning cost estimates and the parallel establishment of decommissioning funds became some of the most important aspects of LCM for nuclear installations and important components of the preliminary decommissioning plan and its revisions. Financial aspects of decommissioning are extensively dealt with in IAEA (2005). A growing awareness of uncertainties about "real" decommissioning costs led to a number of initiatives and studies by the international organizations, focusing on standardization of cost items and better understanding of variations. Among many reports published in this field, it may be worth quoting (ISDC, 2012) (NEA, 2017). The estimate of decommissioning costs and the establishment of a decommissioning fund are fundamental components of the preliminary decommissioning plan and of the principle of sustainability as applied to nuclear decommissioning.

Last but not least, it should never be forgotten that the cost of deferring active decommissioning can be high, due to running costs of a shutdown plant (maintenance, security, surveillance, etc.).

A-1.5 Design and construction features facilitating decommissioning

A-1.5.1 Minimization of activation

The composition of the materials that are exposed to neutron flux during operation should be carefully assessed, especially in regard to trace elements, which can generate considerable quantities of neutron-activated radioisotopes. In light-water reactors (LWRs), this applies specifically to cobalt: Naturally occurring ^{59}Co activates to ^{60}Co, which is the critical gamma-emitter for several decades after final reactor shutdown. The activation of corrosion products in the reactor core is followed by the deposition of radioactive materials in plant systems. While it would be impractical to zero the concentration of trace impurities, optimization will dictate a reduction of their concentrations at an acceptable cost that would allow a significant abatement of radiation levels during inspection and maintenance activities—and final decommissioning. Currently, average cobalt limits as low as 0.02%—0.03% for RPV base material are common targets.

The reader should note that a significant source term of reactor cobalt is given by cobalt-rich alloys (e.g., stellites) used in various reactor components: a cobalt reduction program should include these alloys as well. A study of the dose impacts of stellite was conducted on several German PWRs built by Kraftwerk Union (KWU) (Garbett, 1992). The most recent KWU reactors had entirely removed stellite from the RPV and almost all areas of the primary system circuit. The only system where stellite had remained in use was the control rod drive apparatus. Removal of stellite produced a drastic reduction in radiation levels and related worker doses. Channel head radiation levels were generally lesser than 10 mSv/h compared with 50—100 mSv/h for reactors that have not undertaken stellite reduction programs. NEA (2014a), IAEA (2011a), and IAEA (2019) are reference reports including inter alia cobalt specifications for light-water reactors.

A second benefit resulting from the minimization of activation products is the facilitation of waste management (IAEA, 2007).

A-1.5.2 Control of liquid and airborne contamination

The rooms containing radioactive liquids should be designed to have drip pans and floor curbs with enough capacity for the largest leak assumed. The floors inside the curbs should slope down to address any leak to floor sumps capable to gather all liquid waste. Oil spills should be prevented from mixing with water drainage. The spread of contamination during decommissioning can also be reduced by designing components for easy dismantling (e.g., in modular blocks, see Section A-1.5.7).

To minimize the transfer of contamination to adjacent clean areas (cross-contamination), nuclear facilities are generally categorized into active and inactive zones, with checkpoints at the boundaries for loose materials, components, waste drums, and workers.

The design should include buffer areas for temporary waste storage during operation and for waste transfer, monitoring, and tracking.

Accurate design of the plant ventilation is essential to maintain contamination control and limit the extent of the areas to be decontaminated and the generation of radioactive waste at the decommissioning stage. In view of future decommissioning, the design should also consider the deposition of airborne particulates in ventilation ducts: To reduce deposition, the installation of prefilters has proven beneficial (IAEA, 2007).

A-1.5.3 Spent fuel pools

The design of pools associated with earlier nuclear facilities (spent fuel pools, pool-type research reactors) did not include many features that would simplify decommissioning. Many pools were constructed of unsealed concrete with little attention being given to leak minimization or detection. Bare concrete tends to absorb radioactive contamination. Effective water treatment systems were not common in early plants, and this often resulted in the formation and deposition of sediments. Some fuel cladding materials (e.g., Magnox alloys in UK's gas-cooled reactors) and pool equipment were prone to corrosion when staying underwater for extended periods. Inattention to pools storing degraded spent fuel and hazardous sludge for many decades has caused many problems. In addition to activation products, sludge can contain alpha-emitters from leaking fuel and long-lived fission products. Further complications were due to inadequate records of the pool construction details and its growing radioactive inventory (see Section A-1.5.10).

Pools at NPPs are often kept in better condition than at research installations because of a stricter spent fuel management and more focus on safety culture. Recent types of nuclear fuel offer much higher leak-tightness and are less prone to erosion/corrosion.

Implications of design on pool performance are exemplified by the Apsara reactor (Sasidharan et al., 2006). The Apsara pool was built as a reinforced concrete structure with no lining. Initially, the leakage of pool water through the bare concrete was a major issue. Attempts to stop the leakage included painting the inner surfaces, coating, replastering, and pressure grouting, but all missed the goal. Eventually, the pool was lined with stainless steel 304L plates. The stainless steel lining was positioned over a carbon steel frame that was attached to the concrete. The void between the lining and the pool wall was filled with cement mortar. Following this intervention, the Apsara pool operated for over 30 years with no leakage.

The American Nuclear Society produced design criteria for pool-type independent spent fuel storage installation (ANS, 1988). In addition to construction, safety, and so on, this standard addresses decommissioning. The standard defines details such as design for easy decontamination, floor and wall linings, embedment of pipes, and leak detection and collection. It stresses that attention should be given to interfaces with adjacent installations, security, and the availability of cooling and waste cleanup systems. Shortly after the promulgation of the ANS standard, the IAEA published guidance about design and operational measures to facilitate pool decontamination (IAEA, 1990).

More recently, the IAEA suggested beneficial features for the design and construction of new nuclear installations to facilitate decommissioning (IAEA, 2011a). Specific design features and actions that are pertinent to new pool facilities include the following:

— Double-walled containment
— Preinstallation of leak detection devices
— No construction of structures partly or wholly below ground
— Permanent decontamination facilities
— Sealed surfaces to prevent migration of contaminants inside structures
— Provision of onsite dry waste storage facilities in order not to use pools for waste storage
— Keeping up-to-date records of pool inventory, and its design and construction features
— Preserving archive samples of construction materials

The OECD Nuclear Energy Agency has also published a document relating to pool design that applies specifically to new NPPs (NEA, 2010). By including design features suggested by recent IAEA and NEA reports, there are concrete prospects for improving pools in the new generation of NPPs. For example, the following specific design features to facilitate pool decommissioning were highlighted by Westinghouse for their new generation of NPPs (Wright et al., 2008):

> The walls of the pools are constructed using modular construction techniques and lined with welded stainless steel plates.
> The pools are equipped with leak chases at each weld so that the tank is effectively double walled in the area of plate joints.
> The pool leak detection system is zoned to allow identification of the area of the pool liner that leaked.

The IAEA has published a report on practices for water quality control in pools (IAEA, 2011b), which addresses the general deterioration of materials used underwater in pools. This report addresses water treatment methods and measuring devices for water quality control. Although the report is meant for operating pools, it provides useful guidance on typical circumstances of pools to be decommissioned.

A comprehensive IAEA report on decommissioning of pools including design and construction features relevant to decommissioning can be found in IAEA (2015).

A-1.5.4 Measures to facilitate decontamination

Methods to facilitate surface decontamination should be part of the design (IAEA, 2007). Bare concrete is porous, and water can penetrate, often to depths of 10–20 cm, which would require intrusive dismantling techniques, e.g., scabbling. Concrete floors and walls that are subject to waterborne contamination should be protected by a barrier such as epoxy paint or steel lining.

Stainless steel liners anchored to the underlying concrete are beneficial to decommissioning provided they are not damaged (joints and welds tend to be weak points). Stainless steel slabs can be easily decontaminated. This approach will result in less concrete rubble as radioactive waste. However, the anchors utilized for securing the liner to concrete will make dismantling more difficult.

Plastic coatings will impede or reduce the absorption of radioactive contamination in the concrete; besides, these coatings are far easier to chemically decontaminate than concrete. Floor–wall interface joints should be especially protected to facilitate decontamination. However, protective coatings should maintain their integrity over the entire life of the nuclear facility: They should remain acceptably safe from deterioration caused by radiation, process fluids, and attack from decontamination solutions. Regularly recoating concrete surfaces may turn out to be useful. Plant systems that need to be chemically decontaminated periodically—for example, the primary coolant system in some reactors during outages—need to be designed to simplify filling and drainage after decontamination. In general, all systems transferring radioactive fluids should be properly equipped with connection points, vents, and drains, but these should not impair smooth operation or final dismantling.

A-1.5.5 Layout and egress of components

The plant layout should be primarily conceived to ease radiation and contamination control during operations and maintenance. Facilitation of decommissioning would be an extra bonus. The main objectives of layout design are to minimize radiation doses by component segregation and shielding and to minimize the spread of contamination by containment, zoning, ventilation, and draining. A congested layout will impact on doses both during operation and during decommissioning.

Access measures for the intact replacement of large components (e.g., steam generators) during operation will also facilitate decommissioning. Use can be made of construction pathways (possibly closed during reactor operation). Preplacement of dismantling aids during construction—rather than at the time of dismantling—can facilitate dismantling of selected components and removal of dismantled pieces: These aids may include tracks to guide remotely controlled cutters or lifting equipment. The benefit of intact (aka one-piece) removal consists of reduced dismantling time and associated exposures (see Fig. A-1.3; Section A-1.5.6).

During design, consideration should be given to the use of connectors, latches, and pipe anchors that can be easily removed. However, the integrity of these items during operation or seismic events should not be compromised. A trade-off may be needed as bolted connections (e.g., flanges), while easier to remove than welded connections, are typically subject to higher leakage.

The routing of piping carrying contaminated fluids should not create stagnant flow zones where radioactive crud would preferentially deposit. All piping and tanks should be ideally positioned above ground. This recommendation may be hard to fulfill in practice since belowground piping and tanks benefit from gravity for transfer of fluids. If belowground pipes cannot be avoided, they should be double contained (e.g., within water proof conduits equipped with sumps and water level monitoring) to prevent soil contamination in case of pipe leakage.

Similarly, it is advisable to avoid the installation of embedded piping, which is more difficult to remove than free piping.

To reduce occupational exposures during operations and decommissioning, the piping layout should not overly complicate access. Fig. A-1.4 shows a congested layout where the workers dismantling a given pipe would be exposed to irradiation from surrounding pipes and other components.

Figure A-1.4 Engineering test reactor basement, cubicle interior with tank and sodium-related apparatus. *Credit USDOE, Wiki Commons.*

A-1.5.6 Intact removal of large components

In many decommissioning projects, in situ dismantling of large components (steam generators, primary coolant pumps, pressurizers, and reactor pressure vessel) is the compulsory strategy, if intact removal is made impractical by lack of manoeuvring space, interfering protrusions, or inadequate handling/lifting devices. However, in situ dismantling requires that the dismantling team stay close to the component being segmented and removed in pieces for a longer time than in intact removal. Additionally, the volume of solid waste containers resulting from in situ dismantling will be higher than the intact component alone. Secondary liquid waste, e.g., from decontamination will be minimized in intact removal.

Design provisions should make intact removal possible. The opening of the reactor building to allow transfer of the component to the outside could have been conceived by the design; alternatively, an ad hoc opening can be made during the plant decommissioning (Fig. A-1.5).

Intact removal is not trivial. A full review of the structural strength will be in order, which may require structural reinforcement. Structural issues may concern the component itself, which may have been weakened by prolonged irradiation (embrittlement): The load bearing capacity of the flooring is

FIGURE A-1.5 Trojan NPP intact removal of steam generators. *Courtesy of USNRC.*

another issue. It will be also needed to remove obstacles from the area, install scaffolding and temporary shielding, and cut off the component being removed from attached piping and protrusions. NEA (2012) provides a comprehensive treatment of the subject. Fig A-1.3 is another example of intact removal of large components.

A-1.5.7 Modular design

It has been proposed for newer designs that modular construction would facilitate the dismantling and handling of large structures, thus reducing radiation exposures. However, modular construction can be costlier, lower structural resistance against seismic events, and offer more leakage pathways. The new AP1000 Westinghouse design includes a significant focus on modularization, which had been suggested much earlier for biological shields (NRC, 1979).

Modularization of the biological shield has been a regular feature of the design of small research reactors, e.g., Argonauts. This feature has simplified the dismantling of Argonauts, since the biological shield blocks could be removed manually, with no massive demolition, no airborne dust, and consequently no respiratory protection. Argonaut reactors dismantled include, among others, RB-2 (1987, Montecuccolino Laboratories, near Bologna, Italy Fig. A-1.6), Jason (1999, Greenwich, near London, UK), and Graz University reactor (2006, Graz, Austria). Instead the Moata Argonaut at ANSTO, Lucas Heights near Sydney, Australia had a monolithic bioshield, which had to be fully demolished by wire sawing in 2009 (Fig. A-1.7).

FIGURE A-1.6 RB2 dismantling, lifting bioshield concrete modules. *Photo by M. Laraia, 1987.*

FIGURE A-1.7 Moata reactor dismantling with use of wire sawing. *Photo by M. Laraia.*

A-1.5.8 Design for safe enclosure

If an extended period of safe enclosure (e.g., 60 or more years in some countries) is planned as part of the initial decommissioning strategy for a given facility, it is prudent that the design and construction of that facility take safe enclosure into account. A safe enclosure strategy implies that the entire design life of the plant and buildings is considerably longer than the operational period. The following addresses some of the plant and buildings integrity aspects that should be evaluated by the designers.

The end-of-life radioactive inventory is mostly enclosed within such components as the reactor vessel, steam generator, and piping. Corrosion of the plant systems will reduce their containment functions (think of leaking flanges) or even pose problems of structural resistance (e.g., in case of an earthquake). Corrosion may result from increasing humidity after the reactor is inactive and cold or from in-leakage of rain, surface-, or groundwater. Corrosion allowance or protective coatings are two means to be considered by the designers to protect against corrosion risks during safe enclosure.

The capability to fully drain vessels and piping is another design consideration to reduce hazards during a safe enclosure period. This is not a trivial point since water tends to remain at low points in piping, tanks, and so on.

Consideration should also be given during design to the required extended service of certain facility systems (e.g., radiological monitoring, liquid waste drainage and management systems, ventilation, lighting, etc.). However, the option of retaining these systems active or stand-by during safe enclosure should be weighed against the option of discontinuing these services and installing new systems at the time of final dismantling.

The design should also consider the prolonged integrity of the buildings. Hazards include degradation by atmospheric agents (e.g., industrial fumes from nearby facilities) or water penetration.

A-1.5.9 Operational features facilitating decommissioning

Attention to minimization of operational and decommissioning waste has been raised by the growing volumes of waste in storage at most nuclear sites, the increasing costs of effective waste management, and the shortage of disposal sites. Earlier disposal sites such as Morsleben in the former East Germany have been closed down on safety grounds, and many existing low-level waste (LLW) disposal sites are exhausting their capacity. Therefore, the design capacity of storage facilities needs to cover the whole operating and decommissioning phases pending availability of off-site disposal sites.

A recent US Regulatory Guide (NRC, 2012) is extensively used in the following paragraphs to suggest measures to minimize the production of radioactive waste—with a view to eventual decommissioning. The main requirement for decommissioning is for the nuclear operators to have adequate funds to implement D&ER and terminate their licenses safely and cost-effectively. To be adequate, the decommissioning funds must be sufficient to remove and dispose of all remaining radioactivity that is above the approved waste and site release criteria upon the application for license termination. In other words, the decommissioning fund must cover (1) the costs of packaging, shipping, and disposal for all wastes to be removed from site and (2) site surveys to prove compliance with release criteria. Licensees are expected to adjust the decommissioning fund so that it will be adequate at all times to complete decommissioning. If at any time a licensee identifies residual radioactivity that would require remediation to terminate the license, it should upgrade the fund to cover the additional expenses or do the remediation at once.

The US regulations require licensees to minimize the introduction of significant amounts of radioactivity into the site, including the subsoil, and to carry out radiological surveys to determine the amounts of residual radioactivity at their sites. Regulations also require that the findings of monitoring and surveys be indicated in records relevant to decommissioning (Section A-1.5.10). Because of the wide variety of facilities and site conditions, the US regulations do not specify the monitoring, surveying, or sampling activities in detail. Each licensee is responsible for the following general objectives:

- minimizing the introduction of radiological contamination into the environment, including the subsoil;
- conducting surveys to establish the extent of significant residual contamination in the environment; and
- taking corrective action to avoid further contamination (e.g., repair leaking components and contain, gather, and remove liquid leaks); and
- as needed, assessing the costs to remediate the residual contamination to site release criteria; licensees are then required to use the conclusions of this evaluation in updating total decommissioning costs and financial assurance.

Several studies have identified typical events that could increase D&ER costs by increasing the likelihood of soil contamination. In particular, prolonged leaks of radioactive substances, especially long-lived and mobile radionuclides, into the subsoil may eventually create significant radiological hazards and heighten the cost of D&D.

It can be concluded that the sites with a higher potential of being seriously contaminated are those housing:

- large volumes of slightly contaminated liquids, and other easily dispersible waste (e.g. powders);
- large inventories of long-lived radionuclides;
- production streams held in unlined impoundments, storage ponds (see Section A-1.5.3), and basins; or
- large amounts of solid waste stored outdoors without containerization or protection from environmental agents.

There are several operational aspects that require attention when considered from the viewpoint of eventual decommissioning. For example, allowing scattered waste from operation to build up in plant areas would pose unnecessary hazards to the operating staff and unduly complicate

FIGURE A-1.8 Research reactor hall at the end of operational lifetime. *Photo by M. Laraia.*

FIGURE A-1.9 Solid waste storage at Paldiski Centre, Estonia. *Photo by M. Laraia.*

decommissioning: This is a typical recurrence at research reactors where all too often research apparatuses are left behind after the completion of the research work (Fig. A-1.8).

Another problem, which was common in the early days of the nuclear industry, is the lack of segregation of different categories and types of radioactive waste. Miscellaneous, diverse waste stored together would require complex retrieval operations, identification, and segregation at the time of decommissioning, would generally produce unnecessary amounts of waste packages, and would incur extra occupational doses (or require expensive remote operation) (Fig. A-1.9).

A-1.5.10 Record generation and keeping

Production and management of records are integral components of any organization's activities, processes, and systems. Records enable efficiency, transparency, responsibility, risk management, and business continuity. They also enable organizations to capitalize on the value of their information assets, e.g., for business purposes, to contribute to the preservation of corporate memory and team spirit, and to respond to the challenges of the global (e.g. digital) environment. The International Organization for Standardization (ISO) offers a complete series of internationally accepted standards about record management (e.g., ISO, 2001; ISO, 2011a; ISO, 2011b).

The main source of information for a D&ER project is official company or facility records. Each phase of a facility's life cycle generates information of interest to the various facets of future decommissioning activities that is captured as a record, although not all information captured during the facility's life cycle is relevant to decommissioning activities. The practice of retaining all facility information with the hope that some of it will be useful for decommissioning might result in an inordinate amount of documentation, which is often unmanageable and may overwhelm the portion that is actually valuable information (IAEA, 2002b). Therefore, it is important to select and segregate a subset of key decommissioning information from the earliest phases of data collection (i.e., from the siting, design, and construction phase and throughout operation of a nuclear facility).

While key information will be typically obtained from company or facility records, information from other sources may be also important and need to be identified, captured, and retained in support of future D&ER projects. These other sources may include inter alia the following:

- historical that could have used during construction;
- personal files;
- manufacturer's drawings;
- anecdotes of staff or regulator, public experiences;
- scientific journals and newspaper articles, TV and radio broadcasts, conference proceedings.

Table A-1.1 includes a list of information categories and specific notions that should be collected and maintained for D&ER purposes.

Unknowns in decommissioning cannot be totally eliminated, regardless of the characterization efforts. This is especially applicable to old facilities where documentation may have been lost or where modifications were carried out without updates to the documents. As a result, when planning for decommissioning, it is safer to assume that unexpected events will occur and have contingencies in place to deal with them should such occurrences arise. This approach will not only improve the efficiency of the decommissioning project but will also improve the safety of prior operations.

TABLE A-1.1 Information relevant to decommissioning and environmental remediation planning.

General: Preconstruction characteristics and the bases for D&ER and end-state objectives.
- Facility and site features
- Preliminary decommissioning plan at design and construction stage
- Ongoing decommissioning plans
- Organizational charts that identify key positions and individual staff

Design and configuration: To understand the buildings, components, and structures.
- As-built SSC drawings (including modifications over the life of the facility and in preparation to decommissioning)
- Plant utilities (air, gas, pressurized and drainage water, electrical, ventilation, fire protection, standing radiological monitoring, etc.)
- Purchase orders, equipment details and specifications
- Space allocation and room classifications
- On-site buildings, structures, and activities pertinent to adjacent facilities

Construction/commissioning: To understand the construction sequencing as an aid for reverse dismantling and equipment removal.
- Structural assessments
- Photos showing construction stages
- Alterations to the initial design

Operations: To understand facility's history and radiological inventory, and identify events to the facility during operations insofar as this information could affect D&ER activities.
- Operating, inspection, and maintenance procedures and logs
- Radiological, chemical, and structural characterization
- Equipment room-by-room inventory
- Shipments of materials and equipment
- Survey reports
- Waste transport logs
- Chemical inventory including shipments
- Operational occurrences such as spills, pipe ruptures, and so on, repair/replacement activities, and related occupational and nonoccupational impacts
- Photos of running operations.

Health, safety, and environment: To estimate a facility's end-of-life hazards.
- On-site and off-site emergency plans and their modifications
- (Radiological and nonradiological) hazard analysis, occupational exposures, and assessment reports
- Safety committee minutes
- Annual safety reports
- Inspection reports
- Accident investigation reports

Legal, regulatory, and licensing documents: To understand actual controls, constraints, and provisions, as well as industry standards and practices.
- Plant technical specifications
- Correspondence with regulators
- Legislation and regulations as applicable to the facility
- Permits/consents
- International and national standards and recommendations

Continued

TABLE A-1.1 Information relevant to decommissioning and environmental remediation planning.—cont'd

Decommissioning and remediation: To document the strategic decommissioning decision-making; planning and implementation processes; to record activities; to transfer knowledge.

- Final shutdown plans and reports
- Interim and final site end-state reports
- Detailed D& ER plans
- Turnover documentation and certificates
- Succession planning
- Safe enclosure records, including care and maintenance activities
- Work package documentation
- Project documents (scope, work breakdown, schedule, hold points, status reports, costs, deliverables, compliance with performance indicators, deviations, and closeouts)
- Supporting documents (equipment maintenance and inspections, safety, occupational exposures, plant radiation and contamination levels, quality management, etc.)
- Industrial hygiene and industrial safety (hazards, reports, etc.)
- Annual D&ER reports
- Photos of D&ER activities
- Lessons learned

Waste: To understand the waste generated by D&ER, track waste streams, as well as to estimate the uncertainties, risks, and future waste management activities.

- Waste management plans and reports
- Waste characteristics (physical, chemical, and radiological)
- Waste shipments
- Liabilities resulting from generation of decommissioning waste

D&ER, decommissioning/environmental remediation.

Once a knowledge management (KM) program (including, but not limited to, records) intended for D&ER is established at an early stage of a facility's lifetime, it is expected that it will continue until the end of D&ER (and beyond in some specific cases, e.g., site redevelopment, see Annex A-2). However, KM is not risk free. The loss of knowledge and memory within a D&ER project may be due to different causes. Numerous examples of key factors for the loss of knowledge are given in NEA (2013) and NEA (2014b) and are compiled in Table A-1.2. Clearly, a combination of factors (e.g., lost competence and shaky economy) can play a decisive role, as is readily shown in a national crisis and lack of resources.

A-1.5.11 Conclusions

With the growing experience in nuclear decommissioning, it has become evident that initiatives to ease decommissioning should be taken as early as the siting, design, and construction of nuclear installations. Experience has also

TABLE A-1.2 Key factors for loss of memory in a D&ER project.

1. Technical/environmental factors
1.1 Site degradation and abandonment 1.2 No records/poor archives, missing/insufficient characterization 1.3 No/insufficient update of records 1.4 Deterioration/destruction of files and archives 1.5 Change of record-keeping technology 1.6 Poor identification/collection of incidents and lessons learned
2. Economic factors
2.1 No/insufficient budgets to fulfil KM duties (e.g., due to donors' ceased support)
3. Human factors
4. Organizational factors
5. New regulations/laws

D&ER, decommissioning/environmental remediation; *KM*, knowledge management.

shown that certain operational aspects will be conducive to safer and more cost-effective decommissioning. However, decommissioning-friendly features and activities should not contradict safe and efficient operations. The decision whether to adopt innovative decommissioning-friendly measures is subject to optimization: Techniques such as cost-benefit analysis or MAUA are available to this end. The key conclusions of this annex can be summarized as follows:

- The needs and constraints of eventual decommissioning should be taken care of by the designers, builders, and operators of a nuclear facility. This includes also the return of experience from decommissioning projects. Regulators should also be aware of decommissioning issues at the beginning of a new nuclear project.
- The design features meant to facilitate operation and maintenance should also consider the decommissioning needs: These multipurpose features will enjoy the highest priority.
- Out of all records inherent to design, construction, and operation, those relevant to decommissioning should also identified ("flagged") at the time they are generated and kept safely as long as needed.
- Proper plant layout and access features for people and components can ease decommissioning.
- The design should minimize the production and deposition of activation and fission products.
- If a long period of safe enclosure is foreseen as part of the decommissioning strategy, the design and construction should ensure the needed integrity of structures and components.

References

American Nuclear Society, 1988. Design Criteria for an Independent Spent Fuel Storage Installation (Water Pool Type), ANS-57.7-1988, R 997. ANS, La Grange Park, IL.

Canadian Standards Association, 2014. Decommissioning of Facilities Containing Nuclear Substances, N294-09 (R2014).

Electric Power Research Institute (EPRI), November 2001. Decommissioning Pre-Planning Manual, Final Report. https://www.epri.com/research/products/1003025.

Electric Power Research Institute (EPRI), November 2006. Decommissioning Planning Experiences from U.S. Utilities, Final Report. https://www.epri.com/research/products/1013510.

Garbett, K., 1992. Overview of the Impact of Stellite Removal on Radiation Fields in KWU PWRs, Water Chemistry of Nuclear Reactor Systems. In: Proc. Int. Conf. Bournemouth, 1992, Vol. 2. British Nuclear Energy Society, Ashford Press, Southampton, pp. 31–38.

International Atomic Energy Agency, 1990. Decontamination of transport casks and spent fuel storage facilities. In: Proceedings Technical Committee Meeting, Vienna, 1989, IAEA-TECDOC-556. IAEA, Vienna.

International Atomic Energy Agency, 1995. The Principles of Radioactive Waste Management, Safety Series No. 111-F. IAEA, Vienna.

International Atomic Energy Agency, 1997. Joint Convention on the Safety of Spent Fuel Management and on the Safety of Radioactive Waste Management. http://www-ns.iaea.org/conventions/waste-jointconvention.asp. adopted on 5 September 1997.

International Atomic Energy Agency, 2002b. Record-keeping for the Decommissioning of Nuclear Facilities: Guidelines and Experience. IAEA, Vienna. Technical Reports Series No. 411.

International Atomic Energy Agency, 2005. Financial Aspects of Decommissioning, IAEA-TECDOC-1476. IAEA, Vienna.

International Atomic Energy Agency, 2006. Management of Long Term Radiological Liabilities: Stewardship Challenges. IAEA, Vienna. Technical Reports Series No. 450.

International Atomic Energy Agency, 2007. Considerations for Waste Minimization at the Design Stage of Nuclear Facilities. IAEA, Vienna. Technical Reports Series No 460.

International Atomic Energy Agency, 2008. Long Term Preservation of Information for Decommissioning Projects. IAEA, Vienna. Technical Reports Series No. 467.

International Atomic Energy Agency, 2011. Design Lessons Drawn from the Decommissioning of Nuclear Facilities, IAEA-TECDOC-1657. IAEA, Vienna.

International Atomic Energy Agency, 2011. Good Practices for Water Quality Management in Research Reactors and Spent Fuel Storage Facilities, IAEA Nuclear Energy Series No NP-T-5.2. IAEA, Vienna.

International Atomic Energy Agency, 2014. Decommissioning of Facilities, General Safety Requirements, Safety Standards Series No GSR Part 6. IAEA, Vienna.

International Atomic Energy Agency, 2015. Decommissioning of Pools in Nuclear Facilities, Nuclear Energy Series NW-T-2.6. IAEA, Vienna.

International Organization for Standardization (ISO), 2001. Information and Documentation-Records Management Part 1: General, ISO 15489-1.

International Organization for Standardization (ISO), 2011. Information and Documentation-Management Systems for Records- Fundamentals and Vocabulary, ISO-30300.

International Organization for Standardization (ISO), 2011. Information and Documentation-Management Systems for Records- Requirements, ISO 30301.

International Atomic Energy Agency Methodologies for Assessing the Induced Activation Source Term for Use in Decommissioning Applications, Safety Reports Series No. 95, 2019. IAEA, Vienna.

OECD Nuclear Energy Agency, International Atomic Energy Agency, European Commission, 2012. International Structure for Decommissioning Costing (ISDC) of Nuclear Installations. NEA, Paris. https://www.oecd-nea.org/rwm/reports/2012/ISDC-nuclear-installations.pdf.

OECD Nuclear Energy Agency, 2006. Decommissioning Funding: Ethics, Implementation, Uncertainties. NEA, Paris. https://www.oecd-nea.org/rwm/reports/2006/nea5996-decommissioning.pdf.

OECD Nuclear Energy Agency, 2010. Applying Decommissioning Experience to the Design and Operation of New Nuclear Power Plants. OECD, Paris. https://www.oecd-nea.org/rwm/reports/2010/nea6924-applying-decommissioning.pdf.

OECD Nuclear Energy Agency, September 2012. The Management of Large Components from Decommissioning to Storage and Disposal, NEA/RWM/R, 8. NEA, Paris. https://www.oecd-nea.org/rwm/docs/2012/rwm-r2012-8.pdf.

OECD Nuclear Energy Agency, 2013. In: The Preservation of Records, Knowledge and Memory (RK&M) Across Generations: Improving Our Understanding RK&M, Workshop Proceedings 12-13 September 2012 Issy-les-Moulineaux, 3. NEA/RWM/R, France. https://www.oecd-nea.org/rwm/reports/2013/rwm-r2013-3.pdf.

OECD Nuclear Energy Agency, 2014a. Radiation Protection Aspects of Primary Water Chemistry and Source-term Management, NEA/CRPPH/R, 2. NEA, Paris (2014). https://www.oecd-nea.org/rp/docs/2014/crpph-r2014-2.pdf.

OECD Nuclear Energy Agency, 2014b. In: Preservation of Records, Knowledge and Memory across Generations (RK&M) Loss of Information, Records, Knowledge and Memory — Key Factors in the History of Conventional Waste Disposal, 3. NEA/RWM/R. https://www.oecd-nea.org/rwm/docs/2014/rwm-r2014-3.pdf.

OECD Nuclear Energy Agency, International Atomic Energy Agency, 2017. Addressing Uncertainties in Cost Estimates for Decommissioning Nuclear Facilities. NEA, Paris. https://www.oecd-nea.org/rwm/pubs/2017/7344-uncertainties-decom-cost.pdf.

US Nuclear Regulatory Commission, December 1979. Facilitation of Decommissioning Light Water Reactors, NUREG/CR-0569. USNRC, Washington, D.C. December 1979. https://inis.iaea.org/collection/NCLCollectionStore/_Public/11/532/11532319.pdf?r=1.

US Nuclear Regulatory Commission, 1986. Identification and Evaluation of Facilitation Techniques for Decommissioning Light Water Power Reactors, NUREG/CR-3587. USNRC, Washington, D.C. https://www.osti.gov/servlets/purl/5505702.

US Nuclear Regulatory Commission, 10 CFR 20.1406 Minimization of Contamination https://www.nrc.gov/reading-rm/doc-collections/cfr/part020/part020-1406.html.

US Nuclear Regulatory Commission, 2012. Decommissioning Planning during Operations, Regulatory Guide 4.22. USNRC, Washington, D.C. https://www.nrc.gov/docs/ML1215/ML12158A361.pdf.

Paltemaa, R., 2015. Decommissioning Safety Requirements at Different Stages of a NPP Lifecycle. In: 3rd Int. Nuclear Decommissioning Summit, Berlin, January 28—29, 2015.

Sasidharan, K., et al., 2006. Refurbishment and core conversion of the Apsara reactor. Nuclear Engineering Design 236, 784—795.

Western European Nuclear Regulators Association (WENRA), November 2011. Decommissioning Safety Reference Levels Report version 2.0. http://www.wenra.org/media/filer_public/2012/08/30/wgwd_decommissioning_safety_reference_levels_report_v20.pdf.

Wright, R., Popp, D., Bull, A., 2008. Decommissioning strategy for the AP1000 in the United Kingdom, Westinghouse Electric Co. In: Proc. Nuclear Decommissioning, Liverpool, 22—24 September 2008. Institution of Mechanical Engineers, London.

Annex A-2

Postdecommissioning reuse and redevelopment of nuclear facilities and sites with a view to long-term sustainability

A-2.1 Introduction

Remember that time is money.
Benjamin Franklin 1706–1790

If nuclear decommissioning is viewed in terms of sustainable development, facility and site reuse/redevelopment options play a key role and should not be disregarded. Sustainable development includes a combination of economic development and preservation of natural resources such as grasslands and ponds, and human-made useful structures. Economic development should also ensure social advancement and community integrity. All factors are best served by carefully planned redevelopment of decommissioned sites to provide job security, new productive activities, and/or new services to the public. Experience with redevelopment in the nuclear and nonnuclear industry suggests that a strong commitment by all those involved (e.g., former and new owners, site planners, donors, public institutions) will be vital in promoting options that are both financially profitable and accepted as responsible and sensible by the broader community and other stakeholders.

Traditionally, it was expected that nuclear decommissioning would be aimed at the removal and disposal of waste and the restoration of a site to almost pristine conditions. This concept is not acceptable any longer as the only strategy available and decommissioning should not be viewed as the sad endpoint of a facility or site program but should rather be the starting point of a successful redevelopment and reuse phase. A decommissioning strategy based on total demolition of a facility and unrestricted release and handover of a site long after "close of (decommissioning) business" is not ideal, since it leaves the further use of the site in the dark. Rather, the focus should move to redevelopment options that must be part and parcel of the early planning for decommissioning. As of today, the postdecommissioning redevelopment of nuclear and nonnuclear facilities is not optimized for the very reason that this concept is not integrated with the decommissioning strategy: Traditionally, decommissioning and site release are planned for and implemented first, and redevelopment options are passed on to the new owners for their planning, which results in significant delays.

Redevelopment can also—at least partly—offset the cost of decommissioning. To best use the redevelopment option, however, the redevelopment potential of a nuclear site should be identified at an early stage in the site life cycle since the extent (and the costs) of decommissioning can be influenced by the selected redevelopment strategy. Early reuse and redevelopment plans should ensure that best use is made of the assets, e.g., infrastructure and land resources (roads, railways, etc.) associated with the site. For example, emphasis should be given to the identification of structurally sound buildings not to be demolished. Redevelopment also supports the (ethical, if not legal) requirement that gapless employment needs to be ensured. While this is a direct responsibility toward the workers, it also affects the reactions of the stakeholders (local community, elected officials, land planners, etc.): Both social and economic factors affect progress and final success of the decommissioning project—at least in terms of image. The responsibility of ensuring sustainable social development should not be seen as a burden but an opportunity suggesting proactive action to the site owners.

A-2.2 Redevelopment and sustainability

Four things come not back: the spoken word; the sped arrow; time past; the neglected opportunity.

Omar ibn al-Halif, Aphorisms

Fundamental environmental principles—reduce, recover, recycle, and reuse (the "Four R's")—are inherent to sustainability and successful decommissioning. Applying these principles to decommissioning helps minimize radioactive contamination and—as far as practicable—recover, recycle, and reuse materials, equipment, buildings, and sites. This annex is especially devoted to buildings and sites that will remain after the end of a decommissioning project—and are ready for a new life as the opportunity arises.

Sustainability includes the need to combine economic development with the preservation of natural resources (such as grassy or forested land) and cultural heritage (such as industrial buildings of historic interest). Reuse is a component of sustainability, for example, because it conserves original long-lasting construction materials. Additionally, the reuse of land and buildings for productive use offers a valuable alternative to spoiling more virgin sites, which would be better used for public, undamaging fruition.

The reader should note that for the purposes of this annex, a greenfield site is a former nuclear site that has been granted unrestricted release from radiological control, buildings have been further demolished, and no further use is planned. Extremely restrictive clearance criteria for unrestricted release have become over the years more and more difficult for large sites to comply with and in some cases have made the cost of unrestricted release almost prohibitive. The issue of decreasing decommissioning costs through a different option (brownfield or restricted release) for the postdecommissioning site end state has therefore become more urgent.

Sustainability also associates economic development with maintenance of the community identity and creation of new social opportunities. All of these aspects can best be served by the preplanned, systematic redevelopment of sites to provide continuity of jobs and/or new job opportunities. Additionally, the concept of sustainability suggests a more transparent and more open decision-making process than has been commonly pursued in many nuclear programs to date. The nuclear industry should learn how to incorporate the participation of a wider range of concerned parties (the stakeholders) and be ready to consider their viewpoints as legitimate (which does not mean that **all** viewpoints are right and realistic, and the issue remains for the responsible organization to tackle conflicting views). Regardless of inherent difficulties for different parties to achieve a reasonable agreement, experience of several decades with redevelopment both inside and outside the nuclear industry suggests that successful engagement and commitment of all stakeholders is an important success factor in decommissioning and postdecommissioning projects. The objective of these consultations is to achieve outcomes that are both technically and financially reasonable for the site owner and accepted as responsible and convenient by the wider range of stakeholders.

Finally, site redevelopment incorporates the concept of circular economy (instead of the old-fashioned linear economy), a growing trend in current industrial projects (IAEA, 2019). It is generally accepted that decommissioning be viewed as a component of a facility's life cycle jointly with design, construction, and operations. Site reuse would "close the life cycle loop" and open a new loop. Multiple iterations are then possible in a sequence of subsequent site reuses. Current guidance includes recognition of decommissioning at a very early phase of a facility's lifetime, but not of site redevelopment yet. As site redevelopment is a continuation of decommissioning, it should also be taken account of while planning for decommissioning and even earlier during a facility's siting, construction, and operations.

The diversity of nuclear facilities and their sites and their different settings in terms of traditions, location, culture, and socioeconomic contexts make it impossible to identify generic redevelopment options. Instead, this annex will seek some general guiding principles and will provide examples of how opportunities for redevelopment can orient the decommissioning planning and be conducive to fast, safe, and cost-effective decommissioning and follow-up activities. Specific examples of site reuse within the nuclear decommissioning process will also be discussed.

By describing the variety of uses for nuclear redevelopment, and by reporting on the main factors supporting successful redevelopment, this annex will hopefully encourage those responsible with nuclear operation and decommissioning to incorporate these considerations in their plans.

To date, there is no other systematic, international coverage of redevelopment of nuclear buildings and sites than two IAEA reports (IAEA, 2006) and (IAEA, 2010), and a recent Elsevier book (Laraia, 2019). These publications literally offer hundreds of case studies for the reader's considerations. This annex will only make use of concrete examples that have not been presented in those publications.

A-2.3 Key assets of nuclear sites

Most nuclear facilities enjoy a flat site, good access to power links, transport and other communication infrastructure, and a skilled workforce, all of which are conducive to fast and effective redevelopment. In some cases, nonnuclear activities are already based on the site before the end of nuclear operations (e.g., catering, public transport, and other contractors' activities) and provide the foundation for site rebirth, often in a new capacity.

As an introductory example, it would be unduly expensive and time-consuming to demolish an ex-nuclear building if that building could be readily adapted as part of the site reuse strategy. Similarly, in view of site redevelopment, building or land decontamination should be executed only to a point compatible with further use of a site: In this way, industrial reuse might be consistent with site release criteria less conservative than for residential scenarios. Restricted release (brownfield) would then replace the traditional unrestricted release (greenfield). As one more example, the decommissioning workers can be hired and retained more easily and at a lower cost if continued reemployment within the site redevelopment project is foreseen. Also, redevelopment of a nuclear site can facilitate public acceptance of the nuclear industry as a whole.

An important task in investigating the redevelopment options of a given site is to identify the relevant assets of the site. Typical assets for a nuclear site (sometimes referred to as the "family jewels") could include the following:

- electricity network connections;
- transport infrastructure such as road, rail, or sea accesses;
- a local skilled workforce;
- offices, especially those situated in historical buildings;
- support services (schools, gas stations, etc.)
- machine shops, workshops, and general manufacturing facilities, especially those equipped with powerful machinery;
- a large level land area suitable for a large production factory investment or for a smaller factory with adjacent space for future enlargement; or
- natural features (parks, forests, small lakes, etc.)

Many nuclear power plants (NPPs) have a good connection to a national or international (Fig. A-2.1) electricity network, which makes them attractive for redevelopment either as an alternative electricity generation plant or for an industrial development utilizing large amounts of electric power.

Site accessibility is essential either for industrial uses or for recreational amenities. A nuclear facility's operation needs frequent access of equipment/consumables/specialists for maintenance and refurbishment, visitors, etc. However, several nuclear plants are located in remote areas due to security or radiological factors. This could complicate industrial redevelopment unless their sites are reused for hazardous or secret installations.

The site employees and workers pose a serious challenge to the decommissioning and the site redevelopment. It is a frequent case that some skilled personnel leave the site before completion of decommissioning due to job security concerns. Early planning for redevelopment may persuade some of them not to leave and be associated with redevelopment activities. The availability of skilled personnel from the nuclear facility is an opportunity for site redevelopment, especially if this is compatible with their actual or potential skills (e.g., through dedicated retraining). The Barnwell site, SC, United

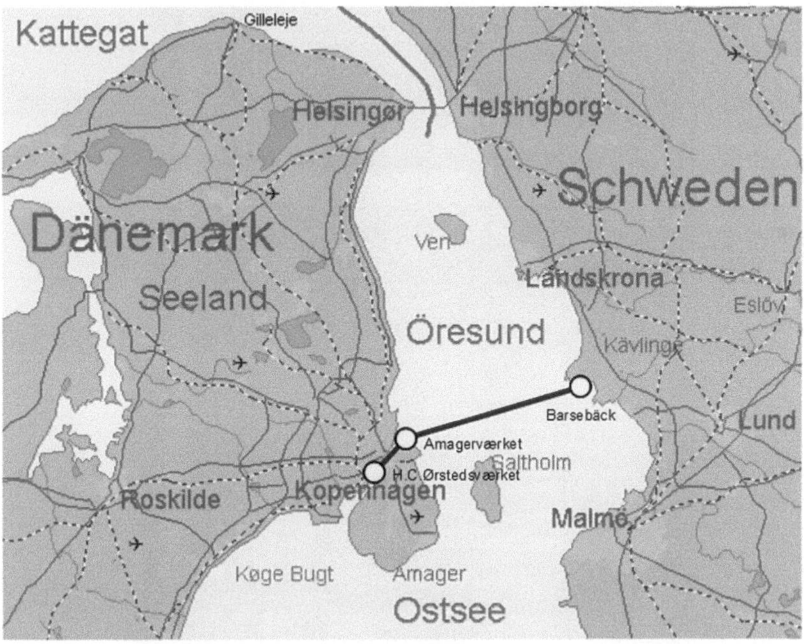

FIGURE A-2.1 Planned 400 kV cable from H.C. Ørsted power station in Denmark via Amager power station to decommissioned Barsebäck NPP in Sweden (Energinet.dk, Wiki commons).

States, is a relevant case in question. At Barnwell, the Government Training Institute is a decommissioned nuclear facility that has been converted into a training site. Fig. A-2.2 shows some training activities at this Barnwell institution. A similar option has been exploited with the support of the Hanford Nuclear Reservation site (Fig. A-2.3).

The former Oak Ridge National Laboratory (whose later offspring is the East Tennessee Technology Park, ETTP) is a large territory where a major transition from nuclear use to other industrial applications has already taken place. A recent case is described in the following. The 27,000 m^2 K-29 was one of five enormous gaseous diffusion buildings situated at ETTP. These installations were used to enrich uranium for defense and civilian purposes until the complex was shut down in the mid-1980s. K-29 was the first building to be demolished (2006). After demolition of the building, the slab was covered with asphalt. Trailers were installed on the slab to house the large ETTP decommissioning workforce.

FIGURE A-2.2 15th civil Support team trains with Army Mountain Warfare School instructors at Barnwell, South Carolina, United States (130417-A-WM282-086, Wiki commons).

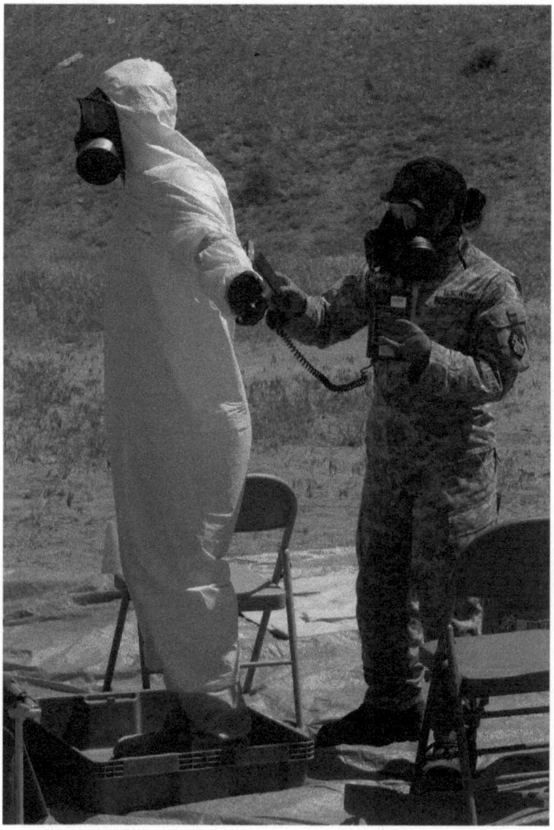

FIGURE A-2.3 A reconnaissance and surveillance chemical, biological, radiological, nuclear (CBRN) soldier assigned to the second chemical Battalion, Fort Hood, Texas, gets inspected for contamination after leaving a live radioactive area. This activity was part of the Defence CBRN Response Force (DCRF) validation training exercises that took place outside the Hanford Nuclear Reservation and the Hazardous Materials Management and Emergency Response (HAMMER) Training and Education Centre test site. *Credit to USDOE, Wiki Commons.*

The slab removal project, which began in October 2018 and ended in June 2019, included the removal of the trailers, asphalt, and slab in view of the economic redevelopment of the site. A regulatory process was required for the 32-ha area to be transferred from government ownership to the community for private sector development.

It is planned to complete all demolitions at ETTP in 2020 and to turn ETTP into a multiuse industrial park. Since decommissioning and site remediation began in 1998, almost 500 installations have been demolished, some 500 ha have been transferred for economic redevelopment, and a 1200-ha conservation area for public use has been created (DOE, 2019).

Offices offer valuable opportunities for practically any forms of redevelopment. Any new site reuse (be it industrial or recreational) will need space for, e.g., administration, reception, coffee bars, etc. Depending on building amplitude, layout, orientation, and so on, reuses such as cinemas, shopping malls, and so on may be of interest. Many nuclear installations house laboratories, warehouses, silos, and so on that can be easily converted to new functions. The challenge here is to harmonize the supply with the demand. Prestigious buildings may be turned into museums, temporary exhibition areas, and so on.

Support services such as schools, restaurants, hotels, or public transport are often available to the workforce during the plant service life and the decommissioning phase. Such services are often viewed as an asset for new organizations planning to settle on the redeveloped site. Long periods of discontinued services should be avoided at all costs, as these interruptions could cause the permanent closure of such services.

Nuclear workshops could be converted to nonnuclear uses, often with the active participation of skilled workers who have ceased their previous functions. In most cases, decontamination of nuclear workshops can be carried out readily. Remaining stocks of spare items or consumables (gas, lubricants, etc.) can be reused for new workshops or other installations on-site.

A large level site is ideal for new industries while keeping enough space for on-site roads, bridges, parking lots, etc. Such infrastructure as electric power, sewage, industrial and potable water is essential. Commonly, all these systems are available during the operation of a nuclear plant, but some retrofitting may be required to adjust to the new functions.

If recreational reuse is planned, trees, lakes, meadows, and other natural features are key assets. Lakes and ponds may help reestablish original flora and fauna: Marinas and nautical clubs could be considered. The return of wildlife can foster ecological/biological initiatives that would make best use of environmental assessments conducted during former plant's operations. Large fields can be returned to agriculture, and the preserved buildings can be reused for the local food industry.

Fig. A-2.4 refers to a site redeveloped for recreational activities. The Rocky Flats site, a former nuclear weapons production facility, has been cleaned up and returned to grasslands and prairies. By the way, this picture emblemizes the great interest of the media (an important category of stakeholders) in redevelopment of former nuclear sites.

The radioactive systems of Trojan NPP were fully dismantled by 2006. The giant cooling tower was demolished by dynamite implosion in the same year. Demolition work on the remaining structures went on through 2008. The central office building and the reactor building were demolished in 2008. Five buildings remained: two warehouses, a small building on the river side, a guard cabin, and offices outside the restricted-access facility. Then the site was returned to unrestricted use (except for the remaining independent spent fuel

FIGURE A-2.4 Journalists at the Rocky Flats site, June 4, 2014. *Credit EPA, Wiki Commons.*

storage installation, ISFSI). Most of the former NPP site is now Trojan Park, a pleasant day-use park offering opportunities for picnicking, trekking, cycling, fishing, rowing, and golfing. The 30-ha park includes a lake, multipurpose footpaths, and a wide sport playing field (Fig. A-2.5).

FIGURE A-2.5 Trojan Park. *Credit to Gary Halvorson, Oregon State Archives, Wiki Commons.*

A-2.4 Factors interfering with redevelopment of buildings

A beneficial reuse is normally easier to identify for a newer structure than for an older one. The long-term use of an older structure could be not cost-effective or could be even impossible due to structural deterioration and lack of compliance with modern building standards. During nuclear decommissioning, at-depth decontamination or removal of large concrete pieces can degrade the integrity of building structures. A financial assessment will be required to ensure that funds are available to retrofit structurally deteriorated areas to modern standards.

Since many of these nuclear structures will be more than 30 years old at the redevelopment stage, compliance with current building requirements could be possible only at high cost and after complex technical reevaluations, thus actually excluding some reuse options.

A number of subsurface/underground features may interfere with the ability to redevelop a nuclear site after completion of decommissioning. Such features include tanks, cables, trenches, water pipes, fire protection systems, sewerage, and other waste collection systems. Some of these systems may require removal or remediation during decommissioning. Depending on the planned reuse of the site, clean underground systems may or may not be beneficial to the redevelopment.

In many environments, full demolition of buildings and reconstruction from scratch remains the preferred choice, although this may conflict with the sustainability principle advocated in this book (e.g., in terms of waste minimization and reduced use of raw materials). Regulatory constraints, timing, and lack of interest by sponsors (all boiling down to costs) may often preclude redevelopment of buildings. These factors explain why many redevelopment projects are subsidized by public funds.

A-2.5 Nuclear reuse

In many instances, to continue the use of a nuclear site for nuclear activities can be a logical decision. This is due to benefits such as reemployment of skilled staff used to working in radiation environments, public acceptance of nuclear operations, available characterization of the site (seismicity, demography, prevailing winds, etc.), local availability of services, etc. To date, to build new NPPs at existing nuclear sites is deemed profitable. In the UK, EDF Energy plans to build four reactors at two sites, with public consultation completed and initial groundwork beginning on the first two reactors, sited at Hinkley Point. EDF applied for a site license for Hinkley Point in mid-2012, and the initial license for Hinkley Point C was granted by UK's regulators by the end of that year. Planning permission and environmental licenses (e.g., for radioactive and nonradioactive discharges) for Hinkley Point C were granted in 2013 (EA, 2013). Horizon Nuclear Power (a UK subsidiary of Japan's Hitachi) has plans for new reactors at the sites of the defunct Wylfa and Oldbury NPPs. An agreement has also been made, which allows in principle for Chinese-designed reactors to be built on the site of the decommissioned Bradwell NPP. Recently, the Nuclear Decommissioning Authority (NDA), as the owner of decommissioning sites in the United Kingdom has established a strategy to optimize the commercial value of those sites—either for nuclear or nonnuclear redevelopment—as a means to mitigate the cleanup expenses (NDA, 2016). Estimating the worth of nuclear sites for new owners is uncertain, but in general the scarcity of new nuclear sites is expected to increase the commercial profitability of existing sites.

In some situations, following decommissioning of a given nuclear facility, a nuclear facility of different characteristics may be installed. For example, a decommissioned facility might be converted to a radioactive waste storage facility (e.g., in a country not having a national waste disposal site, decommissioning waste might be stored at their generation site together with operational waste—see the Würgassen case in Section 7.40). A few cases of nuclear reuse have been arbitrarily selected and are given in the following.

The Hanford Test Reactor (HTR, aka Test Pile) was constructed in late 1943 on the Hanford Site, United States, as a zero-power test facility. For some 30 years, the facility was used in scientific research to test nuclear reactor materials for purity and as a neutron source for reactor experiments in physics and instrumentation. In 1972, HTR was shut down. Advancements had simply made it possible to conduct such testing and experimentation elsewhere leaving the facility obsolete.

UNC Nuclear Industries Inc., then custodian of the facility in Hanford's 300 Area, was requested in 1976 by the US Department of Energy (DOE) to defuel HTR. Work commenced in August of that year and was completed within 2 months. In June 1977, the DOE authorized UNC to accomplish the final phase of the HTR dismantling and release of its host facility, the 305 Building, for use in other R&D programs at Hanford.

Following demolition and waste removal operations, the D&D crew left an open pit of dimensions that would meet the construction needs of the new tenant. Fig. A-2.6 shows the postdemolition status of the 305 Building before the Hot Cell Verification Facility (HCVF) was installed.

Shown in Fig. A-2.7 is the HCVF assembly checkout/qualification area pit constructed in approximately the same location as the open pit in Fig. A-2.6. Other structural modifications to the 305 Building interior are also visible. The steel beam assembly supports a 5-t bridge crane, and the receiving and storage bay access door can be seen at rear right. Flooring was also installed.

In 1985, the 305 Building was converted again to support mechanical development, mockup, and testing for the Plutonium/Uranium Extraction (PUREX) Process Facility Modifications Program. Later, the building was used to test shipment cylinders, casks, and capsules (Gerber, 1993). It had

FIGURE A-2.6 The postdemolition status of 305 Building before construction of the Hot Cell Verification Facility (HCVF). *Credit to USDOE.*

FIGURE A-2.7 HCVF assembly checkout/qualification area. Hot Cell Verification Facility. *Credit to USDOE.*

potential historic status according to the criteria of the National Historic Preservation Act (NHPA), but the fact that the reactor had been removed and the building had been through several alterations made it unlikely that it could be listed in the NHPA. The reader should note that a number of old industrial facilities/sites including some nuclear ones have been given historic recognition and converted into monuments, museums, or cultural centers (among them the Hanford's B Reactor): This option has been often considered a viable redevelopment for decommissioned facilities.

The adaptation of the former circulator halls for buffer waste storage at the decommissioning Bradwell NPP, United Kingdom, is shown in Fig. A-2.8.

In another option, spent fuel could be stored at the decommissioned site. The construction of an ISFSI is currently common to a number of decommissioned US NPP sites, pending availability of a centralized high-level waste repository in the country (Fig. A-2.9). It should be noted that ISFSIs impede the effective redevelopment of the whole site.

The history of the Baker Brothers site in Toledo, OH, United States, can be found in DOE (2020). During WWII, DuPont and the University of Chicago contracted the Baker Brothers Inc. to machine and shape natural uranium slugs from processed uranium metals for use as nuclear fuel in the production reactors sited at Oak Ridge and Hanford. Environmental remediation under the Formerly Utilized Site Remediation Action Program was executed in 1995 and was certified completed in 2001. The property is now privately owned and houses several commercial enterprises (Fig. A-2.10).

FIGURE A.2-8 Bradwell circulator halls being converted into waste stores. *Credit to Magnox Ltd.*

FIGURE A-2.9 Aerial view of decommissioned Yankee Rowe site. Note the white ISFSI casks in the middle of the photo. *ISFSI*, independent spent fuel storage installation. *Courtesy of NRC.*

FIGURE A-2.10 Baker Brothers Inc. site, Toledo, Ohio. *Credit to USDOE, Wiki Commons.*

A truly historic case is described in the following. In late 1943, the air raids on Berlin were becoming so heavy to impede scientific experiments. German physicists started to seek a site, which was relatively safe from air raids. Additionally the site had to be located as far as possible from the advancing Soviet Army in case (as it did happen later) Germany would be occupied by the Allied Forces. During an exploratory visit to Haigerloch, Southwest Germany, the scientists found the rock-sheltered beer cellar of a local inn and rented it for their work.

The reactor was placed in a concrete cylinder. Between the external concrete shell and the internal aluminum shell, ordinary water was filled in as a reactor coolant. The aluminum shell had a diameter of 210 cm and a similar height and contained a magnesium vessel. The space between the two vessels contained a 40 cm layer of graphite bricks. The 664 uranium cubes (edge of 5 cm) attached to the lid were then inserted into the magnesium vessel in a 3-D grid. The lid was bolted on top.

A neutron source was placed into the center of the reactor. There were openings in the top to insert neutron probes for monitoring the neutron flux in all parts of the reactor. Finally, heavy water was poured in slowly, and the growth of the neutron flux was monitored during the filling-up. If the reactor had gone critical, the experiment would have been successfully terminated.

Later calculations demonstrated that the Haigerloch reactor was too small to reach criticality. However, enlarging the reactor was impossible in April 1945 due to the lack of the needed quantities of both heavy water and uranium. Shortly after the last experiment, Haigerloch was occupied by a selected American squad. The Americans disassembled the reactor, and most German scientists moved to other countries to continue their work.

The site is now a public museum: It houses a replica of the original reactor, and items from the time experiments, as well as a number of boards and other media illustrating the beginning and the growth of the nuclear era (Atomkellermuseum undated).

A-2.6 The nonnuclear industry

As the nonnuclear industry has considerably predated the nuclear industry, it is somewhat obvious that reuse/redevelopment considerations appeared first in the nonnuclear industry. Additionally the sheer number of redundant factories demanded an urgent response to the issues posed by deserted, often hazardous sites. While full demolition and reconstruction was the approach taken initially, environmental, socioeconomic, and political arguments dictated by the sustainability principle slowly led to more consideration being given to reuse/redevelopment.

Adaptive reuse usually means the reuse of sites or buildings for purposes other than those they were originally conceived for. When old structures become unfit for their functions and requirements, adaptive reuse comes to hand as a sustainable option for site regeneration.

The urban areas that have mostly become subject to adaptive reuse in the 20th century include uncountable industrial sites: The abandonment of these sites was mostly due to the development of technology and ecological factors that drove the industrial processes away from the city center. Therefore, adaptive reuse becomes a path to urban renewal in which historical industrial sites that have been abandoned reassume a social meaning. Heritage buildings and spaces, once secured, retrofitted, and maintained, remain physical witnesses of collective memory. This contributes to their fascination. It should be noted that the principles and factors driving adaptive reuse remain valid for nonurban sites (such as many nuclear sites): In all cases, site-specific circumstances will dictate the preferred solution.

The following provides a few representative examples of redevelopment projects for industrial factories and sites.

Circus Space is a London-based center of education and exhibition of the traditional circus arts. Once the building was the vestry of St Leonard Shoreditch Electric Light Station, built in 1895 (a vestry was a committee of the local secular and ecclesiastical government for an English parish, who originally met in the vestry or sacristy of the church, consequently named informally as the "vestry"). Rubbish was burned at this Victorian station to generate electricity for the borough. It also provided steam to heat the public baths. The former "generating chamber" and "combustion chamber" now provide facilities for circus training and production. The building was rescued from deterioration in 1994 and redeveloped.

The last relics of its former use can be seen in a Latin inscription above the door: "E pulvere lux et vis," or "Out of the dust, light and power" (AtlasObscura, undated) (Fig. A-2.11).

Also based in the Shoreditch borough of London is the revamped Tramshed restaurant. This building was built as an electricity generating substation and transformer station for the tramway system. As expected, the history of the site gives charm and style to the restaurant (Fig. A-2.12).

A research quoted in Historic England (2015) includes 17 case studies of former industrial sites in the United Kingdom that have been redeveloped, as well as a survey of developing organizations, tenants and owners, local authorities, and workshops with a variety of other stakeholders committed to developing historic buildings. Redeveloping a former industrial site for residential purposes remains a common option for industrial sites: One such case is London Fanshaw Street, where 19th century former industrial buildings have been converted into living spaces and commercial premises (Fig. A-2.13).

In 1891, the family Mendl founded a large bakery in Vienna. An anchor was imprinted on all their products, which became the logo of the company and later on gave the company its current name "Ankerbrotfabrik." The bakery was a U-shaped construction in brickwork and included a bakery dedicated to only baking brown bread, a storeroom, a cart, and a horse shed, as well as a

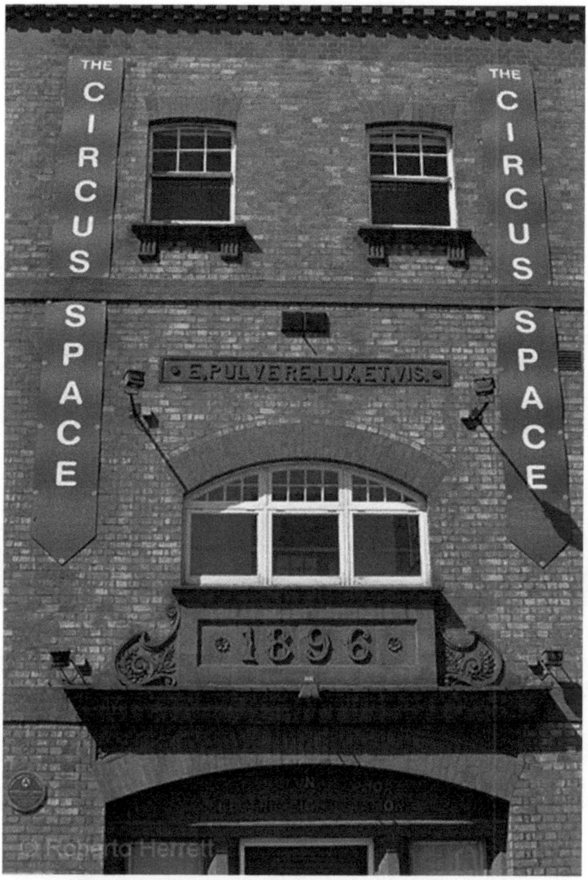

FIGURE A-2.11 Circus Space, London, a former power station. *Photo by Laraia, M., June 2019. Beyond Decommissioning- The Reuse and Redevelopment of Nuclear Installations. Woodhead Publishing (Paperback ISBN: 9780081027905).*

residential and office building. The company became the biggest bakery in Europe in only 10 years. In 1938, the factory was (aryanized) by the Nazi Germans, and the Mendl family fled into exile. During the war, the factory was heavily damaged and nevertheless managed to continue production. Soon after WWII, the Mendls came back to Vienna and the Ankerbrotfabrik was returned to them. Many ownership changes followed. Eventually the family Ostendorf started an extensive restructuring program, which led to the abandonment of the older section of the bakery. This abandoned building was close to demolition but was rescued in 2009 when acquired by "Loft City GmbH & Co KG." The new cultural center Brotfabrik Wien is coming into being as part of the historic building (Fig. A-2.14). It features halls, galleries, studios, offices, lofts, and restaurants. The selected buildings are located around two inner

FIGURE A-2.12 Entrance to Tramshed, London. *Photo by Laraia, M., June 2019. Beyond Decommissioning- The Reuse and Redevelopment of Nuclear Installations. Woodhead Publishing (Paperback ISBN: 9780081027905).*

FIGURE A-2.13 London Fanshaw Street, a former industrial site. *Photo by Laraia, M., June 2019. Beyond Decommissioning- The Reuse and Redevelopment of Nuclear Installations. Woodhead Publishing (Paperback ISBN: 9780081027905).*

FIGURE A-2.14 The "Ankerbrot" factory, Vienna, Austria. *Photo by M. Laraia.*

courtyards of the old factory. In this process, the buildings, legally protected by a preservation order, are not only maintained but partly rebuilt to match their original features (Rapp et al., 2011).

The Centre for Art and Media (ZKM is the German acronym), Karlsruhe, Germany, was founded in the 1980s. Because of the growing media realm and changes in the art field, the local government, the University of Karlsruhe, the University of Music Karlsruhe, the Centre for Nuclear Research Karlsruhe, and various organizations and representatives of the art world established the ZKM Project Group in 1986.

In 1989, the fusion of the arts in one ZKM building became a concrete chance. Initially, the ZKM was spread in various buildings in Karlsruhe. Then for a while, an area to the south of the Karlsruhe Railway Station had been selected for a new building. However, plans for the new construction were dismissed in 1992 on grounds of costs and space in favor of the conversion of an idle factory building. This case exemplifies the constant debate between a new construction or demolition/reconstruction and conversion of an existing building.

The so-called Hall A, an industrial relic constructed in 1914–1918 as a weapons and munitions factory, was selected for the conversion. Hall A was a building of considerable size (312 m long) and enormous volume. Its impressive interior consists of a series of ten glazed atriums.

The conversion began in 1993. Moving to Hall A in 1997, the ZKM had a media theater, concert and events spaces, studios and institutes for research and production, and a media museum, all in the same place. In a second phase, the spaces for the Museum of Contemporary Art (moved in 1999) and the Karlsruhe University of Arts and Design (moved in 2001) were established (Verberne, 2013).

The adaptive reuse strategies applied to the Fiat Lingotto and Van Nelle manufacturing factories (see further in the following) both inaugurate factories of information and creativity while keeping the original identity in line with today's culture. These two projects maintain the overall architecture of the site and visual integrity of the buildings.

Fiat's Lingotto factory, Turin, Italy, was originally completed in 1923; unlike any other car factory, Lingotto was based on a spiral assembly line that climbed up through the building and a concrete banked racetrack on the roof. This structure has five floors and helical ramps at the northern and southern ends, which basically provided the vertical movement of the production process. The structure is supported by a grid of reinforced concrete columns and beams. The columns are some 35 cm wide and are installed on a grid of some 6 m.

In 1983, the owner Fiat opened a competition to seek ideas for the site rehabilitation. The design concept submitted by the famous architect Renzo Piano was the winner.

The proposal aimed at a multiuse complex intended to stimulate collaboration between public and private sectors and assist Turin in its research, education, cultural, and technological priorities. The design solution was not preservation or restoration in the traditional or historical sense, but a form of adaptive reuse that was largely based on the spirit of the site and the structures.

The Fiat's Lingotto factory, like late 19th- and early 20th-century factories in many industrial cities, was located on the then periphery of the city, and when production stopped, workers lost their jobs, and the local economy was disrupted. Consequently, the stigma of the site was an additional issue, which had to be faced in the redevelopment process.

While the original concrete structure was found to match the new functions planned, the original steel windows of the outer shell were heavily corroded and had no insulation, and the stucco, which had been applied to the brick surfaces, was falling apart. The new system has insulated glass that looked like the old windows; while keeping the brickwork openings, the windows are less divided than the old ones, resulting in slightly larger sheets of glass. Colorful roll-down canvas sunblinds were placed on the exterior. The building's exterior remains generally unchanged—only the windows and their arrangement have been modified.

The redevelopment of Fiat Lingotto has been a big architectural hit. Due to its efficient structure, simple outside, and lack of inner finishes, the redeveloped factory had few limitations and offered many different uses (Moshaver, 2011).

The Van Nelle Factory, Rotterdam, Netherlands, was built in 1925–1931. It is a model of *Nieuwe Bouwen* architecture in the country. In 1997, the concept of the new Van Nelle Design Factory was established. It involved the assembly of businesses dealing in design and communication and providing them with a comprehensive infrastructure required by the remote location of the factory. The renovation and adaptive reuse project firstly addressed the three factory buildings and was bound to adhere to the original design rationale.

The building design, including doors, partitions, and windows, maintained the general sense of openness and transparency. In continuity with the character and design of the old buildings, the new buildings tend to look industrial and efficient.

The factory grounds extensively keep their original design. The glass aerial passageways linking the buildings and the shipping office are among the most conspicuous features.

The original factory was designed on the assumption that a modern, transparent, and healthy working environment in a green landscape would be good both for production purposes and for the well-being of the workforce. Its design took in full account the technical uses expected of a factory, while giving a special focus to the development of the lower social classes.

The factory results from the application to architecture of cultural and technical concepts developed in the early 1900s. This new, functional approach enjoyed mass support from the very beginning. Therefore, the spirit of production has been extended to the new building.

The complex has enormous value in the context of architectural heritage.

The main factory buildings were built in reinforced concrete with a flat slab floor system. Its current appearance results from an early curtain wall comprised of steel window assemblies, each one floor high supported on a floor-by-floor basis that runs continuously in front of the structural frame.

The factory's structure has been preserved in its original state, although it no longer contains any machinery, the manufacture of tobacco products having been discontinued in 1998.

When the factory was given a different objective and scope (information and creativity) just before the turn of the millennium, the identity of the old building was carefully preserved. The architect in charge of the redevelopment based his work on the assumption that the factory would be recognized as a world heritage site (Moshaver, 2011).

References

Environment Agency, 2013. Applications by NNB Generation Company Limited for Environmental Permits for a Proposed New Nuclear Power Station at Hinkley Point, Somerset: our decisions. https://assets.publishing.service.gov.uk/government/uploads/system/uploads/attachment_data/file/291291/LIT_7929_d2fe43.pdf.

Gerber, M.S., September 1993. Multiple Missions: The 300 Area in Hanford Site History, WHC-MR-0440. https://www.osti.gov/servlets/purl/10116166.

Historic England, Use of Historic Buildings for Residential Purposes, Scoping Report – Draft, 3 July 2015. https://historicengland.org.uk/content/heritage-counts/pub/2015/use-of-historic-buildings-for-residential-purposes-pdf/.

https://www.atlasobscura.com/places/circus-space.

https://www.haigerloch.de/de/Tourismus/Atomkellermuseum.

International Atomic Energy Agency, 2006. Redevelopment of Nuclear Facilities after Decommissioning. IAEA, Vienna. Technical Reports Series No.444.

International Atomic Energy Agency, 2010. Redevelopment and Reuse of Nuclear Facilities and Sites: Case Histories and Lessons Learned. IAEA, Vienna. Nuclear Energy Series No NW-T-2.2.

International Atomic Enery Agency, IAEA Workshop Highlights Novel Trends in Decommissioning, Jun 28, 2019. https://www.iaea.org/newscenter/news/iaea-workshop-highlights-novel-trends-in-decommissioning#:~:text=Decommissioning%20of%20nuclear%20power%20plants,and%20reduced%20risk%20of%20delays.&text=Circular%20economy%20is%20a%20socio,making%20the%20most%20of%20resources.

International Atomic Enery Agency, 2019, IAEA Workshop Highlights Novel Trends in Decommissioning, https://www.iaea.org/newscenter/news/iaea-workshop-highlights-novel-trends-in-decommissioning#:~:text=Decommissioning%2520of%20nuclear%20power%20plants,and%20reduced%20risk%20of%20delays.&text=Circular%20economy%20is%20a%20socio,making%20the%20most%20of%20resources.

Laraia, M., June 2019. Beyond Decommissioning- The Reuse and Redevelopment of Nuclear Installations. Woodhead Publishing. Paperback ISBN: 9780081027905.

Moshaver, A., 2011. Re Architecture: Old and New In Adaptive Reuse of Modern Industrial Heritage. https://digital.library.ryerson.ca/islandora/object/RULA%3A1778.

Nuclear Decommissioning Authority, 2016. Strategy Effective from April. https://assets.publishing.service.gov.uk/government/uploads/system/uploads/attachment_data/file/512836/Nuclear_Decommissioning_Authority_Strategy_effective_from_April_2016.pdf.

Rapp, C., et al., 2011. The History of a Great Bakery (in German, Ankerbrot – Die Geschichte einer großen Bäckerei). Brandstätter-Verlag.

US Department of Energy, June 4, 2019. Oak Ridge Transforming Former Uranium Enrichment Area for Development. https://www.energy.gov/em/articles/oak-ridge-transforming-former-uranium-enrichment-area-development.

US Department of Energy, May 2020. Site, A FUSRAP site. Toledo, Ohio. https://www.energy.gov/sites/prod/files/2020/05/f74/ToledoFactSheet_0.pdf.

Verberne, K., 2013. From an Old Decommissioned Building to a New Architecture Museum. http://docplayer.net/35748292-From-an-old-decommissioned-building-to-a-new-architecture-museum.html.

Bibliography

Baggio, P.M., 2014. From Industrial Archaeology to Industrial Tourism: the Upper Vicenza Province (in Italian: Dall'archeologia industriale al turismo industriale: Il caso dell'Alto Vicentino). University Ca' Foscari, Venice. http://dspace.unive.it/bitstream/handle/10579/4989/843851-1182188.pdf.

Barasch, D., Thuras, D., 2019. Ruin and Redemption in Architecture. Phaidon Publ, ISBN 9780714878027.

Bullen, P.A., Love, P.E.D., 2010. Rhetoric of Adaptive Reuse or Reality of Demolition: Views from the Field. Cities 27, 215–224.

Edwards, R.I.B., 2017. The Adaptive Reuse of the Former Thesen Island Power Station: a Case Study. Cape Peninsula University of Technology. http://etd.cput.ac.za/handle/20.500.11838/2540.

Giuliani, I., 2010. Industrial Decommissioning and Creative Cities- Two Processes of Urban Alteration between Physical Revitalization and Territorial Upgrading: Zona Tortona e Ventura Lambrate, Milan. In: in Italian: Dismissione industriale e città creative- Due processi di trasformazione urbana tra riqualificazione fisica e strategie di promozione del territorio: i casi di Zona Tortona e Ventura Lambrate a Milano. https://www.politesi.polimi.it/handle/10589/1202.

International Atomic Enery Agency, IAEA Workshop Highlights Novel Trends in Decommissioning, Jun 28, 2019. https://www.iaea.org/newscenter/news/iaea-workshop-highlights-novel-trends-in-decommissioning#:~:text=Decommissioning%20of%20nuclear%20power%20plants,and%20reduced%20risk%20of%20delays.&text=Circular%20economy%20is%20a%20socio,making%20the%20most%20of%20resources.

Tappe, S., 2017. In: Adaptive Reuse of Warehouses in Relation to Neighborhood Cohesion and Identity: a Case Study of New Orleans, Oklahoma City, and Minneapolis, 21. Architecture Undergraduate Honors Theses. http://scholarworks.uark.edu/cgi/viewcontent.cgi?article=1024&context=archuht.

US Department of Energy, 2015. Adaptive Reuse: Reimagining Existing Buildings. Better Buildings Summit May 27-29. https://betterbuildingssolutioncenter.energy.gov/sites/default/files/Thurs%20-%20Adaptive%20Reuse%20Reimagining%20Existing%20Buildings.pdf.

Zuidhof, V., 2009. The Adaptive Reuse Process of the Industrial Heritage of the Former Rotterdam Dry-Dock Company, A Master Thesis. University of Rotterdam. https://thesis.eur.nl/pub/6494.

Glossary

When the source is not specified, definitions are taken from IAEA, Safety Glossary Terminology Used in Nuclear Safety and Radiation Protection, Edition 2018 https://www-pub.iaea.org/MTCD/Publications/PDF/PUB1830_web.pdf. Exceptions include the following notations: (*) IAEA Technical Reports Series No. 444; (**) IAEA Nuclear Energy Series No NW-T-2.2.; (***) English Heritage, Conservation Principles, Policies and Guidance, 2008; other references as annotated

Accident: Any unintended event, including operating errors, equipment failures and other mishaps, the consequences or potential consequences of which are not negligible from the point of view of protection and safety.

Adaptive reuse:** The act of finding a new use for a building (or area). It is often described as a process by which structurally sound older buildings are developed for economically viable new uses.

Aging: General process in which characteristics of a structure, system, or component gradually change with time or use.

ι Although the term aging is defined in a neutral sense—the changes involved in aging may have no effect on protection or safety, or could even have a beneficial effect—it is most commonly used with a connotation of changes that are (or could be) detrimental to protection and safety (i.e., as a synonym of ageing degradation).

Analysis: Often used interchangeably with assessment, especially in more specific terms such as "safety analysis." In general, however, analysis suggests the process and result of a study aimed at understanding the subject of the analysis, while assessment may also include determinations or judgements of acceptability. Analysis is also often associated with the use of a specific technique. Hence, one or more forms of analysis may be used in assessment.

Approval: The granting of consent by a regulatory body.

Area

ι **Controlled area:** A defined area in which specific protection measures and safety provisions are or could be required for controlling exposures or preventing the spread of contamination in normal working conditions, and preventing or limiting the extent of potential exposures.

ι **Operations area:** A geographical area that contains an authorized facility. It is enclosed by a physical barrier (the operations boundary) to prevent unauthorized access, by means of which the management of the authorized facility can exercise direct authority.

Assessment: See Analysis.

Asset: Anything that has value to the organization (International Organization for Standardization [ISO], Information and Documentation—Management Systems for Records—Fundamentals and Vocabulary, ISO-30300 (2011).

ι There are many types of assets, including the following:

- Information
- Software, such as a computer program
- Physical, such as a computer
- Services
- People, and their qualifications, skills, and experience
- Intangibles, such as reputation and image

Barrier: A physical obstruction that prevents or inhibits the movement of people, radionuclides, or some other phenomenon (e.g., fire) or provides shielding against radiation.

Bioassay: Any procedure used to determine the nature, activity, location, or retention of radionuclides in the body by direct (in vivo) measurement or by in vitro analysis of material excreted or otherwise removed from the body.

Biological shield: A mass of absorbing material placed around a reactor or radioactive source to reduce the radiation to a level safe for humans (USNRC Glossary 2020).

Brownfield*: Real property, the expansion, redevelopment, or reuse of which may be complicated by the presence or potential presence of a hazardous substance, pollutant, or contaminant.

Calibration: A set of operations that establish, under specified conditions, the relationship between values of quantities indicated by a measuring instrument or measuring system, or values represented by a material measure or a reference material, and the corresponding values realized by measurement standards.

Canyon: A large heavily shielded concrete building containing a remotely operated plutonium or uranium processing facility (US Department of Energy, National Environmental Policy Act; Glossary of Terms Used in DOE NEPA Documents. September 1998 https://www.energy.gov/sites/prod/files/NEPA_Glossary%2008_2011.pdf).

Cause

ι **Contributing cause:** An event or condition that collectively with other causes increases the likelihood of an accident but which individually did not cause the accident (DOE G 225.1A-1, Implementation Guide for Use with DOE O 225.1 Accident Investigations, 1997 https://www.directives.doe.gov/directives-documents/200-series/0225.1A-EGuide-1/).

ι **Direct cause:** The latent weakness (and the reasons for the latent weakness) that allows or causes the observed cause of an initiating event to happen, including the reasons for the latent weakness. Corrective actions designed to address direct causes are sometimes termed repairs.

ι **Observed cause:** The failure, action, omission, or condition that directly leads to an initiating event.

ι **Root cause:** The fundamental cause of an initiating event, correction of which will prevent recurrence of the initiating event (i.e., the root cause is the failure to detect and correct the relevant latent weakness(es) and the reasons for that failure).

Change management: The process, tools, and techniques to manage the people side of change to achieve the required business results (https://www.prosci.com/change-management/thought-leadership-library/the-what-why-and-how-of-change-management).

Characterization: Determination of the nature and activity of radionuclides present in a specified place.

ι For example, characterization is the determination of the radionuclides present in an area contaminated with radioactive material (e.g., as a first step in planning remediation).

Clearance: Removal of regulatory control by the regulatory body from radioactive material or radioactive objects within notified or authorized facilities and activities.

Clearance level (or clearance criteria): A value established by a regulatory body and expressed in terms of activity concentration and/or total activity, at or below which regulatory control may be removed from a source of radiation.

Cold and dark: A US DOE term for a form of passive safe enclosure. "Cold and dark" ensures a facility is isolated from all external sources of hazardous energy. This includes both identifying potential sources of stored energy as well as isolating electrical and mechanical devices that might pose a hazard to employees" (Lewis, B.K., Gilmour, J.C., and Willis, M.L., The Cold & Dark Process at the Savannah River Site, WM2009 Conference, March 1–5, 2009, Phoenix, AZ, US, Waste Management Symposia https://pdfs.semanticscholar.org/383e/573f2b0501be57bc4eff2fb0abe104548956.pdf).

Committed effective dose E (τ): The sum of the products of the committed organ or tissue equivalent doses and the appropriate tissue weighting factors W_T, where τ is the integration time in years following the intake. The commitment period is taken to be 50 years for adults (ICRP Publication 103).

Confinement: Prevention or control of releases of radioactive material to the environment in operation or in accidents.

Containment: Methods or physical structures designed to prevent or control the release and the dispersion of radioactive substances.

Contamination: Radioactive substances on surfaces or within solids, liquids, or gases (including the human body), where their presence is unintended or undesirable, or the process giving rise to their presence in such places.

Control: The function or power or (usually as controls) means of directing, regulating or restraining.

Cost–benefit analysis: A systematic technical and economic evaluation of the positive effects (benefits) and negative effects (disbenefits, including monetary costs) of undertaking an action.

Criticality: The state of a nuclear chain reacting medium when the chain reaction is just self-sustaining (or critical), i.e., when the reactivity is zero.

ι Often used, slightly more loosely, to refer to states in which the reactivity is greater than zero.

Deactivation: The process of placing a contaminated (nuclear, radiologically, or radioactive), excess facility in a stable condition to minimize existing risks and protect workers, the public, and the environment (US DOE). It is precursor to either long-term safe enclosure or dismantling.

Decommissioning: Administrative and technical actions taken to allow the removal of some or all of the regulatory controls from a facility.

- Decommissioning typically includes dismantling of the facility (or part thereof) to reduce the associated radiation risks, but in the IAEA's usage this needs not be the case. A facility could, for example, be decommissioned without dismantling and the existing structures subsequently put to another use (after decontamination).

Decontamination: The complete or partial removal of contamination by a deliberate physical, chemical, or biological process.

Defense-in-depth: A hierarchical deployment of different levels of diverse equipment and procedures to prevent the escalation of anticipated operational occurrences and to maintain the effectiveness of physical barriers placed between a radiation source or radioactive material and workers, members of the public or the environment, in operational states and, for some barriers, in accident conditions.

Demolition*: Clearance and removal of a structure to achieve greenfield or carry out the redevelopment plan.

Design: The process and the result of developing a concept, detailed plans, supporting calculations, and specifications for a facility and its parts.

Deterministic effect: A radiation induced health effect for which generally a threshold level of dose exists above which the severity of the effect is greater for a higher dose.

- **Contrasting term: Stochastic effect**

Dismantling: The taking apart, disassembling and tearing down of the structures, systems, and components of a facility for the purposes of decommissioning.

- The two main types of dismantling are immediate dismantling and deferred dismantling.

Deferred dismantling: This is deferred after permanent shutdown. For a nuclear installation, the nuclear fuel is first removed. Part or all of a facility containing radioactive material is either processed or put in such a condition that it can be put into storage. The facility is maintained until it can subsequently be decontaminated and/or dismantled. Deferred dismantling can involve the early dismantling of some parts of the facility and the early processing of some radioactive material and its removal from the facility, as preparatory steps for storage of the remaining parts of the facility.

Immediate dismantling: This begins shortly after permanent shutdown. Equipment and the structures, systems, and components of a facility containing radioactive material are removed and/or are decontaminated to a level that permits the removal of regulatory control from the facility and its release, either for unrestricted use or with restrictions on its future use.

Discharge: Planned and controlled release of (usually gaseous or liquid) radioactive substances to the environment.

- Strictly, the act or process of releasing the radioactive substances, but also used to describe the radioactive substances released

Disposal: Emplacement of waste in an appropriate facility without the intention of retrieval.

Disposition: Consigning of, or arrangements for the consigning of, radioactive waste for some specified (interim or final) destination, for example, for the purpose of processing, disposal, or storage.

Emergency: A nonroutine situation that necessitates prompt action, primarily to mitigate a hazard or adverse consequences for human life and health, property, and the environment.

- This includes nuclear and radiological emergencies and conventional emergencies such as fires, release of hazardous chemicals, storms, or earthquakes.
- It includes situations for which prompt action is warranted to mitigate the effects of a perceived hazard.

Emergency response: The performance of actions to mitigate the consequences of an emergency for human life and health, property, and the environment.

Employer: A person or organization with recognized responsibilities, commitments, and duties toward a worker in the employment of the person or organization by virtue of a mutually agreed relationship.

End state: A predetermined criterion defining the point at which a specific task or process is to be considered completed.

ι Used in relation to decommissioning activities as the final state of decommissioning of a facility, and used in relation to remediation as the final status of a site at the end of activities for decommissioning and/or remediation, including approval of the radiological and physical conditions of the site and remaining structures.

Entombment: The encasing of part or all of a facility in a structure of long-lived material for the purposes of decommissioning.

Environment: The conditions under which people, animals, and plants live or develop and which sustain all life and development, especially such conditions as affected by human activities.

Environmental remediation: See Remediation.

Evacuation: The rapid, temporary removal of people from an area to avoid or reduce short-term radiation exposure in an emergency.

Event: In the context of the reporting and analysis of events, an event is any occurrence unintended by the operator, including operating error, equipment failure or other mishap, and deliberate action on the part of others, the consequences, or potential consequences of which are not negligible from the point of view of protection and safety.

Exposure: The state or condition of being subject to irradiation.

ι Exposure should not be used as a synonym for dose. Dose is a measure of the effects of exposure.

ι **External exposure:** Exposure to radiation from a source outside the body. Contrasted with internal exposure.

ι **Internal exposure:** Exposure to radiation from a source within the body. Contrasted with external exposure.

(Note by the author: this book extends the scope of radiation exposure to exposure from toxic or chemical contamination.)

External event: Events unconnected with the operation of a facility or the conduct of an activity that could have an effect on the safety of the facility or activity.

ι Typical examples of external events for nuclear facilities include earthquakes, tornadoes, tsunamis, and aircraft crashes.

Failure (technical): Loss of the ability of a structure, system, or component to function within acceptance criteria.

ι Note that the structure, system, or component is considered to fail when it becomes incapable of functioning, whether or not this is needed at that time.

ι A failure may be the result of, e.g., a hardware fault, a software fault, a system fault, an operator error, or a maintenance error.

First responders: The first members of an emergency service to respond at the scene of an emergency.

Fogger: A device that spreads a chemical in the form of a fog (dictionary.com).

Greater-than-class C (GTCC) waste: In 10 CFR § 61.55, the US NRC categorizes low-level radioactive waste (LLRW) as class A, B, C, or GTCC. The last is LLRW with radionuclide concentrations exceeding the limits for class C waste. Part 61 classification provides for increased physical and administrative controls according to increased hazards and states that GTCC is unsuitable for near-surface disposal (US Nuclear Regulatory Commission, *Code of Federal* Regulations "Licensing Requirements for Land Disposal of Radioactive Waste," 10 CFR 61).

Greenfield*: A condition when the nuclear site has been granted unrestricted release from regulatory control, buildings have been demolished, and no further redevelopment is planned.

Hazard: The potential for harm or other detriment, especially for radiation risks; a factor or condition that might operate against safety.

Human factors: Environmental, organizational and job factors, and human and individual characteristics, which influence behavior at work in a way which can affect health and safety (HSE, UK https://www.hse.gov.uk/humanfactors/introduction.htm).

Incident: Any unintended event, including operating errors, equipment failures, initiating events, accident precursors, near misses or other mishaps, or unauthorized act, malicious or nonmalicious, the consequences or potential consequences of which are not negligible from the point of view of protection and safety. The word incident is used to describe events that are, in effect, minor accidents, i.e., that are distinguished from accidents only in terms of being less severe.

Infrastructure*: Public improvements that support development, including street lighting, sewers, flood control facilities, water lines, gas lines and telephone lines, etc.

In situ: In the natural or original position or place (Merriam-Webster Dictionary)

In situ disposal: See Entombment.

Inspection: An examination, observation, surveillance, measurement, or test undertaken to assess structures, systems, and components and materials, as well as operational activities, technical processes, organizational processes, procedures, and personnel competence.

Institutional control: Control of a radioactive waste site by an authority or institution designated under the laws of a state. This control may be active (monitoring, surveillance, remedial work) or passive (land use control) and may be a factor in the design of a facility (e.g., a near-surface disposal facility).

ι Most commonly used to describe controls over a disposal facility after closure or a facility undergoing decommissioning.

ι Also refers to the controls placed on a site that has been released from regulatory control under the condition of observing specified restrictions on its future use to ensure that these restrictions are complied with.

ι The term institutional control is more general than regulatory control (i.e., regulatory control may be thought of as a special form of institutional control).

ι Institutional control may apply in situations which do not fall within the scope of facilities and activities. As a result, some form of institutional control may be considered more likely to endure further into the future than regulatory control.

Knowledge management: An integrated, systematic approach to identifying, managing, and sharing an organization's knowledge and enabling groups of people to create new knowledge collectively to help in achieving the organization's objectives.

Legacy (waste, site, issue, equipment etc.): In broad usage, a legacy issue is one from the past that still has to be dealt with. A legacy situation has been developed as a result of past actions and decisions.

Legacy has become a cliché term in the nuclear field, but its meaning is not well defined. Different people understand it differently. A better terminology would be "past practices that were never subject to regulatory control" (IAEA Safety Glossary 2018) or were subject to controls, which are nowadays considered inappropriate.

Very often, we hear that site X is a legacy site, as a good deal of legacy facilities was built and operated prior to establishment of current regulations. Following that logic, most existing facilities worldwide could be declared legacy, as the regulations constantly evolve. Which year to put the borderline between legacy and nonlegacy? The same year for all countries or each country should pick a year by itself?

As used in this book, "legacy" loosely refers to practices, tools, installations, etc. that are considered aged, obsolete, semiabandoned or generally unreliable.

Lesson learned: An adverse work practice or experience that is captured and shared to avoid a similar event from occurring again. A lesson learned may also be a good work practice, or innovative approach, that is captured and shared to promote repeat application (slightly rephrased from DOE/RL-97-63).

License: A legal document issued by the regulatory body granting authorization to perform specified activities relating to a facility or activity.

Licensee: The holder of a current license.

Life cycle: As opposed to lifetime, it implies that the life is genuinely cyclical.

Limit: The value of a quantity used in certain specified activities or circumstances that must not be exceeded.

Life cycle management: Life management (or lifetime management) in which due recognition is given to the fact that at all stages in the lifetime there may be effects that need to be taken into consideration.

Maintenance: The organized activity, both administrative and technical, of keeping structures, systems, and components in good operating condition, including both preventive and corrective (or repair) aspects.

Migration: The movement of radionuclides in the environment as a result of natural processes.

Minimization (of waste): The process of reducing the amount and activity of radioactive waste to a level as low as reasonably achievable, at all stages from the design of a facility or activity to decommissioning, by reducing the amount of waste generated and by means such as recycling and reuse, and treatment to reduce its activity, with due consideration for secondary waste as well as primary waste.

Mixed waste: Radioactive waste that also contains nonradioactive toxic or hazardous substances.

Monitoring: Continuous or periodic measurement of radiological or other parameters or determination of the status of a structure, system, or component.

Multiattribute utility analysis: A decision analysis technique that provides rigorous, sound, and demonstrated ways to combine dissimilar measures of costs, risks, and benefits, along with individual preferences, into high-level, aggregated measures that can be used to evaluate alternatives (IAEA-TECDOC-1279 Non-Technical Factors Impacting on the Decision Making Processes in Environmental Remediation, Vienna 2002).

Near miss: A potential significant event that could have occurred as the consequence of a sequence of actual occurrences but did not occur owing to the conditions prevailing at the time.

Nuclear facility: A facility (including associated buildings and equipment) in which nuclear material is produced, processed, used, handled, stored, or disposed of.

Nuclear installation: see Nuclear Facility.

Occupational exposure: Exposure of workers incurred in the course of their work.

Off-site: Outside the site area.

On-site: Within the site area.

Operating organization: Any person or organization applying for authorization or authorized and/or responsible for safety when undertaking activities or in relation to any nuclear facilities or sources of ionizing radiation.

Operator: *This term is not used in this book to avoid confusion with either "operating organization" or "operative."*

Operative: Worker, workman.

Overpack: An enclosure used by a single consignor to contain one or more packages and to form one unit for convenience of handling and stowage during transport.

Packaging: Preparation of radioactive waste for safe handling, transport, storage, and/or disposal by means of enclosing it in a suitable container.

Permanent (final) shutdown: The cessation of operation of a facility with no intention to recommence operation in the future.

Plant: In this book, a plant is a complex of facilities where individual units are located, e.g., multiple generating units, power transmission equipment, fuel processing facilities, auxiliary and service buildings, and other infrastructure. Units refer to individual facilities such as natural gas combustion turbines, coal-fired boilers, or nuclear reactors.

Precursor, precursory event: Event that precedes and signals the potential occurrence of similar events (DOE/RL-97-63).

Procedure: A series of specified actions conducted in a certain order or manner.

ι The set of actions to be taken to conduct an activity or to perform a process is typically specified in a set of instructions.

Process: A course of action or proceeding, especially a series of progressive stages in the manufacture of a product or some other operation.

Program, project: These two terms are often used interchangeably in the technical literature. However, for the purposes of this book, a project is a temporary undertaking to create a unique product or service. A project has a defined start and end point and specific objectives that, when attained, signify completion. A program, on the other hand, is defined as a group of related projects managed in a coordinated way to obtain benefits not available from managing the projects individually. A program may also include elements of ongoing, operational work. So, a program is comprised of multiple projects and is created to obtain broad organizational or technical objectives. There are many differences between a project and a program including scope, benefits realization, time, and other variables. One notable difference is time; for example, a project by definition has a beginning and an end (or at least one hopes so!); certain programs, while having a beginning may not have an end (Dictionary of Project Management Terms, by Ward, J. LeRoy, ISBN: 9781890367459).

Public*:** Of, concerning, done, acting, etc. for people as a whole.

Record(s): Information created, received, and maintained as evidence and as an asset by an organization or person, in pursuit of legal obligations or in the transaction of business (International Organization for Standardization [ISO], Information and Documentation—Management Systems for Records—Fundamentals and Vocabulary, ISO-30,300 (2011)).

Redevelopment*: Planning, development, replanning, redesign, clearance, reconstruction, or rehabilitation of all or part of a project area.

Regulatory body: An authority or a system of authorities designated by the government of a state as having legal authority for conducting the regulatory process, including issuing authorizations, and thereby regulating the safety of nuclear installations, radiation safety, the safety of radioactive waste management, and safety in the transport of radioactive material.

Note: the above definition refers only to nuclear safety and radiation protection. This book recognizes that other regulatory bodies can be involved in a decommissioning or redevelopment project.

Release: The action or process of setting free or being set free, or of allowing or being allowed to move or flow freely. Release is used in both a physical "scientific" sense (discharge) and a "regulatory" sense (clearance).

Remediation: Any measures that may be carried out to reduce the exposure due to existing contamination of land areas through actions applied to the contamination itself (the source) or to the exposure pathways to people.

ι Complete removal of the contamination is not implied.

ι The use of the terms rehabilitation and restoration is discouraged. Such terms may be taken to imply that the conditions that prevailed before the contamination can be achieved again and unconditional use of the land areas can be restored, which is not usually the case (e.g., owing to the effects of the remedial action itself). Often, remediation is used to restore land areas to conditions suitable for limited use under institutional control.

Repair*:** Work beyond the scope of maintenance, to remedy defects caused by decay, damage, or use, including minor adaptation to achieve a sustainable outcome, but not involving restoration or alteration.

Restoration: A restoration is when you bring something back to what it was (vocabulary.com). See comment under Remediation.

Restricted use (or release): The use of an area or of materials subject to restrictions imposed for reasons of radiation protection and safety.

ι Restrictions would typically be expressed in the form of prohibition of particular activities (e.g., house building, growing or harvesting particular foods) or prescription of particular procedures (e.g., materials may only be recycled or reused within a facility).

Reuse:** The use of a facility or building for a purpose other than that for which it was originally intended and/or used, following the termination of its original use has ceased; or the reuse for the original purpose but under new circumstances.

Safe enclosure (during decommissioning): A condition of a nuclear facility during the decommissioning process in which only surveillance and maintenance of the facility take place (IAEA, Radioactive Waste Management Glossary, Vienna 2003).

ι *Also used in this book to refer to the physical structure encasing the nuclear facility in this condition.*

Safety culture: The assembly of characteristics and attitudes in organizations and individuals which establishes that, as an overriding priority, protection and safety issues receive the attention warranted by their significance.

Security: The prevention and detection of, and response to, criminal or intentional unauthorized acts involving nuclear material, other radioactive material, associated facilities, or associated activities.

Segregation: An activity where types of waste or material (radioactive or exempt) are separated or are kept separate on the basis of radiological, chemical, and/or physical properties, to facilitate waste handling and/or processing.

Service life: The period from initial operation to final withdrawal from service of a structure, system, or component.

Site:** The area containing, or under investigation for its suitability for, a facility (e.g., a repository). It is defined by a boundary and is under effective control of the operating organization.

Spent fuel: Nuclear fuel removed from a reactor following irradiation that is no longer usable in its present form because of depletion of fissile material, poison buildup, or radiation damage.

Stakeholder (interested party): A person, company, etc., with a concern or interest in the activities and performance of an organization, business, system, etc.

Stewardship: The physical controls, institutions, information, and other mechanisms needed to ensure protection of people and the environment at sites where the responsible party has completed or plans to complete "cleanup" (e.g., landfill closures, remedial actions, removal actions, and facility stabilization). Long-term stewardship includes, *inter alia*, land-use controls, monitoring, maintenance, and information management (adapted from DOE, Long Term Stewardship Study, October 2001 https://www.energy.gov/sites/prod/files/em/DOELongTermStewardshipStudy-VolumeI-FinalOctober2001.pdf).

Stochastic effect: A radiation-induced health effect, the probability of occurrence of which is greater for a higher radiation dose and the severity of which (if it occurs) is independent of dose.

ι **Contrasting term:** Deterministic effect.

Storage: The holding of radioactive sources, radioactive material, spent fuel, or radioactive waste in a facility that provides for their/its containment, with the intention of retrieval.

ι In many cases, the only element of this definition that is important is the distinction between disposal (with no intent to retrieve) and storage (with intent to retrieve).

Structures, systems, and components (SSCs): A general term encompassing all of the elements (items) of a facility or activity, which contribute to protection and safety, except human factors.

ι **Components:** Discrete elements of a system. ι Examples of components are wires, transistors, integrated circuits, motors, relays, solenoids, pipes, fittings, pumps, tanks, and valves.

ι **Structures:** Passive elements (e.g., buildings, vessels and shielding).

ι **Systems:** Several components assembled in such a way as to perform a specific (active) function.

Surveillance: A type of inspection to verify the integrity of a facility or structure.

Sustainable*:** Capable of meeting present needs without compromising ability to meet future needs.

Transuranics: Elements with an atomic number greater than that of uranium (Merriam-Webster Dictionary).

Traveler: An inspection and certification document. Travelers are forms which accompanied nuclear components or assemblies during the manufacturing process to document that they were produced according to design specifications and program requirements (The Department of Energy's Rocky Flats Plant: a Guide to Record Series Useful for Health-Related Research; Volume IV Production and Materials Handling, August 1, 1995, History Associates Incorporated https://inldigitallibrary.inl.gov/PRR/93922.pdf).

Unrestricted use: The use of an area or of material without any radiologically based restrictions.

ι There may be other restrictions on the use of the area or material, such as planning restrictions on the use of an area of land or restrictions related to the chemical properties of a material.

ι The term "free release" is occasionally, albeit loosely, used in this book.

Validation: The process of determining whether a product or service is adequate to perform its intended function satisfactorily.

Verification: The process of determining whether the quality or performance of a product or service is as stated, as intended, or as required.

Walk-down: Inspection of local areas in a nuclear power plant *(in this book used for any nuclear facility)* where systems and components are physically located to ensure accuracy of procedures and drawings, equipment location, operating status, and environmental effects or system interaction effects on the equipment, which could occur during accident conditions (American National Standard Glossary—May 19, 2016 http://cdn.ans.org/standards/resources/toolkit/docs/glossary-of-definitions.pdf).

Waste: Material for which no further use is foreseen.

Abbreviations, acronyms, initialisms

aka	also known as
A/C	alternating current
ANL	Argonne National Laboratory
Argonaut	an acronym for **ARGOnne** **N**uclear **A**ssembly for **U**niversity **T**raining: a type of small nuclear research and training reactors
ASME	American Society of Mechanical Engineers
ASN	French Autorité de Sûreté Nucléaire (English: Nuclear Safety Authority)
Bioshield	short for Biological Shield
Bq	Becquerel is the SI unit of radioactivity. One Becquerel is one disintegration per second
CSA	Canadian Standards Association
CX	Chapelcross NPP, United Kingdom
CY	Connecticut Yankee NPP, United States
D&D	Decommissioning
D&ER	Decommissioning and environmental remediation
dB	decibel
DOE	US Department of Energy
EA	UK Environment Agency
EBWR	hxperimental boiling water reactor
EC	European Commission
EPA	US Environmental Protection Agency
EPRI	Electric Power Research Institute
ER	Environmental Remediation
ETTP	East Tennessee Technology Park
Gy	Gray, unit of kerma and absorbed dose, equal to 1 J/kg
Ha	hectare (10^4 m^2, 0.01 km^2)
HCA	high contamination area
HEPA	high-efficiency particulate air filter
HLW	high-level (radioactive) waste
HSE	Health and Safety Executive, United Kingdom
HWCTR	Heavy Water Components Test Reactor, SRS, United States
IAEA	International Atomic Energy Agency, Vienna, Austria
ILW	intermediate-level (radioactive) waste
INES	International Nuclear and Radiological Event Scale
INPO	Institute of Nuclear Power Operations, an organization established in 1979 by the US nuclear power industry in the aftermath of the Three Mile Island accident
ISBN	International Standard Book Number
ISFSI	independent spent fuel storage installation

KM	knowledge management
kPa	kilopascal (1000 times the unit of pressure and stress in the SI system)
L	liter (e.g., a 200-L drum)
LCM	life cycle management
LLNL	Lawrence Livermore National Laboratory, United States
LLW	low-level (radioactive) waste
m, km	meters, kilometers
MAUA	multiattribute utility analysis
MW	megawatt
MWe	megawatt (electric)
MWt	megawatt (thermal)
MY	Maine Yankee NPP, United States
NDA (1)	Nuclear Decommissioning Authority, United Kingdom
NDA (2)	nondestructive analysis
NEA	OECD/Nuclear Energy Agency, Paris
No.	number
NPP	nuclear power plant
NRC	US Nuclear Regulatory Commission
ONR	UK's Office for Nuclear Regulation
OSHA	US Occupational Safety and Health Administration
PCB	polychlorinated biphenyl
PFR	prototype fast reactor (at Dounreay, United Kingdom)
PWR	pressurized water reactor
R&D	research and development
RCT	radiological control technician
RP	radiological protection
RPV	reactor pressure vessel
RV	reactor vessel
RWP	radiological work permit
SI	International System of Units, international decimal system of weights and measures building on the metric system of units
SOGIN	Società Gestione Impianti Nucleari (Italy's State-owned company in charge of the decommissioning of Italian nuclear plants and the management of radioactive waste)
SRS	Savannah River Site, United States
SSC	structures, systems, and components
Sv	Sievert, unit of equivalent dose and effective dose, equal to 1 J/kg
U of C	University of Chicago, Argonne, United States
UK	United Kingdom (of Great Britain and Northern Ireland)
UKAEA	United Kingdom Atomic Energy Authority
US	United States
ÚJV Řež	Nuclear Research Institute (ÚJV is the Czech acronym) based at Řež, Czech Republic
VLLW	very-low-level (radioactive) waste
WENRA	Western European Nuclear Regulators Association
WW (I or II)	World War
YR	Yankee (Rowe) NPP

Index

Note: 'Page numbers followed by "f" and "t" indicate figures and tables respectively.'

S

T

U

V

W

Y

Z

FSC
www.fsc.org
MIX
Papier | Fördert
gute Waldnutzung
FSC® C083411

Kolibri 360 GmbH
Ferdinand-Jühlke-Straße 7
99095 Erfurt, Deutschland
produktsicherheit@kolibri360.de